A JOURNEY THROUGH MATH-LAND

A JOURNEY THROUGH MATH-LAND

The Land of Logical Ideas

Reza Noubary

To order additional copies of this book, contact:
Xlibris
844-714-8691
www.Xlibris.com
Orders@Xlibris.com
834002

Contents

INTRODUCTION

I am mathematics, and if I could speak, my friend,
All the reservations and doubts about me would come to an end
I am a higher-order abstraction, I am universal
I am applied to model any regular pattern or trend
I am the go-to subject for complicated or complex situations
Everybody trusts my factual conclusions and what I represent
I offend no one but those who do not care for truth
Bring peace of mind by helping where there is a demand.

MATHEMATICS, THE LARGEST coherent artifact built by civilization, is the craft of creating new knowledge from old. It is a symbolic language to describe complex ideas using abstraction. Mathematics lives in the world of ideal realities, regular patterns, and certainties. It presents the smooth part of the world, what we expect, not what we observe.

WARM UP

1. The poetry of logical ideas

MATHEMATICS HAS BEEN variously described as an ideal reality, a formal game, and the poetry of logical ideas.

2. A supreme example of mathematical beauty (Euler's identity)

$$e^{i\pi} + 1 = 0$$

is likened to a Shakespearean sonnet or a da Vinci picture, Euler's identity is beautiful because it manages to encompass the five neutral constants in mathematics:

0 – the neutral element for addition and subtraction,
1- the neutral element for multiplication and division,
e – Euler's number, the base of natural logarithms,
i – the imaginary unit, which satisfies $i^2 = -1$, and
π - the ratio of the circumference of a circle to its diameter.

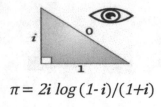

$$\pi = 2i \log(1-i)/(1+i)$$

3. Coincidence?

Einstein was borne on Pi-day and Steven Hockings died on Pi-day.

4. Birthday Surprise

Did you know that it only takes 23 people in a room to give you an evens chance that at least two of them have the same birthday? With 75 people the chances rise to 99 per cent!

5. The universe is not big enough for Googolplex

A googolplex is 10 to the power of a googol, or 10 to the power of 10 to the power of 100. Our known universe does not have enough space to actually write that out on paper. If you try to do that sum on a computer, you'll never get the answer, because it won't have enough memory.

6. Did you know?

- Every odd number has an "e" in it.
- Albert Einstein was born on Pi day and Stephen Hawking died on Pi day
- Zero is not represented in Roman numerals.
- Among all shapes with the same perimeter, a circle has the largest area.
- The easiest way to remember the value of Pi (3.1415926) is by counting each word's letter in 'May I have a large container of coffee'.

- Former US president Garfield discovered a nice proof of the Pythagorean Theorem.
- If a sequence of events occur in random times and random sizes then after the first one the chance that a second one would be bigger is 50% but the expected time for its occurrence in infinite.
- Base 2 (Binary) numeration system is more efficient than decimal system. Trinary (base 3) is more efficient than Binary. The most efficient base is, as expected, Base e.
- The probability that two randomly chosen integers have greatest common divisor equal to 1 is $6/Pi^2$.
- The ratio of the median times for any two successive records tends to e.
- When looking for the best applicant or best lover out of n people one popular strategy is to examine the first m and pick anyone after that who is better than all previous ones. It is shown that the optimal value of m is n/e and the probability that it is actually the best is 1/e. There are some evidence that some birds do follow this strategy quite closely when searching for a mate.
- Isaac Newton did badly in school - he paid no attention to the syllabus or other requirements and studied whatever he felt like studying on his own.

7. Have a question Dad

Why do I need to learn subtraction? Well son, to make a difference.

$$\times - \curlyvee = -$$

8. Three halves

My Mom is half-British, half German, and half Italian. But

little john there are too many halves there. Yes teacher I know but my Mom is a big woman.

9. Which of the three results is correct?

Consider the infinite sum
$$1 - 1 + 1 - 1 + 1 - 1 + 1 - 1 + ...$$
Here is one way to evaluate the sum:
$$(1 - 1) + (1 - 1) + (1 - 1) + (1 - 1) + ... = 0$$
Here is another way:
$$1 - (1 - 1) - (1 - 1) - (1 - 1) - ... = 1$$
A third way, using the identity
$$1 + x + x^2 + x^3 + ... = 1/(1 - x),$$ with $x = -1$, leads to $1/2$.

10. Which is the correct answer?

Think about two young families with children living in a remote neighborhood. One family has a boy and two girls, and the other has two boys and a girl. Suppose that we want to calculate the ratio of boys to girls in that neighborhood.

We may find the average number of boys $((1+2)/2) = 1.5$ and girls $((2+1)/2) = 1.5$ in that neighborhood and calculate the ratio as $1.5/1.5 = 1$.

Alternatively, we may find the ratios of boys to girls in each family, $1/2 = 0.5$ and $2/1 = 2$ respectively, and calculate their average as $(0.5 + 2)/2 = 1.25$.

11. Who Is the Better Free Throw Shooter?

During the last basketball season, Jim attempted one hundred free throws in the first half of the season and made thirty. His free-throw percentage was $30/100 = 0.300$. He also attempted twenty in the second half of the season and made eight ($8/20 = 0.400$). His stats were better than Curt's stats—$5/20$ (0.250) for the first half of the season, and $35/100$ (0.350) for the second half. For the season, however, Curt's free-throw percentage (40/120 = 0.333) was higher than Jim's (38/120 = 0.317).

The above examples are cases of what is known as Simpson's paradox. The paradox occurs because collapsing the data can lead to an inappropriate weighting of the different populations.

12. The world of the numbers

Numbers have an interesting history and a world of their own. Like human beings, some numbers play a more significant role than others. Some represent certain concepts some very complicated. Some get used more often than others. Some never get used. For example, although there are significantly more irrational than rational numbers, irrational numbers are hardly used. Some numbers have a special place in certain cultures and even religions. Some are loved by scientists, artists, and even the general public. But there is one, and only one number, so famous that a particular day is named after it.

13. Pi and Mysteries of Universe

Pi is an irrational number and, as such, it takes an infinite number of digits to give its exact value. This means that we can neither get to the end of it nor can we find the next digit using a pattern in its earlier digits. In other worlds, it is an infinite, non-repeating decimal.

Now, being infinite and non-repeating, it sensible to assume that every possible number combination exists somewhere in Pi. This assumption implies that when converted into numbers (ASCII text), somewhere in that string of digits are answers to all the great questions of the universe, the DNA of every being and everything that has ever been written or will be written. In short, everything trivial or amazing man could think of would be somewhere in the ratio of a circumference and a diameter of a circle.

In fact, right now it is possible to check to see if somebody's birth date or social security number occurs as a string of digits in the first 200 million digits of Pi at the Pi Search Site, http://www.angio.net/pi/.

But could this be the true? Well, nobody is really sure. So at the present we can only say that this is a reasonable possibility.

Pi even enters into Heisenberg's Uncertainty Principle; the equation that defines how precisely we can know the state of the universe.

It also together with Euler's number e Pi appears in the complex equation representing the Bell-Curve.

$$f(x) = \frac{1}{\sqrt{2\pi}} e^{\frac{x^2}{2}}$$

14. Pi and pizzas are linked

You multiply Pi multiplied by the radius squared to find the area and multiply area by height to find the volume, That means the volume of a pizza that has a nominal radius of (z) and height (a) will, of course, be: Pi × z × z × a

15. Piem

Some non-mathematicians also like to memorize digits of Pi. To help them few interesting tricks are developed. Here is an example called a piem (argh!). It is a little poem where the length of each word presents a digit of Pi.

How I need a drink, (3 1 4 1 5)
alcoholic of course, (9 2 6)
after the heavy lectures, (5 3 5 8)
involving quantum mechanics. (9 7 9).

There once was a girl who loved Pi
I never could quite fathom why

To her it's a wonder
To me just a number
Its beauty revealed by and by –<u>Eve Anderson</u>.

Three point one four one five nine
Makes the lazy student whine,
But give this ratio a try–
You'll find that it's as easy as pi! –<u>Fred Russcol</u>.

16. An Icon

- Literary nerds invented a dialect known as Pilish, in which the numbers of letters in successive words match the successive digits of Pi such as these excerpts:
  "May I have a large container of coffee (3.1415926)?"
- There is a music video about Pi that has wizards, robots, and a graffiti artist in it.
- The song "Bye-Bye Miss American Pi" is a parody of Don Mclean's song "American Pie", rewritten as "American Pi".
- Pi is four times the infinite sum and difference of odd number's reciprocals
  $$Pi = 4(1/1 - 1/3 + 1/5 - 1/7 + 1/9 - 1/11 + 1/13 ....,)$$

17. Magic of 73

I DID WRITE THIS POEM WHEN I TURNED 73.
I am 73, I think the number, not me, has a magical power
Although it is made of a spring, summer, fall, and winter like any other
But if you calculate; 73 x your age x 13837, you will see/note the magic
That may make you to believe that numbers can do a wonder
If it does not surprise you just let it go and do not bother
Your friend and family may have fun with it, so pass it all specially to your mother.

I DEDICATE THIS BOOK

To all parents, especially those in heaven
Thanks for all those unconditional love and life lessons
You live in our hearts each and every day we live
(You are desperately missed each and every day our lives)
We think of you so much that we cannot put it in an expression.

Mothers

Most mothers are emotional but are strong like a rock
They are most of the time tired but work around the clock
They worry for their children but are
source of positive energy and joy
Try their best to make children happy,
look for opportunities to knock
Although overwhelmed with responsibilities, they hardly quit
Keep trying, even though they may not be able to absorb the shock
Mothers are prepared for every occasion, even when all is chaos
They try to better our lives every single day by their acts and talk

Fathers

Some of us do not have a living father, but that is okay
Let us show our love and appreciation each and every day
Men are sensitive and needy but do not usually display
They were taught to hide it, that has always been the way
Men hardly talk about their personal problems with others
They keep it inside, suffer, avoid, and
pretend they have nothing to say
They are not certain how to deal with their feelings and needs
So often choose not to express it or do it a wrong way
Though they have invented, fought, and died in millions
They often do not get enough credit, no praise, no pay
The story is long though tough, they badly need love
Let us give them comfort, do whatever we can at least someday
Fathers are their daughter's first love, their son's first hero
Let us wish them a long life and show that we care.

PREFACE

Mathematics, the poetry of logical ideas

THE STORY OF mathematics is fascinating. Its history and philosophy provides an invaluable perspective on human nature and the world around us. Because of its ability to abstract and generalize, mathematics is a unique tool for getting insight into an increasingly large number of disciplines that it is being applied to.

This book plans to trace applications of mathematics. Its goal is to find a way to delight readers about mathematics and open the door for them to see its beauty. It is an attempt to bring a variety of applications within the reach of individuals with some mathematics background or interests. Articles presented constitutes some interesting aspects of life that can be looked upon from mathematical viewpoint. They are made independent from each other whenever possible at the expense of being occasionally repetitive.

Brains Beat Brawn

We are living in the age of technology and innovations. An era that mathematics has situated itself as part of, and essential to, a lived experience. An era that nerds are running away with all the money and mathematics intellect is cool. An era that mathematics has become an undisputable language of science and leadership.

Plan

Other than numbers in totality, we plan to talk about some special numbers such as Pi, Euler's e, and zero. Other mathematics-related topics discussed include faith, randomness, risk, paradoxes, extremes, records, sports, music, natural disasters, diversity, health, coincidences, stock market, and many other topics.

I know that I am not a good writer and that my English is not as good as I would like it to be. However, I decided to challenge myself and hope that readers would give me a break.

An Example: Billion, a Misunderstood Number

We are used to hearing large numbers such as a billion mentioned frequently on the news. People and politicians such as Bernie Sanders are decrying "the millionaires and billionaires" as though these groups are one step apart from one another. Even people who work with numbers sometimes put millionaires and billionaires in the same category without pointing out their huge difference. Let me directly go to some revealing examples.

- If you gave your daughter $1 million and told her to go out and spend $1,000 a day, she would be back in about three years for more money. If you gave her $1 billion, she would not be back for three thousand years.

- If you save $100 per day, it takes 27,397.26 years to reach $1 billion.
- If you and your future children live ninety years, will take more than three hundred generations to save $1 billion.

- If you decide to count to one billion and take just three seconds to say each number, 95.1 years is how long it will take you to count to one billion.
- If you decide to take a billion two-foot-long steps, you would go more than fifteen times around the equator.
- A billion minutes ago, Jesus was alive. A billion days ago, no one walked on the earth on two feet.
- A billionaire is richer than thirteen nations in the world.
- If a billionaire stacks his fortune in hundred-dollar bills vertically, the resulting stack would ascend a staggering 3,585 feet! He could place a belt of dollar bills around Earth.
- One billion pennies stacked on top of each other would make a tower almost 870 miles high.
- The height of a stack of one billion dollar bills measures 358,510 feet or 67.9 miles. This would reach from Earth's surface into the lower portion of the troposphere—one of the major outer layers of Earth's atmosphere.
- The area covered by one billion dollar bills measures four square miles. This is an area equal to the size of 2,555 acres.
- The length of one billion dollar bills laid end to end measures 96,900 miles. This would extend around Earth almost four times.

Summary of Features

This book begins with a warm-up chapter to "walk" the reader through a variety of interesting mathematics-related topics. It strongly emphasizes on interesting applications of mathematics and walks the readers through a variety of mathematical concepts relevant to life. Its goal is to make the book concise and engaging, show how the language of mathematics is related to human culture, and demonstrate its role in comprehending the world.

ACKNOWLEDGMENT

I START WITH MY profound thank-you to people who guided me through the process of completing this book. My gratitude goes to my spouse extraordinaire, Zohreh, who does everything to encourage me to do my work. Hats off to my friends and colleagues who were always encouraging and supportive. I cannot fail to honor the memory of my brother.

I also would like to acknowledge my colleagues JoAnne Growney and Bill Calhoun for their coauthorship of the articles "Can I Be Sensible to Risk," "Changing the Rules of Tennis: An Exercise in Mathematical Modeling," respectively. Special thanks to Megan Mahle, who helped me to organize the book by offering her time and expertise. Her help was invaluable.

ABOUT THE AUTHOR

R EZA D. NOUBARY received his BSc and MSc in Mathematics from Tehran University, and MSc and PhD in Applied Probability and Statistics from Manchester University in England. He has more than fifty years of teaching experience working in several universities in several countries. He has been a visitor in Harvard, Princeton, Penn, UCLA, University of Maryland, University of Kaiserslautern, and Catholic University of Leuven. His research interests include time series analysis, modeling and risk analysis of natural disasters, and applications of mathematics and statistics in sports. He is a fellow of the Alexander von Humboldt and a fellow or member of numerous professional organizations. His outside interests include music, soccer, racquetball, and tennis. He has published several books and more than one hundred research papers in more than ten different disciplines.

ABOUT THE BOOK

In a book, there is wisdom
In a thought, a big dream
In a heart, a burning desir
To find out what is its theme.

THE ESSAYS OF this book are all concerned with the role of mathematics in life and unfolding of intellectuality. It is a book about the uniqueness of mathematics and various contexts it is concerned with. It is intended to be a book not in mathematics but about mathematics, even if some parts of it contain mathematical technicalities.

Mathematics as a Model for Universe

For "is" and "is-not" though with rule and line
And "up-and-down" by logic I define,
Of all that one should care to fathom, I
Was never deep in anything but-wine.

Ah, but my computations, people say,
Reduced the year to better reckoning?—Nay,
'Twas only striking from the calendar
Unborn to-morrow and dead yesterday.
—Omar Khayyam

The various branches of mathematics have, through time, developed as a response to the need for more detailed models to describe new developments, both technological and philosophical. This was true when Newton developed calculus and also true during the late 1800s through the 1920s when a schism developed between the classical mathematicians and some brilliant innovative thinkers, the mathematical crises of the early nineteenth century. One of man's greatest strengths is his ability to question his surroundings and beliefs and, through this questioning, develop new insight and innovation. Most mathematical systems are developed for use in applications. Man's natural inquisitiveness often leads him to develop systems beyond the application and into abstract theory. This theory drives him to investigate the applications and often yields direction for new discoveries that were not previously foreseen or that defy intuition. Georg Cantor (1845–1018) was the most notable of a number of mathematicians who questioned the basic precepts of mathematics and developed the modern methods.

Need to Promote Mathematics

In recent years, we have witnessed many amazing inventions and breakthroughs, mostly by individuals who have received their education here in the United States. Proud of the performance and contributions of the United States' higher-education institutions, we may look ahead and wonder whether this trend will continue and will guarantee our competitiveness in the global world. Unfortunately, this might not be the case, as recent investigations point to a decline in basic knowledge of sciences and mathematics both in our schools and universities. In fact, a large percentage of positions requiring advanced knowledge of sciences and mathematics are now filled by people who have recently moved to the United States from other countries. As a result, in year 2020, for the first time ever, the U.S. Patent and Trademark Office issued more patents to foreigners than to Americans. This trend is likely to continue, since, presently, the two primary sources for graduate students in science and technology at American universities are from China and South Korea.

To see where we stand, let us look at a few more facts and statistics. Very recently, the biggest global school rankings have been published by the OECD (Organization for Economic Cooperation and Development). The study may have some drawbacks considering the facts that (1) some countries have a more homogeneous population and schools than others and (2) students in some countries are better test takers than others. Despite this, the study is considered significant, as it gives access to all countries to compare themselves against the world's education leaders, find out about their relative strengths and weaknesses, and see what long-term economic gains an improved quality in schooling could provide them. As expected the top-five positions were taken by Asian countries—Singapore, Hong Kong, South Korea, Taiwan, and Japan. The five lowest-ranked countries were Oman, Morocco, Honduras, South Africa, and Ghana. Out of the seventy-six countries that were included in the study, United States was ranked twenty-eighth. According to this report, the standard of education is a powerful predictor of the wealth that countries can generate in the long run. Also, last year, among the twenty-nine wealthiest countries, the United States ranked twenty-seventh in regard to the number of college students with degrees in science and engineering. Additionally, among developed countries, the United States was ranked thirty-first in mathematics and twenty-third in science. This is despite the fact that our schools and universities have by far the best facilities in the world. So, considering these, it is logical to think that we may lose ground to foreign rivals unless we find a way to enhance the quality of our mathematics and science education and increase the number of students majoring in this disciplines. This, of course, needs long-term planning to find ways to motivate more students to major in science and mathematics as well as to learn more about other countries' approaches and make changes or adjustments if necessary.

Finally, we also need to combat the culture that is proud of not being good at mathematics. It seems that in our culture, not being good in mathematics implies that the person is socially acceptable or even preferred. According to some statistics, 30 percent of Americans prefer to clean the bathroom than do mathematics. Further, they are not shy to say such a thing to the others.

Something to Smile

A young construction worker has trouble with math, so his foreman tells him to take a philosophy class first in the local college.

"What is that?" asks the worker.

"Here, I will give you an example. Do you own a lawnmower?"

"Yes," replies the worker.

"Then you own a house?"

"Yes . . . how did you know?" asks the worker.

"That's deductive reasoning. Here, I will take it further—do you live with a woman in that house?"

"Yes."

"Then you are a heterosexual."

"Interesting," says the worker. "Sign me up!"

The next week at work, the worker's foreman is talking to him.

"Did you sign up and attend the philosophy class?" what is it? the foreman asked.

"Here, I'll give you an example—do you own a lawnmower?" asks the worker.

"No," replies the foreman.

"Well, then you're a homosexual!" exclaims the worker.

CHAPTER 1

Introduction

Mathematics is like love, a simple idea that can get complicated. This is especially true for mathematicians who are simple people with a complex mind.

1. Quotes about Math

- Mathematics is the most beautiful and most powerful creation of the human spirit.—Stefan Banach
- What is mathematics? It is only a systematic effort of solving puzzles posed by nature.—Shakuntala Devi
- Mathematics is the music of reason.—James Joseph Sylvester
- Mathematics knows no races or geographic boundaries; for mathematics, the cultural world is one country.—David Hilbert
- There should be no such thing as boring mathematics.—Edsger W. Dijkstra

- "Obvious" is the most dangerous word in mathematics.—Eric Temple Bell
- Mathematics allows for no hypocrisy and no vagueness. —Stendhal
- I've always enjoyed mathematics. It is the most precise and concise way of expressing an idea.—N. R. Narayana Murthy
- It is impossible to be a mathematician without being a poet in soul.—Sofya Kovalevskaya
- A mathematician who is not also something of a poet will never be a complete mathematician.—Karl Weierstrass
- Mathematics is not about numbers, equations, computations, or algorithms: it is about understanding.—William Paul Thurston
- Somehow it's okay for people to chuckle about not being good at math. Yet, if I said "I never learned to read," they'd say I was an illiterate dolt.—Neil deGrasse Tyson
- In mathematics the art of proposing a question must be held of higher value than solving it.—Georg Cantor
- It is clear that the chief end of mathematical study must be to make the students think.—John Wesley Young
- Go down deep enough into anything and you will find mathematics.—Dean Schlicter
- Nature is written in mathematical language.—Galileo Galilei
- Mathematics has beauty and romance. It's not a boring place to be, the mathematical world. It's an extraordinary place; it's worth spending time there.—Marcus du Sautoy
- To me, mathematics, computer science, and the arts are insanely related. They're all creative expressions.—Sebastian Thrun
- The essence of mathematics lies in its freedom. —Georg Cantor
- Why do children dread mathematics? Because of the wrong approach. Because it is looked at as a subject. —Shakuntala Devi
- The study of mathematics, like the Nile, begins in minuteness but ends in magnificence. —Charles Caleb Colton

- Wherever there is number, there is beauty.—Proclus
- Life is a math equation. In order to gain the most, you have to know how to convert negatives into positives.—Anonymous
- Mathematics may not teach us to add love or subtract hate, but it gives us hope that every problem has a solution.—Anonymous
- One of the endlessly alluring aspects of mathematics is that its thorniest paradoxes have a way of blooming into beautiful theories.—Philip J. Davis
- The pure mathematician, like the musician, is a free creator of his world of ordered beauty.—Bertrand Russell
- Just because we can't find a solution, it doesn't mean there isn't one.—Andrew Wiles
- Mathematics is a place where you can do things which you can't do in the real world.—Marcus du Sautoy
- Millions saw the apple fall, but Newton asked why.—Bernard Baruch
- The definition of a good mathematical problem is the mathematics it generates rather than the problem itself.—Andrew Wiles
- If I were again beginning my studies, I would follow the advice of Plato and start with mathematics.—Galileo Galilei
- Pure mathematicians just love to try unsolved problems—they love a challenge.—Andrew Wiles
- I've always been interested in using mathematics to make the world work better.—Alvin E. Roth
- The only way to learn mathematics is to do mathematics.—Paul R. Halmos
- Sometimes the questions are complicated and the answers are simple.—Dr. Seuss
- The essence of math is not to make simple things complicated, but to make complicated things simple.—Stan Gudder
- If people do not believe that mathematics is simple, it is only because they do not realize how complicated life is.—John von Neumann

- Mathematics is a game played according to certain simple rules with meaningless marks on paper.—David Hilbert
- Dear Math, please grow up and solve your own problems. I'm tired of solving them for you.—Anonymous
- I am still waiting for the day I'll use mathematics integration in real life.—Derrick Obedgiu
- Arithmetic is numbers you squeeze from your head to your hand to your pencil to your paper till you get the answer.—Carl Sandburg
- That awkward moment when you finish a math problem and your answer isn't even one of the choices.—Ritu Ghatourey
- Mathematics is like love; a simple idea, but it can get complicated.—Anonymous
- If there is a 50-50 chance that something can go wrong, then nine times out of 10 it will.—Paul Harvey
- Math is fun. It teaches you life and death information like when you're cold, you should go to a corner since it's 90 degrees there.—Anonymous
- Pure mathematics is the world's best game. It is more absorbing than chess, more of a gamble than poker, and lasts longer than Monopoly. It's free. It can be played anywhere—Archimedes did it in a bathtub.—Richard J. Trudeau
- Mathematics consists of proving the most obvious thing in the least obvious way.—George Pólya
- In mathematics, you don't understand things. You just get used to them.—John von Neumann
- There are two ways to do great mathematics. The first is to be smarter than everybody else. The second way is to be stupider than everybody else—but persistent.—Raoul Bott
- Five out of four people have trouble with fractions. —Steven Wright
- Mathematics is a hard thing to love. It has the unfortunate habit, like a rude dog, of turning its most unfavorable side towards you when you first make contact with it.—David Whiteland

- In real life, I assure you, there is no such thing as algebra.—Fran Lebowitz
- Mathematics expresses values that reflect the cosmos, including orderliness, balance, harmony, logic, and abstract beauty.—Deepak Chopra

I had a polynomial once. My doctor removed it. —Michael Grant, *Gone*

- The difference between the poet and the mathematician is that the poet tries to get his head into the heavens while the mathematician tries to get the heavens into his head.—G. K. Chesterton
- I know that two and two make four—and should be glad to prove it too if I could—though I must say if by any sort of process I could convert 2 and 2 into five it would give me much greater pleasure.—Lord George Gordon Byron
- But in my opinion, all things in nature occur mathematically.—René Descartes

It's like asking why is Ludwig van Beethoven's Ninth Symphony beautiful. If you don't see why, someone can't tell you. I know numbers are beautiful. If they aren't beautiful, nothing is.—Paul Erdős

This may be called syllogism arithmetical, in which, by combining logic and mathematics, we obtain a double certainty and are twice blessed.—Ambrose Bierce

- They shouldn't be allowed to teach math so early in the morning.—Kendare Blake
- Physics depends on a universe infinitely centered on an equals sign.—Mark Z. Danielewski
- With me, everything turns into mathematics.—René Descartes

- Infinite is a meaningless word: except—it states / The mind is capable of performing / an endless process of addition.—Louis Zukofsky
- A relativist is an individual who doesn't know the difference between an adjective and an adverb.—Bill Gaede
- The important thing to remember about mathematics is not to be frightened.—Richard Dawkins
- One of the endlessly alluring aspects of mathematics is that its thorniest paradoxes have a way of blooming into beautiful theories.—Carl Jung
- All mathematicians share a sense of amazement over the infinite depth and mysterious beauty and usefulness of mathematics.—Martin Gardner
- It is clear that the chief end of mathematical study must be to make the students think.—Martin Gardner
- Mathematics is not about numbers, equations, computations, or algorithms: it is about understanding. – John Wesley Young
- The only way to learn mathematics is to do mathematics. – Shakuntala Devi
- There's no reason to stereotype yourself. Doing math is like going to the gym—it's a workout for your brain and it makes you smarter.—Danica McKellar
- Mathematics is like love; a simple idea, but it can get complicated.—Jarod Kintz
- A man is like a fraction whose numerator is what he is and whose denominator is what he thinks of himself. The larger the denominator, the smaller the fraction.—Leo Tolstoy
- Mathematics, even in its present and most abstract state, is not detached from life. It is just the ideal handling of the problems of life.—Cassius Jackson Keyser
- Mathematics knows no races or geographic boundaries; for mathematics, the cultural world is one country. —Cassius Jackson Keyser

- Out of an infinity of designs a mathematician chooses one pattern for beauty's sake and pulls it down to earth.—Marston Morse
- Mathematics allows for no hypocrisy and no vagueness. —Marston Morse
- Mathematics compares the most diverse phenomena and discovers the secret analogies that unite them.—Joseph Fourier
- Pure mathematics is, in its way, the poetry of logical ideas. —Albert Einstein
- What is mathematics? It is only a systematic effort of solving puzzles posed by nature. —Stefan Banach
- Mathematics is the supreme judge; from its decisions there is no appeal. —Tobias Dantzig
- The laws of nature are but the mathematical thoughts of God. —Euclid
- Mathematics is the music of reason. - Euclid
- Mathematics is the queen of the science. —Carl Friedrich Gauss
- Mathematics compares the most diverse phenomena and discovers the secret analogies that unite them. —Karl Weierstrass
- The essence of mathematics is not to make simple things complicated, but to make complicated things simple. —Stan Gudder
- Geometry is knowledge of the eternally existent.—Pythagoras
- Mathematics is concerned only with the enumeration and comparison of relations.—Carl Friedrich Gauss
- Mathematics is the science of what is clear by itself. —Carl Jacobi
- Nature is written in mathematical language. —Carl Jacobi
- You don't have to be a mathematician to have a feel for numbers. —John Forbes Nash Jr.
- Only two things are infinite, the universe and human stupidity, and I'm not sure about the former. —Albert Einstein

- A mathematician is a blind man in a dark room looking for a black cat which isn't there.—Charles Darwin
- Dear Math, I'm sick and tired of finding your X. Just accept the fact that she's gone.—Unknown
- In the fall of 1972, President Nixon announced that the rate increase of inflation was decreasing. This was the first time a president used the third derivative to advance his case for re-election.—Unknown
- Mathematics is a game played according to certain simple rules with meaningless marks on paper. – Hugo Rossi
- The mathematics that refers to reality, is not certain, and when certain, it is not reality.—Ritu Ghatourey
- No creation of the human is more powerful and beautiful than mathematics.—Reza Noubary
- The poetry of logical ideas is nothing but mathematics.—Reza Noubary

2. Only If Mathematics Could Talk

"The Book of Nature is written in the language of mathematics."—Galileo Galilei

Like most mathematics teachers, I can do some calculations in my head. My friends often see that as a sign of being a good mathematician. I usually tell them that if you consider this an attribute of a good mathematician, then my calculator is an amazing mathematician. I also wonder why people make such an association. Who started this, and why? Anyway, let us talk about one of the many things that makes mathematics a universal communicating tool.

There are over four thousand languages and dialects in the world, and all of them have one thing in common: they are instruments for communication based on the use of sounds or conventional symbols, words, and sentences. Most also have a category for words representing nouns or objects, and a category for words representing verbs or actions. The more developed languages are also described in terms of a vocabulary of symbols or words, a grammar consisting of rules of how they may be used, and a syntax or propositional structure that places the symbols in linear structures.

Taking the commonality of the major languages as a starting point provides an interesting way of looking at the world of mathematics and its language. One model proposed in mid-'90s suggests that we may think about mathematical nouns, or objects, as being numbers, quantities, shapes, functions, patterns, data, and arrangements— items that comfortably map onto commonly accepted mathematics content strands. Mathematical verbs may be regarded as the four major actions that we ascribe to problem solving and reasoning: modeling and formulating, transforming and manipulating, inferring, and communicating. Taken as a whole, these four actions represent the process that we go through to formulate a problem and solve it.

Symbolic Language

As we know, a part of the English language is used for making formal mathematical statements and communicating definitions, theorems, proofs, word problems, and examples. Although the English language is a source of knowledge, it is not designed for doing mathematics and most other hard sciences. Mathematics is usually written in a symbolic form/language that is designed to express mathematical and complex scientific ideas and thoughts. In other words, the symbolic language developed to present and communicate mathematics is a special-purpose language. It has its own symbols and rules of grammar that are quite different from a language such as English. This special-purpose language consists of symbolic expressions written in the way mathematicians traditionally write them. A symbol is a typographical character. The symbolic language also includes symbols that are specific to mathematics. We can usually read expressions in this language in any mathematical article written in almost all languages.

An example of elements of symbolic language include ten digits: 0, 1, 2, . . ., 9. Symbols for operations: +, -, x, /. Symbols that "stand in" for values: x, y, z, ... Special symbols such as π, =, <, $\leq$, ... Here, nouns could be fixed things, such as numbers, or expressions with numbers: 73, 5(3-1/7). The verb could be the equals sign "=," or an inequality such as < or >. Pronouns could be variables like x or y, 5x- 6, x2y, 8/x, which could all be put together into a sentence such as $5x + 14 = 22$. Both English and symbolic language are used in mathematics writing and mathematical lectures.

Let Us Simplify

Recall that language is a type of abstraction (first order) used for communication. It is what enabled humans to pass their knowledge to the future generations and end up controlling the world. Although very useful, most languages have their own shortcoming and limitations in that they can only furnish a finite number of names and words. Additionally, the ordinary languages are not designed for describing, expressing, or explaining the complex scientific ideas and concepts. The symbolic language of mathematics, on the other hand, is equipped with tools for expressing and communicating complex scientific ideas and relationships. For example, unlike words, numbers have no limit. That is why we are better at identifying people by their social security numbers than by their names. In fact, these days, numbers are used more for identification than numeration. Just think about so many different items in a grocery store. Each has their own unique identification that include a great deal of information about them.

3. Mathematics of Humor

Mathematicians never die; they just lose some of their functions.
Many people wonder about a big philosophical questions

concerning the nature of humor. What is humor? What purpose, if any, does it serve? Why are some things funny, and others not? Why something is funny to me but not to you? Or to an American but not to a Japanese? What does humor have to do with mathematics, science, and both global and local logic? I have thought about these questions on and off for ages. After all, people with a good sense of humor are usually popular.

There are various theories of humor, from antiquity to the present day. Lots of eminent thinkers have had something to say here. There is a great deal of diversity, but one theme that keeps coming back is incongruity. Humor arises when two radically different ways of looking at something are juxtaposed. Though incongruity is not enough on its own. There needs to be a point, and timing is very important. This all leads up to an interesting theory. Koestler argues that humor and creativity are closely linked: the patterns of thought are similar in both cases, though the result is different. When you lead listeners' minds in one direction and suddenly make them realize that there is an opposite direction. Or when you talk indirectly about things that are forbidden.

In an interesting book, *Mathematics and Humor*, John Allen Paulos discusses a very imaginative idea! He suggests that catastrophe theory might give us a mathematical tool that lets us understand humor. The idea is not as far-fetched as it may sound. Catastrophe theory, which he introduces in a simple and nontechnical way, is about discontinuous change in dynamic systems, where the system suddenly flips over to a new state as a result of a small change in the input parameters. The effect is irreversible: moving the input parameter back to where it was does not get you back to the previous state. The fundamental theorem of catastrophe theory states that, surprisingly, there are only a very small number of ways in which the discontinuous change can happen.

Paulos thinks this is what happens when we find something funny. We have two potential models for our theory, and as we acquire information, we initially consider one of the models to be the plausible one. In most cases, we do not even think of the other candidate. Suddenly, we get an extra piece of information that pushes us over the edge of the cusp, after which the second model immediately becomes the preferred one. Now we see the story differently, and we cannot go back to our earlier way of seeing it. He argues, reasonably persuasively, that this explains many of the things we notice about humor, including the importance of timing. If the information is presented in the wrong way, you give away the joke by revealing the second model too soon, and there is no discontinuity.

The book is cooler than you may imagine from reading the description above. He illustrates his ideas with plenty of jokes; also, he is well aware of all the things that might be wrong with his account and of the fundamental absurdity inherent in trying to reduce humor to mathematics. Indeed, he suggests that you should think of the book itself as a kind of joke. So read it as a Zen koan, whose purpose is to awaken you to a new view of the world that you had not previously even considered. It has worked on many levels.

In the study of folklore, a folk is defined as any group that has at least one thing in common. Some of these things include nationality, race, or profession. Mathematicians as a folk share a common core of mathematical folklore, which, like other folklore, exists in multiple forms and variations. A part of this folklore consists of different versions of classic jokes such as the absentminded mathematics professor joke that, through history, is attached to various mathematics legends.

The collection of mathematical folklore is often enjoyed not only by mathematicians and their students but by non-mathematicians as well. This is because every joke contains a portion of truth or lie about the mathematicians and mathematical pop culture.

It is important to note that once a joke goes to "public domain," it is often modified, and personalities in the joke are replaced by whoever people find them more interesting or can relate to. As such, it is neither appropriate nor necessary to attach an authorship to them. Furthermore, most of the collected sayings and jokes are repeated in newspapers, radios, TVs, and several webpages, which makes it difficult to credit a particular source. Some people even like to attribute jokes to well-known mathematicians to make them more interesting.

What Makes Us Laugh

Scientists have applied the logic commonly used to explain our sense of humor and have tried to find out how our brain reacts to a joke that makes us laugh. Several hypotheses of humor have been proposed in an attempt to explain this. Take, for example, words that sound like number 8, which can also sound like the verb "ate," with multiple meanings that depend on their context. Ask a five-year-old, "Why was 6 afraid of 7?" and say, "Because 7-8-9!" and watch as you suddenly become their favorite stand-up comic.

Some studies have attempted to explore the core elements of humor and apply a new way of mapping and evaluating the components of humor to determine exactly what makes a joke funny. "Funniness is not a pre-existing 'element of reality' that can be measured; it emerges from an interaction between the underlying nature of the joke, the cognitive state of the listener, and other social and environmental factors," says Liane Gabora from the University of British Columbia. Mathematics cannot describe physical properties of our brains. To test the theories of humor, the researchers broke the construct of a joke down into its components, including the setup, the person who is telling the joke, their relationship with the audience, and the surroundings.

Using the resulting formula, the researchers have applied various scores to weigh a joke's components and have predicted how people might find the overall structure funny. They have then come up with a list of jokes and created a number of variants for each one, such as delivering the punch line without a setup or by presenting it with a modification on its script. The jokes and their variants were tested on people such as undergraduate students, who gave them each a rating based on how funny they thought the joke was. For the most part, the variants weren't considered as funny as the original joke, but they did help the researchers pinpoint what exactly the audience found funny. Previous attempts to understand why puns make our lips pull back into a grin, our diaphragm spasm in laughter, and our brain release endorphins assumed the sudden change in meaning as the joke resolved was to blame. Here is an example of a joke and its variant:

- Mathematicians never die; they just lose their identities.
- Mathematicians never get crazy; they become irrational.
- Algebraic symbols are used when you do not know what you are talking about.
- When a statistician passes the airport security check, they discover a bomb in his bag. He explains, "Statistics shows that the probability of a bomb being on an airplane is 1/1000. However, the chance that there are two bombs at one plane is 1/1000000. So I am much safer . . ."
- Q: What does the zero say to the eight?
  A: Nice belt!
- If a statistician can have his head in an oven and his feet in ice, then, on the average, he will be fine.
- Q: Did you hear the one about the statistician?
  A: Probably . . .
- A mathematician believes nothing until it is proven. A physicist believes everything until it is proven wrong. A chemist doesn't care. A biologist doesn't understand the question.
- Golden rule for math teachers: you must tell the truth and nothing but the truth, but not the whole truth.
- Teacher: Now, suppose the number of sheep is x . . .
  Student: Yes sir, but what happens if the number of sheep is not x?
- This is a one-line proof . . . if we start sufficiently far to the left.
- There are two groups of people in the world: those who believe that the world can be divided into two groups of people, and those who don't.

- The less you know, the more you make.
- Why was the geometry book so adorable? Because it had acute angles.
- I saw my math teacher with a piece of graph paper yesterday. I think he must be plotting something.
- What did the triangle say to the circle? "You're pointless."
- What's a math teacher's favorite kind of tree? Geometry.
- Parallel lines have so much in common . . . It's a shame they'll never meet.
- Did you hear about the mathematician who's afraid of negative numbers? He'll stop at nothing to avoid them.
- I met a math teacher who had twelve children. She really knows how to multiply!
- Do you know what seems odd to me? Numbers that aren't divisible by two.
- Why was six afraid of seven? Because seven, eight, nine!
- What are ten things you can always count on? Your fingers.
- Why did the two fours skip lunch? They already eight!
- There's a fine line between a numerator and a denominator . . . But only a fraction would understand.
- Why did the student get upset when her teacher called her average? It was a "mean" thing to say.
- I poured root beer into a square cup. Now I have beer.
- Why was the math book so sad? Because it had so many problems.
- Why can't a nose be twelve inches long? Because, then, it would be a foot.
- What did one algebra book say to the other? "Don't bother me, I've got my own problems."
- Mathematics and alcohol do not mix, so please do not drink and derive.
- The problems for the exam will be similar to the ones discussed in the class. Of course, the numbers will be different. But not all of them. Pi will still be 3.14159 . . .

About the Book by Paulos

I've always felt that one of the really big philosophical questions that concerns the nature of humor is, what is humor? What purpose, if any, does it serve? Why are some things funny, and others not? I've thought about this stuff on and off for ages.

The author starts by giving you a quick tour through various theories of humor, from antiquity to the present day. Lots of eminent thinkers have had something to say here: Plato, Hobbes, Kant, Hazlitt, Schopenhauer, and Bergson all get quoted. There is a great deal of diversity, but one theme that keeps coming back is incongruity. Humor arises when two radically different ways of looking at something are juxtaposed. Though, as the author immediately notes, incongruity isn't enough on its own. There needs to be a point, and timing is very important. This all leads up to his own theory, whose immediate predecessor is Koestler, in *The Act of Creation*. Koestler argues that humor and creativity are closely linked: the patterns of thought are similar in both cases, though the result is different.

He tells you about formal theories and models, and how a formal theory can have models that are very different. He also talks about self-reference and grammar. The grammar is necessary, however, when he wants to discuss puns and plays on words.

Finally, he gets to the point, and a very imaginative one it is too! He suggests that catastrophe theory might give you a mathematical tool that lets you understand humor. The idea isn't as far-fetched as it may sound. Catastrophe theory, which he introduces in a simple and nontechnical way, is about discontinuous change in dynamic systems, where the system suddenly flips over to a new state as a result of a small change in the input parameters. The effect is irreversible: moving the input parameter back to where it was doesn't get you back to the previous state. The fundamental theorem of catastrophe theory states that, surprisingly, there are only a very small number of ways in which the discontinuous change can happen.

4. Mathematicians Laugh Too

To the mathematician who discovered the zero, thanks for nothing.
A mathematician who is afraid of negative numbers would stop at
nothing to avoid them.
Calculus jokes are all derivative trigonometry jokes are too graphic.
Arithmetic jokes are pretty basic algebra jokes are usually formulaic.

It is generally recognized that once a joke goes to "public domain," it is modified or is changed to fit to a beloved teacher or to legendary figures. As such, there should be no authorship in it. In addition, majority of the jokes are repeated in a number of web pages, which makes it difficult to credit a particular Internet source. So thanks to all Internet collectors of mathematics jokes.

Some Beginner and Intermediate Math Jokes

I will do algebra, I will do trig
I will solve any problem, small or big
I will do geometry; I will do computation for sure
Only graphing is where I draw the line. That is the trick.
(However, graphing is where I draw the line you know what for.)

- Why was the math book sad? It had many problems.
- Spelling book to math book? "I know I can count on you!"
- Why did the two fours skip lunch? They already eight!
- Why was the fraction worried about marrying the decimal? Because she would have to convert.
- Be humble like equal sign. You are not greater than or less than others.
- What do you call couple 7 and 9? The odd couple.
- Why do plants hate math? It gives them square roots.
- There are three kinds of people in the world: those who can count and those who cannot.

- How do you stay warm in any room? Just huddle in the corner, where it is always ninety degrees.
- Why does nobody talk to circles? Because there is no point.
- Did you hear about the mathematician who is afraid of negative numbers? She would stop at nothing to avoid them.
- After a sheepdog chased all the sheep into the pen, he told the farmer, "All forty accounted for."
  "But I only have thirty-six sheep," the farmer replied.
  "I know," said the sheepdog. "But I rounded them up."
- Why did the triangle make the basketball team over the square? He always made three-pointers.
- Why did the kid always wear glasses during math class? They improve di-vision.
- Why cannot you trust a math teacher holding graphing paper? They must be plotting something.
- What do baby parabolas drink? Quadratic formula.
- Some statistics joke are outlier.
- Pi and an imaginary number were fighting:
  "Get real," pi said.
  "Be rational," the imaginary number said.
- Why did the mathematician spill all of his food in the oven? The directions said, "Put it in the oven at 180."
- An engineer thinks that his equations approximate reality. A physicist thinks reality approximates his equations. A mathematician does not care about reality.
- "Mathematics is the art of giving the same name to different things."—J. H. Poincare
- I do not think—therefore, I am not.
- "A mathematician is a blind man in a dark room looking for a black cat, which is not there."—Charles R. Darwin
- Algebraic symbols are used when you do not know what you are talking about.
- Philosophy is a game with objectives and no rules. Mathematics is a game with rules and no objectives.

- A mathematician believes nothing until it is proven. A physicist believes everything until it is proven wrong. A chemist does not care. A biologist does not understand the question.
- Golden rule for math teachers: you must tell the truth and nothing but the truth, but not the whole truth.
- Q: What do you get when you add two apples to three apples? A: A senior high school math problem.
- Teacher: Now, suppose the number of sheep is x . . .
  Student: Yes, sir, but what happens if the number of sheep is not x?
- "There are four airplanes flying, and then two more airplanes join them. How many airplanes are flying now?"
  The child says, "I know, of course, that $4 + 2 = 6$, but I cannot figure out what the airplanes have do with this!"
- "This is a one-line proof . . . if we start sufficiently far to the left."
- Of course, the numbers will be different. But not all of them. Pi will still be 3.14159 . . ."
- Sex and drugs? They are nothing compared with a good proof!
- A lecturer: "Now we will prove the theorem. In fact, I will prove it all by myself."
- The highest moments in the life of a mathematician are the first few moments after one has proved the result, but before one finds the mistake.
- Golden rule of deriving: never trust any result that was proved after 11:00 p.m.
- Relations between pure and applied mathematicians are based on trust and understanding—pure mathematicians do not trust applied mathematicians, and applied mathematicians do not understand pure mathematicians.
- Some mathematicians become so tense these days that they that they do not go to sleep during seminars.

- "If I have seen further than others, it is because I was standing on the shoulders of giants."—Isaac Newton
- These days, even the most pure and abstract mathematics is in danger to be applied.
- The reason that every major university maintains a department of mathematics is that it is cheaper to do this than to institutionalize all those people.
- Interesting theorem: All positive integers are interesting.
  Proof: Assume the contrary. Then there is a lowest non-interesting positive integer. But, hey, that is interesting! A contradiction.
- Theorem: There are two groups of people in the world: those who believe that the world can be divided into two groups of people, and those who do not.
- There are three kinds of people in the world: those who can count and those who cannot.
- There really are only two types of people in the world: those that do not do math and those that take care of them.
- "The world is everywhere dense with idiots."
- Cat Theorem: A cat has nine tails.
  Proof: No cat has eight tails. A cat has one tail more than no cat. Therefore, a cat has nine tails.
- Q: How do you tell that you are in thehands of the Mathematical Mafia?
  A: They make you an offer that you cannot understand.
- "The number you have dialed is imaginary. Please rotate your phone ninety degrees and try again."
  Q: How many times can you subtract 7 from 83, and what is left afterwards?
  A: I can subtract it as many times as I want, and it leaves 76 every time.
- Pope has settled the continuum hypothesis! He has declared that cardinals above eighty have no powers.
- A circle is a round straight line with a hole in the middle.
- Q: Why did the mathematician name his dog Cauchy?

- A: Because he left a residue at every pole.
- Q: What does the zero say to the eight?
  A: Nice belt!
- Life is complex: it has both real and imaginary components.
- "Divide fourteen sugar cubes into three cups of coffee so that each cup has an odd number of sugar cubes in it."
  "That's easy: one, one, and twelve."
  "|But twelve is not odd!"
  "Twelve is an odd number of cubes to put in a cup of coffee . . ."
- A statistician can have his head in an oven and his feet in ice, and he will say that, on the average, he feels fine.
- Q: Did you hear the one about the statistician?
  A: Probably . . .
- Q: What is a dilemma?
  A: A lemma that proves two results.
- Some say the pope is the greatest cardinal. But others insist this cannot be so, as every pope has a successor.
- Was General Calculus a Roman war hero?
- "What is your favorite thing about mathematics?"
  "Knot theory."
  "Yeah, me neither."
- In Alaska, where it gets very cold, pi is only 3.00. As you know, everything shrinks in the cold. They call it Eskimo pi.
- A famous mathematician was to give a keynote speech at a conference. Asked for an advance summary, he said he would present a proof of Fermat's Last Theorem—but they should keep it under their hats. When he arrived, though, he spoke on a much more prosaic topic. Afterward, the conference organizers asked why he said he would talk about the theorem and then did not. He replied this was his standard practice, just in case he was killed on the way to the conference.

Note. The followings few poems were posted by different people in different sites so that it was not possible to find the original sources.

'Tis a favorite project of mine
A new value of pi to assign.
I would fix it at 3, for its simpler, you see.
Than 3 point 1 4 1 5 9

If inside a circle a line
Hits the center and goes spine to spine
And the line's length is "d"
The circumference will be d times 3.14159

A challenge for many long ages.
Had baffled the savants and sages.
Yet at last came the light:
Seems old Fermat was right—to the margin add 200
pages.

Integral z-squared dz
from 1 to the cube root of 3
Times the cosine of three pi over 9
equals log of the cube root of "e."

The poem below was written by John Saxon to explain this equality:

$$((12 + 144 + 20 + (3 * 4^{(1/2)})) / 7) + (5 * 11) = 9^2 + 0$$

A Dozen, a Gross and a Score,
plus three times the square root of four,
Divided by seven plus five times eleven,
equals nine squared and not a bit more.

A graduate student from Trinity
Computed the cube of infinity;
But it gave him the fidgets to write down all those
digits,
So he dropped math and took up divinity.

A conjecture both deep and profound
Is whether the circle is round;
in a paper by Erdos written in Kurdish,
A counterexample is found.

There once was a number named pi.
Who frequently liked to get high.
All he did every day was sit in his room and play.
With his imaginary friend named i.

There once was a number named e.
Who took way too much LSD.
She thought she was great.
We know she was not greater than 3.

There once was a log named Lynn.
Whose life was devoted to sin.
She came from a tree whose base was shaped like an e.
She is the most natural log I have seen.

More Jokes

- It is only two weeks into the term that, in a calculus class, a student raises his hand and asks, "Will we ever need this stuff in real life?"
  The professor gently smiles at him and says, "Of course not—if your real life will consist of flipping hamburgers at McDonald's!"
- Q: What does the PhD in math with a job say to the PhD in math without a job?
  A: Paper or plastic?
- What is the difference between an argument and a proof?
  An argument will convince a reasonable man, but a proof is needed to convince an unreasonable one.

There are also parodies of the logic utilized in mathematical proofs:

- Theorem. A cat has nine tails.
Proof. No cat has eight tails. A cat has one more tail than no cat. Therefore, a cat has nine tails.
- An engineer thinks that his equations are an approximation to reality. A physicist thinks reality is an approximation to his equations. A mathematician doesn't care.
- A mathematician, a physicist, and an engineer were traveling through Scotland when they saw a black sheep through the window of the train. "Aha," says the engineer, "I see that Scottish sheep are black." "Hmm," says the physicist, "you mean that some Scottish sheep are black." "No," says the mathematician, "all we know is that there is at least one sheep in Scotland, and that at least one side of that one sheep is black!"
- Here, the mathematician again gives impractical answers to real-world questions: Three men are in a hot-air balloon. Soon, they find themselves lost in a canyon somewhere. One of the three men says, "I've got an idea. We can call for help in this canyon, and the echo will carry our voices far." So he leans over the basket and yells out, "Helloooooo! Where are we?" (They hear the echo several times.)
Fifteen minutes later, they hear this echoing voice: "Hellooooo! You're lost!!"
One of the men says, "That must have been a mathematician."
Puzzled, one of the other men asks, "Why do you say that?" The reply: "For three reasons: (1) He took a long time to answer, (2) he was absolutely correct, and (3) his answer was absolutely useless."
- Three statisticians go out hunting together. After a while, they spot a solitary rabbit.
The first statistician takes aim and overshoots. The second aims and undershoots.

The third shouts out, "We got him!" (Source: chjilloutdamnit / Reddit)

- There was a statistician that drowned crossing a river . . . It was three feet deep on average. (Source: anatiferous_outlaw / reddit)
- I often use this joke in the class to assess their alertness: A man went to a pizza place and asked the waiter to bring a plain pizza. "Do you want me to cut it for you, sir?" "Yes." "Eight pieces or six?" "Make it six, I do not think I can eat eight."
- In a true-false test, the teacher asked the following question: The answer to this question is false: True or false?
- Husband: Research shows that a woman speaks twice as much as a man.
  Wife: Well, that is because, for you to listen, we need to repeat everything twice.
  Husband: What did you say?
- Patient: Doc. you are amazing. You cured my hearing problem.
  Doctor: Thanks. Your bill comes to $500.
  Patient: What? I cannot hear you.

5. My Poems about Mathematics

Mathematics is a beautiful and well-defined symbolic language
Symbols are its building block, together with well-
thought definitions they form a package
It plays a key role in abstractions higher than
the ordinary language we speak
A powerful communication tool that science, especially
hard sciences, utilize to their advantage.

As far as the laws of mathematics refer to reality,
they are not certain, that is for sure

As far as they are certain, they do not represent
reality only its perceived image, nothing more
Mathematics models only the smooth part of the real
world, it deals and presents theoretical expectation
Real-world observation always deviates from expectation,
that is the part we need to study deeper, that is the core.

Like many others I think poetry has some similarities with math
They follow parallel structures and expressions with a similar stand
Poetry utilizes the least number of words,
math the least number of symbols
They respectively describe feelings and concepts
using as much elements as they demand.

Like others, I learned addition and subtraction when I was a child
I thought I knew a great deal of math, I could
do it all, did not need any help or guide
I did not know that it was such a wide subject, something
that could make even logical minds go wild
A subject that many have some problems with,
they find it useful but very hard.

Most people I know separate mathematics from the daily life
But to me it has a lot to do with it, I can
picture it easily unlike my wife
I cannot imagine a life separate from math even
though it may not be directly evident
If you use it in your daily life, you will see its
usefulness, it works like a sharp knife

Math helps you to become a problem solver,
it helps you to develop a logical mind
It provides you with a tool to deal with this complex
issue, feel confident, and be able to rewind.

For some people mathematics is feared for its complexity
For others, it is enjoyed for its exactness and explicitly
I enjoy its beauty, elegance, as well as its truth and simplicity
The world likes it for its helpfulness, usefulness, and necessity.

When I watch elegant mathematics in action
It is hard for me not to admire its beauty and sophistication
A symbolic language useful to all human beings regardless
A universal tool familiar to all that
wherever you go it will function.

Hey Math, hope you see or feel my crazy love
You are what I always think about, you are my real drive
I admire you for being truthful and straightforward
You never change color, you ever cheat
or fool, yet you always evolve.

I am terrible in math and I say that proudly
I think that makes me more acceptable socially
Modern life is more brain than body as you know
That gives nerds an upper hand, they are
not losers any longer, hardly.

Hi Math, I wonder why people do not understand you
You help them every single day, yet those
who really respect you are only a few
Some are unaware, some ungrateful, and do
not value what you really offer and do
They do not realize that you are the mother
of science, you help us to go through.

People often do not understand you my dear friend, Math
You help them every day, yet they do not appreciate it. That is sad

They do not realize how hard life would
be without you and your help
That is the altitude that make most math lovers like me really mad.

It seems that brilliant mathematicians of the last
centuries pushed the math to its peak
The contribution of recent mathematicians, though
significantly more, is relatively weak
Does this mean that we humans have arrived
at our possible limit in math?
An interesting question, hard to answer, I do
not see much hope, future looks bleak.

It is odd that half of the counting numbers are odd
Even and odd numbers are like the main colors, black and white
Like stars in the sky, some numbers appear brighter than others
Some are made up of smaller numbers and they can always divide.

Why is eight afraid of seven even though it is one unit bigger?
Because seven ate nine even though nine is a greater figure
Now I see that ten is shaking even though
it's a double-digit number
I guess, because it is sitting next to the nine
and it is almost time for dinner.

My friend, have you ever liked to study some basic math?
I think some see math like a cat and themselves as a rat
I always liked mathematics though some of
my friends did not consider that cool
For me math always looked like a beautiful
world, I am grateful and glad.

My friend looked at the pictures of his past classmates and said hi
That reminded him of high school algebra that often made him cry

He had such bad memories after he failed it
twice, he was so sad, he wanted to die
Now that he is older, it comes very easy to him,
I do not understand, I wonder why?

Would you like to be a number in the
multiplication table of numbers?
That means to be treated the same as other numbers as members
Do you think this is a good question to
pose, yes or no? not easy to decide
It is one of those questions, the answer
lives in the world of wonders.

Before you judge others, walk in their shoes, try to walk their path
Think twice, be fair, do research, use your mind, and have class
Others like you are trying to do their best to go through their lives
Nothing is always true or false, right or wrong, life is not like math.

I like to do some addition, but you clearly prefer subtraction
I want to do some multiplication, but I see that you prefer division
I talk about problems like dissatisfaction,
but you prefer rationalization
I like to do some scientific experimentation,
but you clearly lack ambition.

Where should I look for truth; in church or in a math book?
Where should I look for love; in a bar or search for a romantic look?
Where should I look for work; Internet,
newspaper, or bulletin board?
Where should I look for a friend; in
school, at work, or on Facebook?

Mathematics, you are certainly far more than numbers
Why do people often associate you to them, I wonder?
You always speak the truth and nothing else but truth

Education about you should get more
attention, you are a great partner.

There is a secret in everything named x
Seeking to reveal itself to the wisdom
It is a closed door waiting to be opened
Wishing for a smart mind to enter its kingdom.

In the end of a proof, there is a big earning
In the processes, there is a lesson for learning
The action possesses more joy than one
It will make you feel that you are in the point of turning.

In a math book, there is wisdom
In a theorem, there is a kingdom
No solution can be made up or faked
The reasoning buys you your freedom

Mathematics does not have a season
It is valued for a its special reason
It shows you the direction when you are lost
It opens your mind and takes you out of prison.

Mathematics has its own special kingdom
The key to enter it is called brain or wisdom
Membership requires love, passion, and a desire to learn
What it offers is beyond a simple mental freedom.

There is a secret in every equation that provokes curiosity
and passion
To uncover it we need to see its beauty and
realize it is worth to complete the mission
The outcome is refreshing, it always results
in a long-term compassion

It is satisfying as participants understand its
logic and its beautiful expression.

1, 2, 3, 4, 5, 6, 7, 8, 9
They are building block of the decimal system, they all look fine
Other than digits representing numbers,
they sometimes are used as a sign
They like each other, together they have
fun, they frequently wine and dine.

9, 8, 7, 6, 5, 4, 3, 2, 1
They are all of the same age neither of them is young
7 is the boss, because of that famous joke I told my young son
7 ate 9 is the reason, that is what people say that was done
Numbers stay together as if they are family, they are one
None of them hate others, none tries to hit the others and run.

Why can people in math problems have 60
bananas and a huge watermelon
Yet, me drinking fifteen beers at night is considered
a problem, I may be considered a felon?
I am the black sheep of the family, everybody
wants to give me a lecture and advice
But nobody cares to complain to math teachers
or make a story about it my son.

Binary and I
When I work with binary numbers, I feel
at home, somewhere I have been
When I do mathematical manipulations, I feel
comfortable, something I have seen

When I think about binary numbers, I feel I
am watching a black and white movie

It brings back both bad and good memories,
the times I felt depressed or groovy

I feel that I am living in a world with my dear
friends known as And, Or, and Not
Where everything has two faces, just present
or absent, things between are cut.

0
10
001
01011
1100001
010110011
11000101010101

Probability

The probability theory investigates the laws
concerning phenomena influenced by chance
It is the branch of science whose goal is to do things
like quantitative inference for instance
It is also a measure of chances or certainties
about occurrence of the event of interest
It has applications in all real-life problems and decisions
that involve uncertainty, such as investing and insurance
In recent years it is used extensively as
our understanding is enhanced
The results in new ways of looking at the
world, a significant advance.

Statistics

Statistics is the science of learning about the whole
from a part known as data or sample
For instance blood in your entire body is the whole,
a drop of it is the part that is a nice example

Elements of statistics are population, sample,
inference, and the margin of error
The methods of analysis use inductive reasoning,
plus some form of art that is ample.

Central Tendency

There are sets of numbers known as measures for central tendency
The popular ones are mean, median and
mode values for expectancy
Each has its advantages and shortcomings, as they should
If you pay attention you find out what
they are, you certainly would.

Mean

To find the mean simply add all the
measurements in your sample first
Then take the sum and divide it by the
number of measurements next
Mean is greatly affected by outliers, so be
careful when you are applying it.
It is misleading, as it is not the 50-50 point
that you see it if you just test.

Median

To find median, order the measurements in ascending order first
Pick the measurement in the middle of the list that comes next
Unlike mean, median is not affected by the outliners, that is a good
But then it does not utilize all the available
information as it should.

Mode

To find the mode, just look at a measurement
with the highest frequency
The one that you may consider a typical value
that occupies the biggest vacancy

You may have more than one sample mode that may be a problem
Again, we are not using all the information,
especially those with high infrequency.

Trimmed Mean

Trimmed mean combines the ideas from mean and median
It arranges the measurements in ascending
order and drops extremes then
The number of the measurements dropped
is often applied to both ends of data
It eliminates the effects of outliers and finds
the mean of the rest that is the plan.

Variability

There are two frequently used measures for spread or variation
One is known as range and the other is
well-known standard deviation
Range is the difference between the largest
and smallest measurement
Simple but not a good measure as it ignores
all the details it needs for betterment
Standard deviation is average deviation
from the mean or expectation
The deviations of all measurements from the
mean are squared and added for calculation
Next the square root of the sum is divided by
the sample size minus one then is taken
The final number, which is zero or larger,
is the value of standard deviation
The samples with a larger standard deviation have a larger spread
Data with large standard deviations creates
problem for inference, a bad situation.

Math puzzles

Critical thinking and skill to logic are important and even necessary. Puzzles in general and math puzzles in particular challenge one's mind to understand and learn logical thinking. They also strengthen the capacity of abstract thinking. The following is a typical example.

- There are twelve coins. One of them is false; it weighs differently. It is not known if the false coin is heavier or lighter than the right coins. How to find the false coin by three weighs on a simple scale?

Brainteasers

While brainteasers do not always deal directly with math skills, they are important tools in the development for critical thinking. Brainteasers are more than just simple puzzles and riddles, as they often involve lateral thinking, which means being creative. Below are some examples:

- You are in a room that has three switches and a closed door. The switches control three light bulbs on the other side of the door. Once you open the door, you may never touch the switches again. How can you definitively tell which switch is connected to each of the light bulbs? Answer: Turn on the first two switches. Leave them on for five minutes. Once five minutes have passed, turn off the second switch, leaving one switch on. Now go through the door. The light that is still on is connected to the first switch. Whichever of the other two is warm to the touch is connected to the second switch. The bulb that is cold is connected to the switch that was never turned on.

- A man is looking at a photograph of someone. His friend asks who it is. The man replies, "Brothers and sisters, I have none. But that man's father is my father's son." Who was in the photograph?
  Answer: His son.
- A man stands on one side of a river, his dog on the other. The man calls his dog, who immediately crosses the river without getting wet and without using a bridge or a boat. How did the dog do it?
  Answer: The river was frozen.
- What makes this number unique: 8,549,176,320?
  Answer: It has each number, zero through nine, listed in alphabetical order.
- What five-letter word becomes shorter when you add two letters to it?
  Answer: Short.
- Your parents have six sons including you, and each son has one sister. How many people are in the family?
  Answer: Nine—two parents, six sons, and one daughter.
- An Arab sheik is old and must leave his fortune to one of his two sons. He makes a proposition: both sons will ride their camels in a race, and whichever camel crosses the finish line *last* will win the fortune for its owner. During the race, the two brothers wander aimlessly for days, neither willing to cross the finish line. In desperation, they ask a wise man for advice. He tells them something; then the brothers leap onto the camels and charge toward the finish line. What did the wise man say?
  Answer: The rules of the race were that the owner of the camel that crosses the finish line last wins the fortune. The wise man simply told them to switch camels.
- Fourteen of the kids in the class are girls. Eight of the kids wear blue shirts. Two of the kids are neither girls or wear a

blue shirt. If five of the kids are girls who wear blue shirts, how many kids are in the class?

Answer: Nineteen.

- How can $8 + 8 = 4$?

  Answer: When you think in terms of time. 8:00 a.m. + 8 hours= 4:00 p.m.

- The water level in a reservoir is low but doubles every day. It takes sixty days to fill the reservoir. How long does it take the reservoir to become half full?

  Answer: Fifty-nine days. If the water level doubles every day, the reservoir on any given day was half the size the day prior. If the reservoir is full on day 60, that means it was half full on day 59, not on day 30.

- A farmer needs to take a fox, a chicken, and a sack of grain across a river. The only way across the river is by a small boat, which can only hold the farmer and one of the three items. Left unsupervised, the chicken will eat the grain, and the fox will eat the chicken. However, the fox will not try to eat the grain, and neither the fox nor the chicken will wander off. How does the farmer get everything across the river?

  Answer: The farmer must follow these steps. Take the chicken across the river. Come back with an empty boat. Take the grain across the river. Bring the chicken back. Take the fox across the river. Come back with an empty boat. Take the chicken across the river.

- A man describes his daughters, saying, "They are all blonde but two; all brunette but two; and all redheaded but two." How many daughters does he have?

  Answer: Three. A blonde, a brunette, and a redhead.

- There is a word in the English language in which the first two letters signify a male, the first three letters signify a female, the first four signify a great man, and the whole word, a great woman. What is the word?

  Answer: Heroine.

- What is next in this sequence of numbers: 1, 11, 21, 1211, 111221, 312211, _____?
Answer: 13112221. Each sequence of numbers is a verbal representation of the sequence before it. Thus, starting with 1, the next sequence would be "two ones," or "11." That sequence is followed by "two one," or "21," and so on and so forth.

- You are in a place called Wally's World, and there is only one law. There is a mirror, but no reflection. There is pizza with cheese, but not sausage. There is pepper, but no salt. There is a door, yet no entrance or exit. What is the law?
Answer: Each word in Wally's World must contain double letters.

- Four people arrive at a river with a narrow bridge that can only hold two people at a time. It's nighttime, and they have one torch that has to be used when crossing the bridge. Person A can cross the bridge in one minute, B in two minutes, C in five minutes, and D in eight minutes. When two people cross the bridge together, they must move at the slower person's pace. Can they all get across the bridge in fifteen minutes or less?
Answer: Yes, they can cross in exactly fifteen minutes.
The group of four must follow these three steps.
First, A and B cross the bridge, and A brings the light back. This takes three minutes.
Next, C and D cross, and B brings the light back. This takes another ten minutes.
Finally, A and B cross again. This takes another two minutes.

- In my hand, I have two coins that are newly minted. Together, they total thirty cents. One is not a nickel. What are the coins?
Answer: A quarter and a nickel.

- Find a number less than one hundred that is increased by one-fifth of its value when its digits are reversed.
Answer: Forty-five (1/5 of 45 = 9, 9 + 45 = 54).

- I have a large money box, ten inches wide and five inches tall. Roughly how many coins can I place until my moneybox is no longer empty?
  Answer: Just one, after which it will no longer be empty.

- An elevator is on the ground floor. There are five people in the elevator including me. When the lift reaches the first floor, one person gets out and two people get in. The lift goes up to the second floor, three people get out, five people get in. It then goes up to the next floor up, no one gets out, but twelve people get in. Halfway up to the next floor, the elevator cable snaps; it crashes to the floor. Everyone else dies in the elevator except me. How did I survive?
  Answer: I got off on the first floor.

- What four-letter word can be written forward, backward, or upside down, and can still be read from left to right?
  Answer: Noon.

- In reply to an inquiry about the animals on his farm, the farmer says, "I only ever keep sheep, goats, and horses. In fact, at the moment they are all sheep bar three, all goats bar four, and all horses bar five." How many does he have of each animal?
  Answer: The farmer has three sheep, two goats, and one horse. You can solve this easy math riddle with a quick hypothetical. Take sheep: we know that there are three animals that are goats and horses, so we suppose there are two goats and one horse. Checking this hypothesis gives us three sheep, which works out because there are four non-goats: three sheep, and one horse!

- One brother says of his younger brother, "Two years ago, I was three times as old as my brother was. In three years' time, I will be twice as old as my brother." How old are they each now?
  Answer: One way to solve this math riddle is to use even numbers: The older brother will be twice as

old as his younger brother in three years' time. This immediately rules out the older brother currently being eight, eleven, and fourteen, so he must be seventeen, and the younger brother seven. Two years ago, they were fifteen and five, respectively, and in three years' time, they will be twenty and ten.

- Old Granny Adams left half her money to her granddaughter and half that amount to her grandson. She left a sixth to her brother, and the remainder, $1,000, to the dogs' home. How much did she leave altogether?
  Answer: This one might have tripped you up! But the trick is not to focus on the hypothetical amounts but on the fractions: Adding one-half, one-quarter, and one-sixth tells us that the total is a fraction of twelfths (2 + 4 + 6 = 12). You can also think about it as 6/12, 3/12, 2/12, which equals 11/12. If the remainder is $1,000, that must be one-twelfth, so the total is $12,000.

- John noticed that the amount he was paying for his lunch was a rearrangement of the digits of the amount of money he had in his pocket, and that the money he had left over was yet another rearrangement of the same three digits! How much money did John start with?
  Answer: John started with $9.54. The money can be written with just three digits, so it must be between $1.01 and $9.99. Trial and error shows that there is only one set of numbers that fits this question: $9.54 = $4.59 + $4.95.

- What is the smallest number that increases by twelve when it is flipped and turned upside down?
  Answer: The answer is eighty-six. When it is turned upside down and flipped, it becomes ninety-eight, which is twelve more than eighty-six.

CHAPTER 2

Mathematics

What language other than mathematics can
explain the complexity of this universe?
Mathematics, a discipline where nothing is fake.
Mathematics, like journalism, is about interesting stories.
The only difference is that the stories have to be true.

FOR MANY, MATHEMATICS is anything but exciting or beautiful. For mathematics lovers, on the other hand, these are some of the many reasons to study the subject. According to G. H. Hardy, "The mathematician's patterns, like the painters paints or poets poems, is beautiful. In fact, the beauty is the first test." Paul Erdős expressed his views on the ineffability of mathematics when he said, "Why are numbers beautiful? It is like asking why is Beethoven's Ninth Symphony beautiful. If you do not see why, no one can tell you." Of course, not all the mathematics is beautiful. For most mathematicians, elegant mathematics is the limit of its beauty.

For some, the beauty of mathematics lies in the search for truth even more than discovering it.

> I am mathematics; some see me as a hero
> Some dislike me they see me of no zero
> I am the foundation of science and logical thinking
> Provide assistance to fact seekers that is true

1. What Is Mathematics After All?

Mathematics is an abstract science. It can be studied in its own right (pure mathematics) or as is applied to other disciplines (applied mathematics). It is known among mathematicians that "the relations between pure and applied mathematicians are based on trust and understanding." Pure mathematicians do not trust applied mathematicians, and applied mathematicians do not understand pure mathematicians. Mathematicians often seek out patterns and use them to formulate new conjectures. They then resolve the truth or falsity of them by mathematical proof. When outcomes seem to be good models of real phenomena, mathematical reasoning is used to provide insight about the situation. A mathematical model is an abstract model that uses symbolic language to describe the behavior of a system. It is a representation of the essential aspects of a system in usable form. They are used for simulating real-life situations to forecast their future behavior. Of course, models are not the same as the real thing.

For more than two thousand years, mathematics has been a part of the search for understanding the world. Mathematical discoveries have come both from the attempt to describe the natural world and from the desire to arrive at a form of inescapable truth from careful reasoning. Meaningful mathematics is like journalism—it tells an interesting story. But, unlike some journalism, the story has to be true. It is the study of quantity, structure, space, and change. The truth is established by rigorous deduction from appropriately chosen axioms and definitions.

Confusion

Most people relate to mathematics through numbers and often refer to their manipulation as mathematics and the manipulators as mathematician. Numbers manipulation, in fact, has not much to do with mathematics. If so, my calculator would be a great mathematician. Mathematics is partly a symbolic language for expressing and communicating complex ideas and relationships, and partly a perfect world arrived at by smoothing the rough world. Most of the smooth curves in calculus book only represent the ideal world or our expectations, but not the real world.

Recall that the ordinary language is incapable of describing, expressing, or explaining the complex scientific ideas and concepts. Additionally, no language is universal. Since mathematics is a man-made science, it is unique and free of the uncertainties of the real world. In fact, it is the only discipline where theorems and proofs and deductive reasoning are used with zero margin of error. Other disciplines are mostly based on observation or experimentations where generalization takes place using inductive reasoning. As such, there is always a margin of error. Physicist and engineers use mathematics to model specific problem, much like a tailor who makes suit. Mathematicians are like tailors who make all types of suits and let the user pick the one that fits the situation the best.

Mathematics is a concise language with well-defined rules for manipulations. Mathematical model is a description of a system using this language. The modeling includes method of simulating real-life situations with mathematical equations to forecast their future behavior. This definition suggests that modeling is a cognitive activity in which we think about and make models to describe devices, objects, entities, or state of affairs. It is representation of the essential aspects of systems to represent knowledge of that system in usable form.

In general, mathematical modeling has three potential uses, not necessarily independent:

1. A model may be *descriptive* in the sense that it synthesizes the available information on a process with no real attempt to explain the underlying mechanism.
2. A model may be also *explanatory* in the sense that it makes certain underlying assumptions about the process under study and derives the logical implications of those assumptions.
3. A model may be *predictive*—that is, it may be constructed for the purpose of predicting the future values or the response of the system to factors that have not been observed.

There is no such thing as the best model for a given phenomenon. Models are neither right nor wrong, but in the final analysis, a model is judged using a single, quite pragmatic factor, the model's usefulness. Models can, however, be judged as appropriate or inappropriate. Such a judgment, which must take into account the goals of the study and whatever qualitative information is available, benefits considerably from a clear statement of the model.

1. Multiplication Our Parents Learned in Schools

Multiplication and division as practiced in centuries ago had little in common with modern operations bearing the same names. Multiplication, for instance, was a succession of duplations, which was the name given to the doubling of a number. In the same way, division was reduced to mediation (i.e., "halving" a number). A clearer insight into the status of reckoning in the Middle Ages can be obtained from an example. Using modern nations:

Today	Thirteenth Century
46 x	46 x 2 = 92
13	46 x 4 = 92 x 2 = 184
138	
46	46 x 8 = 184 x 2 = 368
598	368 + 184 + 46 = 598

2. Mathematics of Dating

Although it may be hard to believe, mathematics could help you with your dating and mating problems. How? Consider, for example, the following question asked by someone who is seeking to meet the "right" one:

1. How many people should I date before committing to a long-term relationship?
2. How can I maximize my chance of finding the "right one"?

Here are some tips to find an answer. Suppose that you have a list of potential partners that you can potentially date and you are able to rank according to your personal taste and preferences. Also, suppose that after each date, your options are to commit or break up permanently (never date that person again).

Strategy 1

Let us assume that potential dates appear in a random order and you rank them according to your personal criterion. Here, you are looking for the date with highest rank (record rank) among potential dates. Here are a few details. Your first date is the first record as she/he is the best so far. At this point, you can either commit or date more, hoping that a future date would be even better. Both decisions involve risk. You may miss a better future date by not dating more or miss a the best by breaking up with a past date. Let us see how mathematics can help you. As mentioned, my first date is the first record, as it is the best so far. Any future date better than the first date is the second record, and that could be just your second date or third or even the very last one. If the first date is the best, you have encountered only one record. If your second date is the best, two records, and so on. If you date ten people, you could see one to ten records. Given the number of the potential dates, the probabilities of encountering certain number of records can be mathematically calculated. For ten potential dates, these probabilities are presented in the table below. For example, the chance of encountering only one record—that is, the chance that your first date would be the best—is 10 percent. The chance of encountering two records—that is, the chance that there would be only one future date better than your first date—is 28.29 percent. As we see, occurrence of three records in ten potential dates has the highest chance (31 percent).

This implies that if you keep dating until you meet the third record date, you maximize your chance of finding the best. In other words, your chance of meeting "the best" is largest if you wait to meet the third record date.

No. of Records	Probability
1	0.10000
2	0.28290
3	0.31316
4	0.19943
5	0.07422
6	0.01744
7	0.00260
8	0.00024
9	0.00001
10	0.0000

For one hundred dates, your chance is maximized (21 percent) if you wait to meet the fifth record.

Strategy 2

Here, you first date a certain number of people (initial group) and break up with them permanently. Then date more until you find a date better than the best in the initial group. If there was no such a date, you commit to your last date.

For ten potential dates, this strategy maximizes your chance of finding the best (77.4 percent) if you include three people in the initial group. For five and fifteen, your chance is maximized if you include two and five people in the initial group, respectively.

3. Is It My Anniversary? Should I Buy Flowers?

Men are known to forget "important" dates (e.g., their anniversary). Here is an example where men may be helped by mathematics.

What Is Expected Value?

Suppose that you flip a fair coin and win a dollar if head comes up and lose a dollar if tail comes up. If you play this game for few hours, at least theoretically you expected to break even—that is, neither win nor lose. This is because you expect to win half of the time and lose half of the time. Mathematically, this is calculated as $1/2 (+1) + 1/2 (-1) = 0$. These types of games are called fair games.

Now, suppose that John is not sure if today is his anniversary and is wondering whether he should buy flowers in the way home or not. He remembers his statistics class and decides to treat it as an expected value problem. He formulates the problem as follows: let A be the event that it is his anniversary and B be the event of buying flowers. Suppose that the probability that A is p. These are the four possibilities: is anniversary, buy flowers; is anniversary, do not buy flowers; is not anniversary, buy flowers; and is not anniversary, do not buy flowers. Knowing his wife and all that has happened in the past, he comes up with the following monetary payoffs for each attention and constructs the following payoff table:

	Anniversary(A)	No Anniversary(A')
Flowers(B)	10	-1
No Flowers(B')	-10,000	10

Now, if he buys flowers, then the expected return or payoff will be $10p - (1-p)$. This is because the possible values are 10 and -1 with respective probabilities p and 1 - p. If he does not buy flowers, the expected return or payoff will be $-10,000p + 10(1-p)$. To see for which values of p buying flowers result in a greater payoff, we should solve the following inequality:

$$10p - (1-p) > -10,000p + 10(1-p) \text{ or}$$
$$10,021p > 11 \text{ or } p > 1/911.$$

So the expected payoff for buying flower is larger if $p > 1/911$. Now, let us analyze this further.

1. First, there are 365 days in a year. Thus, the probability that any given day is his anniversary is $1/365 > 1/911$. So a man who has been married for several years should know this without any calculation.
2. 911 is a well-known number and has association with a sad and unexpected event that changed America.

5. Billion, a Misunderstood Number

We are used to hearing large numbers such as a billion mentioned frequently on the news. People and politicians such as Bernie Sanders are decrying "the millionaires and billionaires" as though these groups are one step apart from one another. Even people who work with numbers sometimes put millionaires and billionaires in the same category without pointing out their huge difference. Let me directly go to some revealing examples.

- If you gave your daughter $1 million and told her to go out and spend $1,000 dollars a day, she would be back in about three years for more money. If you gave her $1 billion, she would not be back for three thousand years.
- If you save $100 per day, it takes 27,397.26 years to reach $1 billion.
- If you and your future children live ninety years, it will take more than three hundred generations to save $1 billion.
- If you decide to count to one billion and take just three seconds to say each number, 95.1 years is how long it will take you to count to one billion.
- If you decide to take a billion two-foot-long steps, you would go more than fifteen times around the equator.
- A billion minutes ago, Jesus was alive. A billion days ago, no one walked on the earth on two feet.
- A billionaire is richer than thirteen nations in the world.
- If a billionaire stacks his fortune in hundred-dollar bills vertically, the resulting stack would ascend a staggering 3,585 feet! He could place a belt of dollar bills around Earth.
- One billion pennies stacked on top of each other would make a tower almost 870 miles high.
- The height of a stack of one billion dollar bills measures 358,510 feet or 67.9 miles. This would reach from Earth's surface into the lower portion of the troposphere— one of the major outer layers of Earth's atmosphere.

- The area covered by one billion dollar bills measures four square miles. This is an area equal to the size of 2,555 acres.
- The length of one billion dollar bills laid end to end measures 96,900 miles. This would extend around Earth almost four times.

If Only Billionaires Help

Now, think about issues such as health insurance. Suppose that the cost of a group insurance for a family is $10,000 per year. For ten thousand families, the cost comes to 10,000 x $10,000 = $100,000,000 per year, or $1 billion for the next ten years. This respectively is 0.55 percent, 0.9 percent, and 1.25 percent of the wealth of the top billionaires—Jeff Bezos, Bill Gates, and Warren Buffet. Remember that these people sometimes express their concerns for the poor and middle class.

Final Words

Despite their heavy presence in our societal psyche, there are only around 550 billionaires in the United States. Surprisingly, the word "billionaire" has become unremarkable in the American vernacular. There are about fifteen million people with net worth over $1 million. This group is often referenced alongside billionaires. Of course, some of us understand that there is a massive difference between these two parties. Using some examples, we tried to make sense of the gap in wealth of a person with $1 million and a person with$1 billion.

6. Pascal's Argument regarding God

This interesting application is based on Blaise Pascal's philosophical argument (1660) regarding God. Let E denote the event that God exists, and B denote the event that a person believes that God exists. Suppose that p, the probability that God exists, is a number between zero and one. Now, each person has options of becoming a believer or nonbeliever. This leads to four possibilities: God exists and person is believer, God exists and person is nonbeliever, God does not exist and person is believer, God does not exist and person is nonbeliever. Suppose that the payoffs of these options can be expressed in dollars as follows:

	God Exists	God Does Not Exists
Believer	100,000	- 10,000
Non-Believer	- 100,000.000.00	100,000

Now, in this case, before discussing further, let us repeat the example used in the previous section in case you did not read it. Suppose that you flip a fair coin and win a dollar if head comes up and lose a dollar if tail comes up. If you play this game for few hours, you expect to break even—that is neither win nor lose. This is because you expect to win half of the time and lose half of the time. Mathematically, this is calculated as $1/2 (+1) + 1/2 (- 1)=0$.

Now, the expected payoffs are $10^5 p - 10^4 (1 - p)$ for a believer and $-10^{10} p + 10^5 (1 - p)$ for a non-believer. The two expected payoffs are equal if $p=1/90911=0.000109$. That implies that the expected payoff is higher for believers if the chance that God exists is bigger than just 0.000109, almost one in ten thousands. In fact, Pascal's original argument uses infinite rather than -10^{10} and concludes that believing in God has a higher payoff for any value of p, no matter how small.

7. Bell Curve, a Law of the Nature

Where the four most famous numbers of
mathematics—0, 1, pi, and Euler's e—meet.

With its symmetrical shape, central peak, and gracefully sloping sides, the bell curve is one of the most important and best-known graphs to both scientists and the general public. It presents distribution of any measurable attribute that has common values around the center and rarer values to either side. From stock market jitters to human heights and IQ, many real-life phenomena follow, at least approximately its pattern. The standard normal often denoted as N (0, 1) has the four most amazing numbers—namely zero, one, Pi (the ratio of circles circumference and its radars), and the Euler number e (the base of the natural logarithm)—in its representation. The area under the curve of a normal distribution represents the probabilities of obtaining possible value of a variable.

Why Special?

We typically observe large-scale phenomena that arise from the interactions of many hidden small-scale processes. It is like looking down on Earth from an aircraft flying far above it. We only see the big items and green spaces without noticing details and movements. The macroscopic patterns we see often vary around a characteristic shape of the area. This is because most observable patterns of nature arise from aggregation of numerous small-scale processes. Self-similar patterns like coldly flowers are called fractals. We might say that as the number of entities contributing to the aggregate increases, they converge in the limit to distributions that define the common patterns of nature.

The best known of the limiting distributions, the Gaussian (normal/bell curve) distribution, follows from a celebrated theorem known as the central limit theorem. The observed outcome of atrocities such as height, weight, or yield arises from the summing up of many small-scale processes, and as such, the distribution typically approaches the bell curve. In fact, based on the central limit distribution, any aggregation of processes that preserves information only about the mean and variance attracts to the normal pattern. Further examples include variations in cancer onset that arise from variable failures in the many individual checks and balances on DNA repair, cell cycle control, and tissue homeostasis. Variations in the ecological distribution of species follow the myriad local differences in the birth and death rates of species and in the small-scale interactions between particular species.

The normal curve has many other names including the error curve. The bell curve achieved notoriety in 1994 as the title of a book by Richard Herrnstein and Charles Murray, which was accused of lending support to the view that certain races are inherently intellectually inferior to others, for which the authors received heavy criticism.

Note that in all complex systems, we wish to understand how large-scale pattern arises from the aggregation of small-scale processes. The central limit theorem is widely known, and the Gaussian distribution is widely recognized as an aggregate pattern. This limiting distribution is so important that one could hardly begin to understand patterns of nature without an instinctive recognition of the relation between aggregation and the Gaussian curve. The equation presenting the normal distribution is complex with some undesirable mathematical properties.

Final Words

The widespread use of normal distributions arises for two reasons. First, many measurements concern fluctuations about a central location caused by perturbing factors or by errors in measurement. Second, in formulating a theoretical analysis of measurement and information, an assumption of Gaussian fluctuations is the best choice when one has information only about the precision or error in observations with regard to the average value of the population under observation.

Bell Curve, a Different Account

With its symmetrical shape, central peak, and gracefully sloping sides, the bell curve is one of the most important, well-studied, and best-known graphs for scientists, businesspeople, and the general public. It presents the distribution of any measurable attribute that has its more likely values around the center, less likely values to either side, and least likely values in the tails. It is used frequently partly because of the fact that distribution of many real-life variables follow its pattern, at least approximately. The standard normal curve N(0, 1), with mean 0 and standard deviation 1, has the four most amazing numbers—zero; one; Pi, the ratio of circle's circumference and radius; and the Euler number e, the base of the natural logarithm—in its representation. The area under the normal curve represents the probabilities of the variable falling in a certain range. The total area under the curve equals one.

What Is Special?

As S. A. Frank explains in "The Common Patterns of Nature," we typically observe large-scale phenomena arising from the interactions of many hidden small-scale processes. It's like looking down on Earth from an aircraft flying far above. We see the big items and big green spaces without observing details and the movements. According to Frank, in all complex systems, our wish is to understand how large-scale patterns arise from the aggregation of small-scale processes. This is where the bell curve plays an important role.

Why Is It Important?

Have you ever wondered why, to estimate our true blood pressure, doctors recommend measuring it several times a day and calculating the average? Like most of us, they know that the average provides a better estimate. A single measurement could be affected by excitement, fatigue, and other sources of variation. People have been aware of this problem for a long time, but they couldn't figure out how much improvement is achieved by adding one more measurement. Now, the central limit theorem allows us to do inference about the sample average when inference cannot be done for a single measurement. The theorem states that under certain conditions, any observed variables in nature that arise from the summing up of many values will have a distribution close to bell curve, and as we increase the sample size, the distribution will approach a complete bell curve.

More on Central Limit Distribution

Why is the central limit theorem so important? Let's look at an example. Suppose that a fair die is rolled once, and your job is to predict the observed number/outcome. Obviously, in this situation, any guess is as good as any other guess. Your chance of being right is 1/6 and of being wrong is 5/6, as there are six possible outcomes. Now, suppose that two dice are rolled or one die is rolled twice, and your job is to guess the average of the two numbers/outcomes. Now, there are thirty-six possible pairs. If you guess average 1, then the chance that you are right is only 1/36, because average of one only occurs if 1 and 1 is observed or 1 is observed twice (1, 1). But if your guess is 3.5, then the chance that you are right is 6/36 because any of the following possible pairs—(1, 6), (2, 5), (3, 4), (4, 3), (5, 2), (6, 1)—will yield an average of 3.5. It's easy to show that 3.5 is the best guess. As the number of rolls is increased and averaged, the average 1 or 6 or any numbers close to them become less likely, the center values (numbers close to 3.5) become more likely, and our guesses become better and better. Moreover, not only does the shape of the distribution tend to the bell curve, but the variability of the sample mean will also decrease and lead to a better inference.

Consider an example in medical science. Since the distribution of most measurements such as blood pressure is unknown, rather than inference about a single measurement, scientists use the central limit, do inference about the average of several measurements, and make a statement about the average. This makes sense since medical science is concerned with each person's average for the purpose of decision-making.

Final Word

According to Frank, there are two main reasons for the widespread use of normal distributions. First, many measurements fluctuate about a central location because of perturbing factors or errors in measurement. Second, when analyzing the measurement, an assumption of Gaussian fluctuations is the best (maximum entropy) choice when the only information available is about the precision or error in observations with regard to the average value of the population of measurements.

References

Frank, S. A. "The Common Patterns of Nature." *Journal of Evolutionary Biology* 22 (2009): 1563–1585. https://stevefrank.org/reprints-pdf/09JEBmaxent.pdf.

Central Limit Theorem and My Shape

This section is about a silly joke. If you are familiar with Roman numerals, you probably know that XL = 40. You may have also heard that people who get to their forties use that as an excuse to be large size and overweight. Here is a story I have made up for being overweight. It is based on theorem known as central limit theorem, which plays the key role in science in general and in statistics in particular.

Central limit theorem is probably the most important and celebrated theorem of statistics. It states that if you have a random sample of measurements from a population and add them up, the distribution of sum will tend to bell-curve (normal distribution) and approximation gets better and better by increasing the sample size.

I used the theorem to make up the following story. Thirty laborers and a foreman, who is in charge of hiring and firing, work in a remote farm. Every evening, meals are delivered the foreman by an outside contractor. The amount of calorie per package varies between 1,200 and 1,800. Before distributing, the foreman steals around 10 percent of each worker's share for his personal consumption.

Let X_i be the amount of calories consumed by a worker. Then the amount of calories consumed by the foreman is $Y = 0.10 (X_1 + X_2 + ... + X_{30})$. Having a hard job and eating only once a day, workers had a very flat belly, like the picture on the left. This is like uniform distribution. The question is, what would be the shape of the foreman's belly? Since X_i's are independent random variables with uniform distribution, according to the central limit theorem, distribution of Y is approximately normal, as in the picture on the right. Foreman, of course, blames the theorem for his problem.

It is interesting to ask which individual we should name normal.

8. There Was More Snow When I Was Young

Although some of my younger friends who have lived in Bloomsburg for a few years consider the winter of 2018 one of the coldest and snowiest, most of my older friends see it as just another typical winter. Could it be because my older friends are forgetful, or is there another explanation for it? I think, like most other things, there are reasons behind this. I hope you will find this explanation convincing.

Let us simplify the problem and consider only the annual snowfall. Think of a group of babies who will be born on January 1, 2019, here and several other places where winters are similar to ours. What I would like to do is find a reasonable answer to the following question: How many personal (not historical) record snowfalls would members of this group experience during their lifetimes?

Clearly, for whole group, the first year's snowfall (2019) will be a record—that is, by age one, they all will have seen one personal record in their short life. But what about the second year (2020)? This will depend on whether in places they live 2020's snowfall was more or less than that of 2019. Since there is no significant trend in the amount of snowfall, the chances of having no (zero) new record and having one new record in 2020 are both 50 percent. So, averaging the zeros and ones, we may count this as $1/2$, which is same as saying that half of the babies are expected to see a record. Using the same logic, the chance that the third year (2021) produces a new personal record is $1/3$ and could be counted as $1/3$ record. The same applies to years 4, 5, . . ., and we get the following formula:

Expected Number of Records $= 1 + 1/2 + 1/3 + 1/4 + \ldots$

For example, for a baby or group of babies who live for one hundred years, we get

$$1+1/2+1/3+\ldots+1/100=5.19$$

That is five records on average. We can also calculate odds for experiencing a specific number of records such as three or four. For example, for one hundred years, the probability of experiencing five records is 21 percent and is, in fact, the most likely scenario.

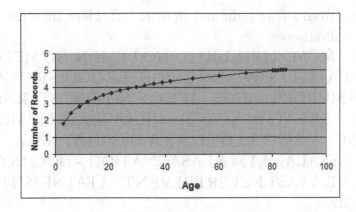

Using the above formula, by age one, all babies will see one record. By age four, the expected or average number of records experienced is two (1 + 1/2 + 1/3 + 1/4 = 2.083). By age eleven, three (1 + 1/2 + 1/3 + 1/4 + ... 1/11 = 3.02). By age thirty-one, four (1 + 1/2 + 1/3 + 1/4 + ... +1/31 = 4.02). And at age eighty-three, five (1 + 1/2 + 1/3 + 1/4 + ... 1/83 = 5). Of course, not all babies will experience the same number of personal records; but if we average number of records each of them experience in their life, we get very close to what the above formula predicts.

Now, what do all these have to do with our original question? Well, isn't this a key to the fact that in our youth, winters were colder with more snow? Just think about my friends, aged thirty-two to eighty-two. Most of them remember the fourth record they have experienced at around age thirty-one. As a result, for them, all winters since then may feel just average or typical.

Madam I M Adam

Did you know that Inauguration Day, January 20, 2021 (1-20-2021) was the first palindrome-number Inauguration Day in American history? Surprisingly the next palindrome Inauguration Day will not occur until 1/20/3021; that is a thousand years from now!

A palindrome is a number, word, phrase, or sentence that reads the same from left to right and right to left. Here are a few well-known palindromes:

323 – MOM - I DID, DID I? - NO LEMON, NO MELON - WAS IT A CAT I SAW? - DO GEESE SEE GOD? - RED RUM, SIR, IS MURDER. - RATS LIVE ON NO EVIL STAR. -EVA, CAN I STAB BATS IN A CAVE? - MR. OWL ATE MY METAL WORM. - A SANTA LIVED AS A DEVIL AT NASA. - GO HANG A SALAMI, I'M A LASAGNA HOG. - DOC, NOTE: I DISSENT. A FAST NEVER PREVENTS A FATNESS. I DIET ON COD.

There is a beauty in symmetry and most humans find symmetrical patterns more attractive than asymmetrical ones. Our preference for symmetry might be a byproduct of our need to recognize objects irrespective of their position and orientation in the visual field. Much can be said about symmetry and its effects on human life. For example, many people find palindromic words and numbers interesting and memorable.

Palindromic numbers are the same whether read from left to right or right to left. The first few palindromic numbers are 0, 1, 2, 3, 4, 5, 6, 7, 8, 9, 11, 22, 33, 44, 55, 66, 77, 88, 99, 101, 111, and 121.

A palindrome can be formed from a number that is not a palindrome by adding the original number to the number formed by reversing the digits and repeating the process if necessary. For example, $17 + 71 = 88$ and $123 + 321 = 444$.

Palindromic Dates

It is interesting to note that 2021 affords 22 palindromic dates including this stretch in December:

12/1/21, 12/2/21, 12/3/21, 12/4/21, 12/5/21, 12/6/21, 12/7/21, 12/8/21, 12/9/21

"The only two years in a century that contain 22 palindromic dates are the ones ending with 11 and 21," "The year 2011 had 22, and in the next century, they will be found in the years 2111 and 2121."

Palindrome weeks will happen every year through 2029 then will not be seen for a century.

Inauguration Day fell on a palindrome date for the first time, starting a 10-day stretch beginning with 1-20-21 and ending with 1-29-21.

Palindromic Poems

A line palindrome is when the individual lines of a text make a palindromic sequence. An example is the poem "Doppelgänger," below by James A. Lindon, which reads identically from the first line to the last, and from the last to the first. It is amazing how an identical line changes its meaning completely from one end of the poem to the other.

"Doppelgänger"
Entering the lonely house with my wife
I saw him for the first time
Peering furtively from behind a bush—
Blackness that moved,
…
Blackness that moved.
Peering furtively from behind a bush,
I saw him, for the first time
Entering the lonely house with my wife.

Susan Stewart has a terrific line palindrome poem in her recent book *Columbarium*, where the form mirrors the action of the poem (a journey into and out of hell). Here is an excerpt:
"Two Brief Views of Hell"
Leaving the fringe of light at the edge of the leaves, deeper, then deeper, the rocking back and forth movement forward through

the ever-narrowing circle that never, in truth, narrowed beyond the bending going in, not knowing whether a turn or impasse would lie at the place … not knowing whether a turn or impasse would lie at the place that never, in truth, narrowed beyond the bending going in, the rocking back and forth movement forward through the ever-narrowing circle.

Leaving the fringe of light at the edge of the leaves, deeper, then deeper.

If you have a half-written poem, try to rewrite it in reverse and sell it as a palindrome.

9. Geometries We Did Not Learn in School

Geometry is the realm of mathematics in which we talk about things like points, lines, angles, triangles, circles, squares, and other shapes, as well as the properties and relationships between them.

For centuries, it was widely believed that the universe worked according to the principles of what most of us learned in school known as Euclidean geometry, where parallel lines never cross. This was the geometry we learned and was the only kind of geometry available. The decisive steps in the creation of non-Euclidean geometry took place in the beginning of the nineteenth century.

Non-Euclidean geometry arises when the parallel postulate is replaced with an alternative one. Doing so, one obtains hyperbolic geometry and elliptic geometry, both referred to as non-Euclidean geometries. The essential difference between the geometries is the nature of parallel lines. Euclid's fifth postulate, the parallel postulate, which, as we know, states that within a two-dimensional plane, for any given line ℓ and a point A, which is not on ℓ, there is exactly one line through A that does not intersect ℓ. In hyperbolic geometry, by contrast, there are infinitely many lines through A not intersecting ℓ; while in elliptic geometry, any line through A intersects ℓ.

Another way to describe the differences is to consider two straight lines indefinitely extended in a two-dimensional plane that are both perpendicular to a third line:

- In Euclidean geometry, the lines remain at a constant distance from each other (meaning that a line drawn perpendicular to one line at any point will intersect the other line, and the length of the line segment joining the points of intersection remains constant) and are known as parallels.
- In hyperbolic geometry, they "curve away" from each other, increasing in distance as one moves farther from the points of intersection with the common perpendicular.
- In elliptic geometry, the lines "curve toward" each other and intersect.

A Summary

One of the frequently cited arguments in favor of revision of mathematics in light of empirical discoveries is the general theory of relativity and its adoption of non-Euclidean geometry. Recall that Euclid developed the idea of geometry at around 300 BC. Most books on this subject start with five main postulates, axioms, or assumptions from which they drive theorems of geometry. These are postulates:

- Given two points, there is a straight line that joins them.
- A straight-line segment can be prolonged indefinitely.
- A circle can be constructed when its center point and a distance for its radius are given.
- All right angles are equal.
- From a point outside a line, only one line can be drown parallel to it.

The last postulate is more complicated than the other four. Over the years, many mathematicians tried to derive it from the first four. However, so far, nobody had yet come up with such a proof.

10. Fractals Geometry

Mathematics and Art. Mathematics and art intersect in our world in beautiful ways. Many mathematicians draw upon art, and many artists draw upon mathematics. When mathematics and art come together, people are often inspired and they can start to see mathematics as a beautiful and creative subject. Poetry is also very mathematical and filled with rich patterns.

Da Vinci used the mathematical principles of linear perspective— parallel lines, the horizon line, and a vanishing point—to create the illusion of depth on a flat surface. In the *Annunciation,* for example, he uses perspective to emphasis the corner of a building, a walled garden, and a path.

According to the professor Annalisa Crannell, "Most realistic art aims to depict a three-dimensional world on a two-dimensional canvas. The geometry of the situation becomes clear if we think of the canvas as a plane and the artist's eye as a point, and remember that light travels in straight lines. The artist's lines of sight intersect the (temporarily transparent) canvas, mapping each point the artist sees in the real world to a corresponding point on the canvas. The picture looks realistic if this mapping is accomplished accurately."

History

Mathematics and art have a long historical relationship. Artists have used mathematics since the fourth century BC when the Greek sculptor Polykleitos wrote his Canon, prescribing proportions conjectured to have been based on the ratio $1:\sqrt{2}$ for the ideal male nude. Persistent popular claims have been made for the use of the golden ratio in ancient art and architecture, without reliable evidence. In the Italian Renaissance, Luca Pacioli wrote the influential treatise *De divina proportione* (1509), illustrated with woodcuts by Leonardo da Vinci, on the use of the golden ratio in art. Another Italian painter, Piero della Francesca, developed Euclid's ideas on perspective in treatises such as De prospectiva pingendi, and in his paintings. The engraver Albrecht Dürer made many references to mathematics in his work *Melencolia I.* In modern times, the graphic artist M. C. Escher made intensive use of tessellation and hyperbolic geometry with the help of the mathematician H. S. M. Coxeter, while the De Stijl movement led by Theo van Doesburg and Piet Mondrian explicitly embraced geometrical forms. Mathematics has inspired textile arts such as quilting, knitting, cross-stitch, crochet, embroidery, weaving, Turkish and other carpet-making, as well as kilim. In Islamic art, symmetries are evident in forms as varied as Persian girih and Moroccan zellige tile work, Mughal jali pierced stone screens, and widespread muqarnas vaulting.

Mathematics has directly influenced art with conceptual tools such as linear perspective, the analysis of symmetry, and mathematical objects such as polyhedra and the Möbius strip. Magnus Wenninger creates colorful stellated polyhedra, originally as models for teaching. Mathematical concepts such as recursion and logical paradox can be seen in paintings by René Magritte and in engravings by M. C. Escher. Computer art often makes use of fractals including the Mandelbrot set, and sometimes explores other mathematical objects such as cellular automata. Controversially, the artist David Hockney has argued that artists from the Renaissance onward made use of the camera lucida to draw precise representations of scenes; the architect Philip Steadman similarly argued that Vermeer used the camera obscura in his distinctively observed paintings.

But if you ask mathematicians to describe what mathematics is all about, their answer might surprise you. They often compare it to art. Or music. Or philosophy. Or any of the seemingly more "arts-oriented" pursuits in the arts and sciences canon. And despite how concrete and absolute mathematics appears on the surface, mathematicians say their field is about as abstract a subject as you can imagine.

"The idea of abstraction is so natural to a mathematician," says Prof. Nitu Kitchloo. "Math begins with intuition. We take something completely concrete, and then we make it so abstract it might be unrecognizable. We play with toys we can't even picture. But this abstraction allows us to draw connections that may not have been evident before."

Kitchloo, who served as chair of the mathematics department at the Krieger School from 2014 until last summer, doesn't spend his days in front of a blackboard scribbling. Rather, when he isn't teaching or meeting with students, you can find Kitchloo at a cafe. He admits he'll spend hours there, nursing a coffee or tea, and staring off into space, deep in thought, just like a philosopher working on an argument or a poet thinking up his next line. What he's doing is trying to get a feeling for the object he's studying. In his case, Kitchloo studies topology, the intersection of geometric forms, and spatial relations.

Why Fractals?

Clouds are not spheres, mountains are not cones, and lightning does not travel in a straight line. The complexity of nature's shapes differs in kind, not merely degree, from that of the shapes of ordinary geometry, the geometry of fractal shapes.

Now that the field has expanded greatly with many active researchers, Mandelbrot presents the definitive overview of the origins of his ideas and their new applications. *The Fractal Geometry of Nature* is based on his highly acclaimed earlier work but has much broader and deeper coverage and more extensive illustrations.

"Why is geometry often described as 'cold' and 'dry'? One reason lies in its inability to describe the shape of a cloud, a mountain, a coastline, or a tree. Clouds are not spheres, mountains are not cones, coastlines are not circles, and bark is not smooth, nor does lightning travel in a straight line" (Mandelbrot 1982, p. 1).

These are the words of the Polish French mathematician Benoit Mandelbrot, who famously introduced the concept of fractals and its applications. He named the phenomenon fractal, derived from the Latin word *fra'ctus*, meaning "broken." In the introduction of his book *The Fractal Geometry of Nature* (Mandelbrot 1982), he states that most fractals tend to have fragmentation and statistical regularities or irregularities occurring at all scales.

Mandlebrot's Finding

One man who saw a need for a new geometry was Benoit B. Mandelbrot. He felt that Euclidean geometry was not satisfactory as a model for natural objects. To anyone who has tried to draw a picture of a non-regular object (such as a tree) on a computer graphics screen using the Euclidean drawing primitives usually provided, this is an obvious statement. The strength of Mandlebrot's finding was his research into the findings of the earlier mathematicians and the development of a practical application of their theory. He showed how these functions yield valuable insight into the creation of models for natural objects such as coastlines and mountains. Mandelbrot popularized the notion of a fractal geometry for these types of objects. Although he did not invent the ideas he presents, Mandelbrot must be considered important because of his synthesis of the theory at a time when science was reaching out for new more accurate models to describe its processes.

The golden ratio and Fibonacci's sequence is also a prominent feature in nature. Fibonacci's sequence in the bones of a human hand is an example. Even our own galaxy that we live in is made up of Fibonacci's sequence.

Fractal Curves

Fractals differ from Euclidian geometry and its straight lines and smooth curves by being fractioned and not having a tangent at any given point. Fractals have been used since the end of the seventeenth century. There are numerous fractal curves discovered and named by different mathematicians—for instance, the Koch curve, the Minkowski curve, and the Peano curve.

Self-Similarity

Self-similarity is another important concept within the field of fractals. The fragmentation tends to recur identically at all scales. Although, when discussing natural occurring fractals, the concept of infinite iterations becomes less useful. Naturally, there are physical limitations such as molecule size and, finally, atom size. Mathematical fractal structures, on the other hand, reach no such limit.

"Fractals are objects with roughness at all scales or, for natural fractals, over at least several orders of magnitude of scales" (West, Deering 1994, p. 9).

Fractals in Nature

In nature, energy efficiency is crucial, and high-performing structures are created with simple material and the information itself being the key to success. To minimize the energy and material spent on this information, nature has conceived an impressive ratio between amount of information put into the system and the complexity of the outcome. Self-similar structures require only one rule, which applies in all scales, and by being information-efficient, fractal structures are created (Gruber 2011, p. 97).

Natural-occurring fractals can be found in the branching of a tree, veins of a leaf, mountain ridges, rivers, vegetables, and bronchial structure of lungs, to name a few.

Joye draws the conclusion that our love for certain fractal dimension can be derived from our biological urge to find suitable habitats. Environments with high fractal dimension offers too much coverage for possible danger, whereas environments with lower fractal dimension do not provide resources enough to prove sustaining, implying that midrange fractal environments are suitable. This biophilic trait is the reason that architecture containing fractal geometry is beneficial for humans.

Which Calculation Is Correct?

I clearly recall my teacher's distinct voice pointing out a fact about fractions, "Remember that a/b + c/d is not equal to (a + c)/ (b + d)." I also remember a classmate questioning its importance and the teacher's response that most of us did not find very convincing. Now, as a mathematics teacher for almost fifty years, I still get the same questions from students who feel that knowing such little mathematical things are not important. Here is usually my response to students who care to listen. I start with a simple example that goes like this:

Think about two young families with children living in a remote neighborhood. One family has a boy and two girls, and the other two boys and a girl. Suppose that we want to calculate the ratio of boys to girls in that neighborhood. To do this, we may find the average number of boys ((1 + 2)/2) = 1.5) and girls ((2 + 1)/2) = 1.5) in that neighborhood and calculate the ratio as 1.5/1.5 = 1. Or, alternatively, find the ratios of boys to girls in each family, 1/2 = 0.5 and 2/1 = 2 separately and calculate their average as (0.5 + 2)/2 = 1.25. As you see, we do not get the same answer. Why? Simply because the average of ratios is not the same as the ratio of the averages.

Who Is a Better Free-Throw Shooter?

During the last basketball season, Jim attempted one hundred free throws in the first half of the season and made thirty. So his free-throw percentage was 30/100 = 0.300. He also attempted twenty in the second half of the season and made eight (8/20 = 0.400). His stat was better than Curt's stat, 5/20 (0.250) for the first half of the season and 35/100 (0.350) for the second half of the season. For the season, however, Curt's free-throw percentage, (5 + 35)/(20 + 100) = 40/120 = 0.333, was higher than Jim's, (30 + 8)/(100 + 2) = 38/120 = 0.317. So, who do you think was a better free-throw shooter in that season? Here, mathematics helps you to decide. How? The conclusion based on denial should always be preferred. Two dimensions versus one. So Jim.

Which Treatment Should I Choose?

Consider a disease for which there are two methods of treatment, A and B, and it is up to the patients to choose one. Suppose that in the past, out of 100 male patients who chose A, 20 recovered (recovery rate = 20/100 = 0.20). Also, out of 210 male patients who chose B, 50 recovered (recovery rate = 50/ 210 = 0.24). These rates suggest that a male patient should prefer treatment B to A (24 percent versus 20 percent). Suppose also that, so far, out of 60 female patients who chose A, 40 recovered (recovery rate = 40/60 = 0.67). Also, out of 20 female patients who chose B, 15 recovered (recovery rate = 15/20 =0.75). These rates suggest that a female patient should prefer treatment B to A (75 percent versus 67 percent) as well.

Now, combining the data, we see that the total number of people (regardless of their gender) who chose A was 160, of which 60 recovered (recovery rate = 60/160 = 0.38). Also, the total number of people (regardless of their gender) who chose B was 230, of which 65 recovered (recovery rate = 65/230 = 0.28). These rates suggest that it is wise to use or choose treatment A over B (38 percent versus 28 percent). This, unlike expectation, contradicts our earlier conclusion based on patient's gender.

The lesson here is that with only two numbers, as in this example, drawing a conclusion is not always straightforward. From a mathematical point of view, the conclusion based on details (higher dimensions/gender) should be preferred. In fact, breaking the data gives a two-dimensional view of it, which is better than a one-dimensional view of it based on collapsed data.

The situation described is an example of what is known as Simpson's paradox. The paradox occurs because collapsing the data can lead to an inappropriate weighting of the different populations. We will return to Simpson's paradox later in this chapter.

Who Is Discriminated Against?

Consider a business with one hundred employees, fifty "type A" and fifty "type B", with comparable duties and average pay of $16,000 (total pay = $800,000) and $14,000 (total pay = $700,000) per person, per year, respectively. Here, the average pay to type A employees is more than that of type B employees, suggesting a pay discrimination in favor of type A employees. But by considering the details, one finds that the data contain other relevant information. Consider, for example, a classifier such as length of employment.

Suppose that ten of type A employees have been working for that business for less than five years with an average pay of $10,000 (total pay = $100,000), and forty for more than five years with average pay of $17,500 (total pay = $700,000); 100,000 + 700,000 = 800,000.

In type B group, forty employees have been working there for less than five years with average pay of $12,500 (total pay = $500,000), and ten for more than five years with average pay of $20,000 (total pay = $200,000); 500,000 + 200,000 = 700,000.

Now, comparison suggests that type B employees' average pay is higher than that of type A employees' in both categories, suggesting a pay discrimination in favor of type B employees. This, again, contradicts our earlier conclusion.

So here is the lesson: it is not straightforward to argue that there has been a discrimination in the matter of pay—that is, one must be careful in reporting averages unless the groups under consideration are homogeneous. The problem gets even more complicated/interesting if we consider more than one classifiers (e.g., education, experience).

In sum, these examples demonstrate how different analyses of the same data may lead to apparently conflicting results and sometimes incorrect decisions.

An Exceptional Family

Many exceptional individuals have contributed to the progress of mathematics, naturally some more than others. The Bernoulli family in mathematics, like their contemporaries, the Bachs in music, is an unusual example of talent in one field appearing in successive generations. The Bernoullis' work helped probability to grow from its birthplace in the gambling hall to a respectable tool with worldwide applications. No fewer than seven Bernoullis, over three generations spanning the years between 1680 and 1800, were distinguished mathematicians. Five of them helped build the mathematics of probability.

Jakob (1654–1705) and Johann (1667–1748) were sons of a prosperous Swiss merchant, but they studied mathematics against the will of their practical father. Both were among the finest mathematicians of their times, but it was Jakob who concentrated on probability. He was the first to see clearly the idea of a long-run proportion as a way of measuring chance.

Johann's son Daniel (1700–1782) and Jakob and Johann's nephew Nicholas (1687–1759) studied probability. Nicholas saw that the pattern of births of male and female children could be described by probability. Despite his own rebellion against his father's strictures, Johann tried to make his son Daniel a merchant or a doctor. Daniel, undeterred, became yet another Bernoulli mathematician. In the field of probability, he worked to fairly price games of chance and gave evidence for the effectiveness of inoculation against smallpox.

An Exceptional Athlete in Numbers

Let me, for a little while, move away from mathematicians and talk about an exceptional runner. Consider the records for men's one-hundred-meter rum. For this event, twenty records are set since 1912. The last three records, 9.58, 9.69, and 9.72, were set in years

2009, 2008, and 2008, respectively, all by an exceptional athlete, Usain Bolt, in his statistics.

1. Unlike other events, here, the time between the last and penultimate records is one year, and the time the last record has held to date is zero.
2. Excluding the last record (9.58), the probability of setting a record 9.58 or less was only 1.8 percent, which, again, indicates how extraordinary the new record is.
3. The probability of setting a record 9.55 or less was only 1.7 percent.
4. The probability of setting a record 9.50 or less was only 1 percent.

Data for one-hundred-and two-hundred-meter runs exhibit long tail, as the present records for these events, 9.58 seconds and 19.19 seconds, respectively, are significantly lower than the previous records. Application to the men's one-hundred-meter data for the period starting January 1, 1977, when IAAF required fully automatic timing, to September 1, 2009, shows that the probabilities of setting a new record such as 9.55 seconds or less and 9.5 seconds or less are respectively

A. 0.0102 and 0.0052, when Bolt's records are included;
B. 0.0043 and 0.0023, when Bolt's records are excluded.

Also, excluding Bolt's records, the probability of setting a record of 9.58 seconds or less by other runners is only 0.0064.

Application of the method to Bolt's individual performance prior to the 2008 Olympics reveals the following:

A. For him, the probability of running the two hundred meters in the 19.30 seconds or less was only 0.00257, indicating that his Olympic record, 19.30 seconds, was completely unexpected.

B. The probability of breaking his own best record, 19.75 seconds, was only 0.0738, indicating that his Olympic performance was exceptional.

Also, application of the method to his individual performance including his 2008 Olympic record reveals that his new record, 19.19 seconds, was even more astonishing.

Application of the method for estimation of ultimate record produces the following 90 percent prediction intervals:

A. (9.40, 9.58) when Bolt's t records are included;
B. (9.62, 9.71) when Bolt's records are excluded.

Note that Bolt's last record 9.58, falls outside the interval B. This demonstrates that Bolt was in a different league.

Pi and Mysteries of Universe

We all know about Pi. In fact, it is probably the only irrational number we know and remember. This is what is amazing about Pi. In seventeenth century, mathematicians proved that pi is an irrational number—that is, it takes an infinite number of digits to give its exact value. This means that we can neither get to the end of it nor find/predict its next digit, as digits of pi follow no pattern. In short, Pi is has infinite nonrepeating decimal.

Having infinite and nonrepeating decimals, it is reasonable to assume that every possible number combination could be found somewhere in pi's decimal representation. This assumption implies that, when converted into numbers (ASCII text), somewhere in the string of digits, one could find answers to all the great questions of the universe, the DNA of every being, and every book or article that has ever been written or will be written. In fact, right now, it is possible to check to see if somebody's birth date or social security number occurs as a string of digits in the first two hundred million digits of pi at the Pi Search Site, http://www.angio.net/pi/. Anything trivial or amazing could be found somewhere in the ratio of a circumference and a diameter of a circle.

A Presidential Coincidence

Most people find coincidences fascinating. One strange historical "coincidence" concerns the early presidents of the United States. Among the first five presidents, Washington, Adams, Jefferson, Madison, and Monroe, three of them, Adams, Jefferson, and Monroe, died on the same day of the year. And the date was none other than the *fourth of July*. Of all the dates to die on, that must surely be the most significant to any American. The probability that three people out of five die in the same day of the year is about one in five million. The fact that the day was July 4 and the five people were presidents make it even less likely and more surprising. To the early presidents, the anniversary of independence meant so much, and, as such, they were really keen to hang on until they had reached it. This is what happened to Thomas Jefferson, the third president. John Adams, the second president, actually died a few hours after Thomas Jefferson.

Does this historical event support the theory that we could postpone our death day by looking forward to something important such as a milestone? I usually ask students to express their views about this and the related issues. What I have noticed during the years is that this example surprises more students than the famous Abraham Lincoln and John F. Kennedy connections.

Technology and Mathematics Education

In today's world, technology plays an important role. It has changed everything including the way we learn, see, and apply mathematics. An interesting question is how technology has affected mathematics education and whether to encourage use of technology in mathematics classes. To learn pure mathematics, one should be able to think abstract. Some believe that using computer often works against this.

A number of educators have tried to study the role and possible effects of technology and, specifically, computers in teaching and learning mathematics.

Such studies compare "mathematicians" who are trained using traditional methods (abstract thinkers) with "mathematicians" who are trained utilizing computers (computer-assisted learners). As of today, there are different opinions among experts and educators about the outcome, validity, interpretations, and usefulness of such comparison.

As expected, there is a general agreement that extensive use of computer modifies student's problem-solving approach and ability, which are key to utilizing mathematics. This, in turn, makes the direct comparison difficult. As a result, the query about effective teaching and learning should probably be replaced by the following questions: which type of "mathematicians" will our teaching method produce, and should we name them the same?

Recall that, compared to capabilities of the human brain, even the most advanced supercomputers are not yet "intelligent." The brain's vast capacity derives from the fact that all its nerve cells operate in mutual association as a "neural network." Given appropriate examples, a network of this nature can learn to store and retrieve information independently. In fact, few of us have difficulty recognizing the face of someone we know as we rush through a busy shopping mall. In a fraction of a second, we recognize the person, notice his hat, and note any movements, size and orientation, light or shade. At the same time, we recall memories and many other items of information about this person. Our brain achieves all this without any effort. Even a child can do it. A computer, on the other hand, although can record images electronically, is able to understand their content only in certain cases and after a long calculation.

So the fundamental question is the following: should we educate our students, especially the mathematics majors, to use computers to learn, or rely mostly on their brain power, or both? And if both, to what extent?

This depends, of course, on how mathematics is learned. Computers allow large numbers of users to retrieve the same or similar information at the same time. The human brain, on the other hand, is more portable and infinitely more powerful. Its powerful neural network permits people to use their imagination, see patterns and associations, and think "creatively" in ways that computers are not yet cable of.

Would the use of computers make the progress of mathematics faster or slower, change its direction, etc.? This may not be the proper question. Computers have and will certainly change the course of progress. They would lead to new areas such as computational mathematics and computer graphics that are important in their own right. However, they would not help to produce mathematicians like Newton or Euler, who could develop extremely complicated theories in their heads.

Some ways of using a computer in teaching mathematics are considered proper; some are not. For example, students may use computer as a tool to construct their own understanding of mathematical concepts. Generally, using a computer as tool to simulate the real world, or at least the quantifiable part of it, is appropriate. Using it merely as a super calculator is improper. The first application would help learning and understanding mathematics. The second would only make the life easier and will save time.

Mathematical Modeling

Human beings has been always curious about how things work and, if possible, model it. This includes smoothing, idealizing, and expressing the real world using symbolic language so that future behaviors could be predicted. Mathematical modeling is an ever-developing part of mathematics, simply because of our desire to formulate the phenomenon around us. Let us present it by an introductory example.

For businesses, one important problem is related to the positioning of stores and deciding on delivery routes. This requires information on the distances of the road, y, between different places. Where a large number of such places are involved, finding these distances by direct measurement is either not easy or time-consuming.

To avoid this problem, the usual approach is to relate the road distances to the straight line distance, denoted by x as measured using a scale map. This relationship will enable us to predict a value of y given a corresponding value of x.

The fundamental question is, how do we obtain this relationship or the model? There is a whole spectrum of differing approaches. Extremes of this spectrum are *conceptual approach* and *empirical approach*.

Remainder of the spectrum usually consists of a combination of the extremes, known as *eclectic approach*.

1) Conceptual Approach

Derives the form of the model on the basis of our understanding of the situation. Uses logical reasoning, "known theory," to obtain the model. How? Here is the example:

a. When $x = 0$, the two points coincide, so $y = 0$.
b. If there is a straight road between the two points, then $y = x$; otherwise, y will be greater than x.
c. The distance y will generally increase with x. However, the randomness in the pattern of roads will mean that even if different pairs of places have the same x, there will be random differences in the y values.

y = Part predictable from x + unpredictable random component

Thus, each y will consist of a part predictable from the x value and an unpredictable, random component.

d. Provided we keep to relatively similar situations (e.g., urban roads), then the form of the relationship should not depend strongly on the distances involved. Thus, if the straight line distances are doubled, we would expect in most cases the road distances also to be approximately doubled.

Consider a few possibilities (simplest relationships):

i. $y = x$: satisfies (a) and (d) but not (b) or (c).
ii. $y = x +$ random component [error]: this now allows (c) but not (b).
iii. $y =$ constant $+ x +$ random component: this helps with (b) but (a) now fails.
iv. $y =$ constant times $x +$ random component: This now satisfies all four provided that the constant is greater than one and that the random component, e, is constrained so that $y \geq x$.

We can write our "conceptual" model as

$$y = Bx + e, B \geq 1, y \geq x.$$

Notice that this model is derived without any data about the actual situation.

2) Empirical Approach

Considers only empirical approach and ignores all the past evidences from previous experiences or knowledge. To apply this approach, we need to begin with the available data. First, we plot it to get some idea about possible pattern to relationship.

Suppose that the data suggests a straight line passing through the origin for the main trend. If the observation (x,y) represents one of the points, the relation can be expressed as

$$y = Bx + e,$$

where e is the error in predicting y from the line. If this works, we could say that we arrived at the same model by both approaches. The conceptual approach used what we might term prior information (i.e., the information that we possessed prior to obtaining the data). The empirical approach ignored this information and used only the empirical information contained in the data.

In practice, we should seek to use both sources of information, the eclectic approach.

To see this, consider the following question: how do we obtain or calculate B?

To answer this, we have to take an empirical approach and use our data to provide a numerical value for B. This value is referred to as an estimate of B. There are some methods and formulae for obtaining B. The question that arises now is, how do we test validity of our model?

The answer is, by testing on more data, we compare the actual measurements with predictions from the model. Alternatively, we can simulate data from a known model and develop a model using empirical approach. Finally, compare that with original model and alter our modeling approach to achieve improvements.

Note that the model, as it has been defined, concentrates on the deterministic (here line) and provided no information about the random component, e. As a further stage in the model construction, we must consider how to model the properties of this random component. For that, we examine the evidence of the observations.

We make, for example, "frequency" table and use known statistical methods to analyze the errors. We thus can see that to discuss the random component of the model, a different language is needed from the mathematical statements of the deterministic component.

This is the language of expectations and probabilities, the language of statistics.

Benefits of Mathematical Modeling

There are many benefits to mathematical modeling. Here are some examples:

1. They are clearly defined and thus easily communicated so that their strengths and weaknesses may be analyzed.
2. They can be manipulated according to the clear rules of mathematics and investigated either analytically or, more experimentally, using a computer.

In fact, with the huge growth in computing power, modelling of all kinds has boomed. It is now possible to model complex engineering structures and test them inside the computer before an actual prototype is built. Now, can we expect our models to be true descriptions of reality?

The answer is no. Models are but approximation to certain aspects of complex reality.

In fact, a clearly stated model provides us with a constant reminder of what is real and what is modeled—that is, what is observed versus what is expected. Is this a weakness?

The answer is no. In fact, one cannot analyze the real world (by definition of analyze). One can only analyze a picture of the world (conceptual model) that is in one's head. The main requirements on such models are that they be accurate enough for the purposes at hand and that they be tractable enough for the needed accuracy.

A Model for Population Growth

Suppose that the goal is to model the population of rabbits in a certain part of Pennsylvania or the population of Philadelphia.

Let y_n be the number of individuals at time n (year or generation n), and b (y) and d (y) denote the number of individuals who were born and died during this period, respectively. Then

$$y_{n+1} = y_n + by_n - dy_n = (1+b-d)\, y_n = ry_n$$

represents the number of individuals at a time $n+1$. For realistic models of growth certain restrictions on r are necessary. Suppose that the action of previous generations determine the growth of a population—that is, $r = r(y_n)$, a function of y_n. Thus, r is a density dependent growth rate.

One of the simplest density dependent models that contains a formulation to represent effects of overcrowding is when r is a linear function of population size—that is,

$$y_{n+1} - y_n = r^* \, y_n \left(1-\tfrac{y_n}{k}\right) = F(y_n)$$

$$y_{n+1} - y_n = r^* y_n \left(1-y_n/k\right) = F(y_n)$$

$r^* =$ intrinsic rate of growth, $k =$ environmental carrying capacity.

The behavior of the logistic equation for some typical values of r^* and y_0, the initial size of population, is studied extensively. It varies from smooth to oscillatory and even chaos. One of the most important questions often asked is, do biological populations exhibit chaotic behavior? This question is unresolved at present. But it is known that if the models with chaotic behavior are relevant to populations, then wild oscillations of populations need not necessarily be the consequence of random environmental fluctuations but might be intrinsic to the population. Another question of interest is, can we use this model for human population (e.g., population of England)? According to experts, since population is changing instantly, continuous models are more appropriate.

The continuous version of classical logistic equation is the a first-order Bernoulli differential equation. Note that there is a big difference between discrete and continuous logistic models. In fact, there is an important distinction between first-order difference and differential equations with constant coefficients. The latter allows no oscillation.

Key to Modeling

In general, mathematical modeling takes the data as a message and seeks models for them using the basic decomposition.

$$\text{Datum} = \text{Systematic Part} + \text{Random Part}$$
$$\text{or}$$
$$\text{Message} = \text{Signal} + \text{Noise}$$

Mathematical modeling is normally applied to the first component. It is the attention given to the second component that perhaps most distinguishes statistical modeling from other kinds. If the statistical properties of our observations remain unchanged as time passes (stationary process), we further have what is known as decomposition theorem:

$$\text{Stationary Process} = \text{Deterministic Part} + \text{Indeterministic Part}$$

Numbers and Their Story

Every day, we go about our lives without appreciating the mathematics. Where did all this comforts come from, and why is it important to know?

Thousands of years ago, people began to use numbers for elementary purposes. Using a tally stick, numbers were used as a representation for something. In Africa, bones were found with notches in them. Remarkably, the notches were a representation of days until the moon cycle, similar to the modern-day calendar. Also, the tally sticks represented the hunting schedule and how long they have been in that location. The tally stick used base numbers and was imperative for the survival of both hunting and gatherers.

Empires such as the Romans were very big in commerce and trade, and with such a growing dynasty, they needed numbers. I am sure everyone has used Roman numerals before; that's just how important numbers are and how powerful they can be. What is great about numbers is that using them enables other things to come about, similar to the Romans using numbers for expressing numerically in Latin. They also used numbers for keeping track of ships and other items important to them.

The Italian Leonardo Fibonacci realized how strenuous and time consuming Roman numbers were to use and began to make his own numbering system. Daring to go against the "prestigious" Roman numerals, Leonardo wrote a book called *Liber Abaci* in AD 1202. Although looked down on compared to Roman numerals, people began to realize just how imperative Leonardo's number system was in the fifteenth century because it was much less time consuming and problematic.

Another big culture with a big impact are the Babylonians. They took the numbers and began to show forms of what we know today to being geometry and algebra. In the 1600s, Isaac Newton took the knowledge passed down from the Babylonians and began to work on numbers with all different possible variable ranges. In doing so, his work brought misunderstanding to original problems and numbers. The Babylonians sometimes are looked down upon for the mere fact that they never incorporated a zero into their numbering system; however, the system was used for over a thousand years and never created a problem for them.

Georg Cantor was also a big help with the numerical system as he cleared up some of the confusion people had with Newton. Although his work cleared up some confusion, many people found his work "mathematical curiosities" instead of practical mathematics because of other findings such as the chaos theory and fractal geometry.

Thinking about it, numbers has been around before writing has. In the Fertile Crescent of southwest Asia, where farming may have developed first, many farmers had a surplus of crop. To store these bins of crop, they created records of what bin and what type of crop they had. Then they would make clay tokens or coins, similar to how people make pottery today, and they would keep a record of what they had.

After researching the beginning of mathematics and numbers, I really have gained a new appreciation for it. It appears that many cultures developed their own form of numbering, and they all can be useful. I guess I never really took the time to respect where it all came from; it's ignorant to think that it's just always been here. I mean, numbers are all over the place; I can't imagine my life without them. I praise the person who started numbering.

Roman Numerals

The system of Roman numerals was slightly modified in the Middle Ages to produce the system we use today. It is based on certain letters that are given values as numerals: Roman numerals are a system of numerical notations used by the Romans. They are an additive (and subtractive) system in which letters are used to denote certain "base" numbers, and arbitrary numbers are then denoted using combinations of symbols. Unfortunately, little is known about the origin of the Roman numeral system.

The following gives the Latin letters used in Roman numerals and the corresponding numerical values they represent. For example: I 1, V 5, X 10, L 50, C 100, D 500, and M 1000. Although the Roman numerals are now written with letters of the Roman alphabet, they were originally separate symbols. The Etruscans, for example, used I Λ? X? 8 ⊕? for IV XL CM. They appear to derive from notches on tally sticks, such as those used by Italian and Dalmatian shepherds into the nineteenth century. Thus, the I descends from a notch scored across the stick. Every fifth notch was double cut, and every tenth was cross cut (X), much like European tally marks today. This produced a positional system: eight on a counting stick was eight tallies, IIIIA?III, but this could be written A?III (or VIII), as the A? implies the four prior notches. Likewise, number four on the stick was the I-notch that could be felt just before the cut of the V, so it could be written as either IIII or IV. Thus, the system was neither additive nor subtractive in its conception, but ordinal. When the tallies were later transferred to writing, the marks were easily identified with the existing Roman letters I, V, X. (A folk etymology has it that the V represented a hand, and that the X was made by placing two Vs on top of each other, one inverted.)

The tenth V or X along the stick received an extra stroke. Thus, 50 was written variously as N, H?, K, 1:P?, ?, etc., but perhaps most often as a chicken track shaped like a superimposed V and I. This had flattened to _1_? (an inverted T) by the time of Augustus, and soon thereafter became identified with the graphically similar letter L. Likewise, 100 was variously)K?, ?, l><l?, H, or as any of the symbols for 50 above plus an extra stroke. The form)K? (that is, a superimposed X and I) came to predominate, was written variously as >I< or ?IC, was then shortened to? or C, with C finally winning out because, as a letter, it stood for *centum* (Latin for "hundred").

The hundredth V or X was marked with a box or circle. Thus, 500 was like a ? superposed on a? or f–7 (that is, like a I> with a cross bar), becoming a struck-through D or a E> by the time of Augustus, under the graphic influence of the letter D. It was later identified as the letter D. Meanwhile, 1000 was a circled X: ?, @?,®?, and by Augustinian times was partially identified with the Greek letter <I>F. It then evolved along several independent routes. Some variants, such as 1:P? and CD (more accurately a reversed D adjacent to a regular D), were historical dead ends (although folk etymology later identified D for 500 as half of <I>F for 1000 because of the CD variant), while two variants of ? survive to this day. One, CI?, led to the convention of using parentheses to indicate multiplication by 1000 (later extended to double parentheses as in ?, ?, etc.); in the other, ? became 00 8 and 1><1?, eventually changing to M under the influence of the word *mille* ("thousand").

The number 1732 would be denoted MDCCXXXII in Roman

numerals. However, Roman numerals are not a purely additive number system. In particular, instead of using four symbols to represent a 4, 40, 9, 90, etc. (i.e., IIII, XXX:X, VIIII, LXXX:X, etc.), such numbers are instead denoted by preceding the symbol for 5, 50, 10, 100, etc., with a symbol indicating subtraction. For example, 4 is denoted IV, 9 as IX, 40 as XL, etc. However, this rule is generally not followed on the faces of clocks, where IIII is usually encountered instead of IV. Furthermore, the practice of placing smaller digits before large ones to indicate subtraction of value was hardly ever used by Romans and came into popularity in Europe after the invention of the printing press.

The history of Roman numerals is not well documented, and written accounts are contradictory. It is likely that counting began on the fingers, and that is why we count in tens. A single stroke I represents one finger, five or a handful could possibly be represented by V, and the X may have been used because if you stretch out two handfuls of fingers and place them close, the two little fingers cross in an X. Alternatively, an X is like two Vs, one upside down. Although the Latin for 100 is *centum* and for 1000 is *mille*, scholars generally do not think that C is 100 and M is 1000 because they are the initial letters of centum and mille. The use of D could be a representation of a C with a vertical line through it representing half. My own views is that M arose out of the use of () symbols to multiply by 1000. This theory is supported by the use of (I) for 1000 and I) for 500. These could easily become corrupted or abbreviated into M or D, which they resemble.

Probability and Mathematics

It is well-known that the ideas of randomness are central to much of modern physics and have overthrown the "clockwork universe" conceptions of earlier centuries. The laws of probability and statistics were developed by such mathematicians as Fermat, Pascal, and

Gauss, and received their first major application in physics in the kinetic theory of gases developed by Maxwell and Boltzmann.

Here, the use of probability is necessary because the number of particles involved is too great for a deterministic/mathematical calculation. With the advent of quantum theory, physics seemed to be based on an essential randomness, whose reality was debated by Bohr and Einstein until the end of their lives. Only later, in the experiments of Alain Aspect, has a convincing demonstration been given that the inescapable randomness of quantum theory is a fact of nature.

Since the molecules and their collisions are so numerous, and the velocities so varied, and since our ignorance of the initial conditions is almost total, Maxwell postulated that positions and velocities are distributed at random; and he was confident that this assumption would describe the gas adequately and would allow one to calculate the mean values of the macroscopic variables. His breathtaking intuition was confirmed half a century later by the work of Albert Einstein (1905) and of Jean Perrin (1908) on Brownian motion.

To clarify, think about one mole of gas. One mole of any molecular substance such as $O2$ contains 6.02×10^{23} molecules. To construct the theory governing a deterministic system of 10^{23} molecules, physicists exploited probabilities through ignorance, and with complete success. The reason for this success deserves discussion. The Soviet physicist Lev Landau has shown that a classical system requiring infinitely many parameters would behave in a totally random fashion; in other words, it would be random unavoidably and not merely by reason of our ignorance.

Boltzmann bases his analysis on the fallowing observation: the molecules are so fast and their collisions so frequent that the system rapidly loses or at least appears to lose track of the initial conditions. Typically, this leads us into the realm of probabilities through ignorance. His based his theory on the following postulates:

1. Every molecule has equal a priori probability of being in region A.

2. The system evolves spontaneously from the less toward the most probable state.

This type of studies led to model for phenomenon such as Brownian motion. The phenomenon was discovered and studied first by the botanist Robert Brown in connection with the erratic motion of pollen grain suspended in fluids. Einstein first presented a quantitative theory of the Brownian motion in 1905 based on kinematic theory and statistical mechanics. He showed that the motion could be explained by assuming that the immersed particle was continually being subjected to bombardment by the molecules of the surrounding medium. Wiener developed rigorous mathematical explanation in 1918 based on stochastic process called Wiener-Levy process.

Demonstration

- A dust particle suspended in water moves around randomly, executing what is called Brownian motion. This stems from molecular agitation, through the impacts of water molecules on the dust particle. Every molecule is a direct or indirect cause of the motion, and we can say that the Brownian motion of the dust particle is governed by very many variables. In such cases, one speaks of a random process; to treat it mathematically, we use the calculus of probabilities described.
- A compass needle acted on simultaneously by a fixed and by a rotating field constitutes a very simple physical system depending on only three variables. However, we shall see that one can choose experimental conditions under which the motion of the magnetized needle is so unsystematic that prediction seems totally impossible. In such very simple cases whose evolution is nevertheless unpredictable, one speaks of chaos and of chaotic processes; these are the terms we use whenever the variables characterizing the system are few.

The method allows one to predict the future exactly, from initial conditions that are likewise exact. The best example of a deterministic theory is classical mechanics. However, this definition of determinism is based on the tacit assumption that the deviations on arrival diminish roughly in proportion to the deviations at departure. In that case, the idealized limit can be envisaged quite clearly; and in fact, gunners can realize excellent approximations to it. But what would happen if initial and final deviations were connected by a relation more complex than simple proportionality?

Fibonacci, Art, Music, and Nature

Where did the decimal numbering system we use come from? When did we change from Roman numerals to a decimal numbering system? Well, it was in the thirteenth century when Fibonacci published his *Liber Abaci*.

According to "Who was Fibonacci?" by J. J. O'Connor and E. F. Robertson, Leonardo Pisano, better known by his nickname Fibonacci, was born in Pisa in AD 1175. He was the son of Guilielmo, a Pisan merchant and member of the Bonacci family. Fibonacci, who sometimes used the name Bigollo, was educated in North Africa. His father's job was to represent merchants trading in Bugia. Fibonacci was taught mathematics in Bugia and traveled widely with his father to Egypt, Syria, Greece, Sicily, and Provence. He could see the colossal advantages of the mathematical systems used in the countries they visited.

In 1200, Fibonacci returned to Pisa and used the knowledge he had gained on his travels to write *Liber Abaci*, which translates as "The Book of Calculations." In *Liber Abaci*, he introduced the Latin-speaking world to the decimal system, as well as the Fibonacci numbers and Fibonacci sequence for which he is best remembered today. The resulting sequence is 1, 1, 2, 3, 5, 8, 13, 21, 34, 55, . . . This sequence in which each number is the sum of the two preceding numbers has proved extremely useful and appears in many different areas of mathematics and science. For example, some plants branch in such a way that they have a Fibonacci number of growing points. Flowers often have a Fibonacci number of petals, and sunflowers have a Fibonacci number of spirals in the arrangement of the seeds. By taking the ratio of successive terms in the Fibonacci series, the special value of 1.61803, called the golden ratio, is obtained. The ratio was used in Greek architecture, such as the Parthenon in Athens, and in geometry to form a star used in many flags of the world.

Fibonacci then wrote a number of important texts that played an important role in reviving ancient mathematical skills and making significant contributions of his own. Of his books, we still have copies of *Liber Abaci* (1202), *Practica Geometriae* (1220), *Flos* (1225), and *Liber Quadratorum* (1225), although he wrote other texts as well. His text on commercial arithmetic, *Di minor guisa*, and his commentary on Book X of Euclid's Elements unfortunately are lost.

Fibonacci was a sophisticated mathematician, and his achievements were clearly recognized. His practical applications rather than the abstract theorems made him famous among his contemporaries. His most impressive work overall is *Liber Quadratorum*, which translates as "The Book of Squares," according to O'Connor and Robertson. This number theory book examines methods for finding Pythogorean triples.

According to O'Connor and Robertson, "After 1228 there is only one known document which refers to Fibonacci. This is a decree made by the Republic of Pisa in 1240 in which a salary is awarded to: 'the serious and learned Master Leonardo Bigollo.' This salary was given to Fibonacci in recognition for the services that he had given to the city, advising on matters of accounting and teaching the citizens."

Fibonacci's contribution to mathematics has been largely overlooked, and his work in number theory was almost wholly ignored and virtually unknown during the Middle Ages. At the age of seventy-five, Fibonacci passed away, but his theories and applications of mathematics are still used today.

Music

It is well known that the Fibonacci sequence of numbers and the associated "golden ratio" are manifested in nature and certain works of art. It is less well known that these numbers also underlie certain musical intervals and compositions. As is presented in numerous publications, music has a foundation in the mathematical study of sequences and series. There are thirteen notes in the span of any note through its octave. A scale is composed of eight notes, of which the fifth and third notes create the basic foundation of all chords, which are based on a tone that is a combination of two steps and one step from the root tone, which is the first note of the scale. In a scale, the dominant note is the fifth note of the major scale, which is also the eighth note of all thirteen notes that comprise the octave. This provides an added instance of Fibonacci numbers in key musical relationships. Interestingly, $8/13 = 0.61538$, which approximates the golden ratio minus one.

Here is another view of the Fibonacci relationship presented by Gerben Schwab in his YouTube video. First, number the eight notes of the octave scale. Next, number the thirteen notes of the chromatic scale. The Fibonacci numbers, in red on both scales, fall on the same keys in both methods (C, D, E, G, and C). This creates the Fibonacci ratios of 1:1, 2:3, 3:5, 5:8 and 8:13.

The golden ratio φ is an irrational number (infinite decimal representation with no pattern) equal to $(1 + \sqrt{5})/2 = 1.61803 \ldots$ It has many properties such as $1.61803 \ldots = 1 + 1/1.61803$. A golden rectangle is a rectangle so that the ratio of the longer side to the shorter side is φ. Credit cards are a typical example. The golden angle is about 137.5 degrees and is related to the golden ratio. It is $1/\phi^2$ of a circle. Also,

$$\phi = 1 + (1/(1+(1/(1+(1/(1+\ldots))))))$$

$$\varphi = 1 + \cfrac{1}{1 + \cfrac{1}{1 + \cfrac{1}{1 + \dots}}}$$

Many plants display Fibonacci phyllotaxis, featuring Fibonacci numbers and the golden angle. Based on a survey of the literature encompassing 650 species and 12,500 specimens, R. Jean (1994) estimated that among plants displaying spiral or multijugate phyllotaxis, about 92 percent of them have Fibonacci phyllotaxis.

References

Meisner, Gary. "Music and the Fibonacci Sequence and Phi." *Phi: The Golden Number*, May 4, 2012. https://www.goldennumber. net/music/.

Mathematics and Art, a Summary

There are many different aspects of math within art. For example, Fibonacci sequence starts with 0, and is when the two previous numbers add up to the following number, so it goes: 0, 1, 1, 2, 3, 5, 8, 13, 21, 34, 55, 89, . . ., etc. When this sequence is drawn out in squares, it makes spiral, also known as the "golden ratio."

The golden ratio is the number 1.618033, also known as phi (Φ). This number is found by the formula $(a + b)/a = a/b$. This relates to Fibonacci's sequence, as ratio of each successive pair of numbers in the sequence approximates phi (1.618 . . .), as 5 divided by 3 is 1.666 . . ., and 8 divided by 5 is 1.60. Phi was first used by Greek sculptor and mathematician. It was Euclid that first talked about Phi as taking a line and separating it into two in a way that the ratio of the shortest segment to the longest will be the same ratio of the longest to the original line.

The golden ratio was first called the "divine proportion" in the

1500s. Many people believed that by taking measurements of your body and then using the formula of phi, you were beautiful if the result was 1.60.

Artists recognize that the Fibonacci spiral is an expression of an aesthetically pleasing principle—the rule of thirds. This is used in the composition of a picture; by balancing the features of the image by thirds rather than strictly centering them, a more pleasing flow to the picture is achieved.

Many Renaissance artists used the golden ratio in their paintings to achieve beauty. For example, Leonardo Da Vinci used it to define all the fundamental proportions of his painting *The Last Supper,* from the dimensions of the table at which Christ and the disciples sat to the proportions of the walls and windows in the background. Da Vinci has used the golden ratio in two of his most famous paintings.

Mathematics and Music, a Different Account

From the beat of a tribal drum to a choir of crickets, music is everywhere. Other than enjoying it throughout the ages, mathematicians have tried to explore the connections between mathematics and music. Their efforts showed how mathematics can be used to analyze musical rhythms, to study the sound waves that produce musical notes, to explain why instruments are tuned, and to compose music. Some have written books that explore the relationship between mathematics and music through proportions, patterns, Fibonacci numbers or the golden ratio, geometric transformations, trigonometric functions, fractals, and other mathematical concepts. One of the first to do so was the famous mathematician Pythagoras (sixth century BC), who first noted the fractional pitch relationships in the lengths of strings (i.e., if one halves a string and plucks it, the pitch is an octave higher). Other mathematicians have noticed particular patterns in music that have mathematical properties. In the field of cognitive research, the mind-body connections between music and mathematics have fueled continuing debate surrounding the so-called Mozart effect, which was first popularized in the early 1990s. This might be explained by the fact that the same parts of the brain are active when listening to Mozart as when engaged in spatial-temporal reasoning.

Several other books are also written for those who wish to understand more about the dynamics of sound and music. Some trace the history of music and its origins in math. Around the year 1200 BC, the famous mathematician Leonardo of Pisa, who was called Fibonacci, wrote a book that promoted the use of the number system that is used today. He also discovered an interesting number pattern: 1, 1, 2, 3, 5, 8, 13, 21, 34, 55, 89, 144, 233, 377, 610, . . .The pattern continues forever. It is generated by adding two preceding numbers to find the third. Through the centuries, mathematicians have noted that the Fibonacci pattern has many connections to science, art, literature, and music.

The Relationship

Music educator Chad Criswell offers easy cross-curricular teaching ideas for math or music teachers. Included: links to more music resources and worksheets. Music teachers are well aware of the very close link between math and music. Without mathematics, music simply would not exist.

The rhythms of a piece of music are based on a standard unit of time (known as a *measure*) that can be subdivided in many different ways. *Drawing parallels between pop music and math concepts* is a great way to reinforce those cross-curricular concepts while using a strong mental hook to keep students' attention. Try these simple ideas to incorporate music or mathematics into your next lesson.

The beats in a piece of music are the pulses you tap your feet to while listening to it. Begin by proposing the notion that those beats are *durations* rather than instances in time. Draw a long horizontal rectangle on the board. That represents a full measure of music.

- Cut a measure in half, and each of the two "chunks" becomes a "half note" that lasts for two beats.
- Cut each half note into two parts, and you get quarter notes. Quarter notes are the part of the music that is most often felt as the beat we tap our foot to. Each quarter note is equal to one beat of the music.
- Each of these four beats can be cut in half again, leaving eight parts to the measure (known as eighth notes in music). Each eighth note lasts for one-half of a beat.
- Divide the eighth notes in half, and you have sixteenth notes! Each sixteenth note lasts for one-fourth of a beat.
- Remember that the names of the rhythms relate to a full measure rather than to the beats themselves. An eighth note is one-eighth of a measure, not one eighth of a beat.

Probability and Music

Wolfgang Amadeus Mozart (1756–1791), an Austrian musician and composer, was truly a child prodigy: he wrote his first symphony at age nine. His music endures more than two hundred years after his death as some of the most popular and beautiful classical music ever composed. He was one of the first composers to write in what became known as the "classical" style, along with Haydn and, later, Beethoven.

Mozart put an interesting spin on one of his works, the Musikalisches Würfelspiel, or Musical Dice Game, by introducing the laws of probability into its composition. For every sixteen bars of music in this piece, Mozart offers two choices for the eighth and sixteenth bars and eleven choices for every other bar. Any combination of choices results in a lovely minuet conforming to harmonic and compositional requirements for the Viennese minuets of his time. Mozart suggests the use of a pair of dice to make the choices: throw the dice and take the sum of the resulting numbers as the choice. More melodies can be made from this piece than there are people on Earth today!

Teaching Aspect

Music teachers are well aware of the very close link between math and music. Without mathematics, music simply would not exist. The rhythms of a piece of music are based on a standard unit of time (known as a measure) that can be subdivided in many different ways. Drawing parallels between pop music and math concepts is a great way to reinforce those cross-curricular concepts while using a strong mental hook to keep students attention.

Once students understand the concept of beats and measures, the discussion can be taken to the next level, and common musical symbols can be used as a way to do fractional math.

Draw a diagram with four quarter notes in place of the four subdivided boxes mentioned previously. Put plus signs between each quarter note and an equals sign at the end. How many beats are in this math problem? Four!

- Make things more challenging and get into fractions by adding eighth notes and sixteenth notes. Two eighth notes plus a sixteenth note equals one and one-half beats (0.5 +0.5 +0.25 = 1.25 beats)
- Use the rectangle diagram to reinforce the math concepts at work here.

Music can be a lot more complicated than this, but the examples above relate to the most common time signature used in music (4/4 time) and the most commonly heard time signature in popular music. Finding a piece of music to help demonstrate those concepts is as easy as listening to the radio or borrowing a student's iPod. Further, many great free music worksheets related to these and other musical concepts are available for use in the classroom. Try to incorporate these tricks into either a math or a music class to show just how closely related these two subjects truly are.

Simpson's Paradox

Many misleading uses of statistical methods can be found in newspapers, magazines, or even journal articles summarizing experimental research. One source of this is poor statistical reasoning. However, it is also possible for perfectly correct statistical reasoning to give misleading or puzzling results. It is often difficult to decide which of several ways of analyzing data is most appropriate. The fact is that there is more to the wise use of statistics than a knowledge of classical statistical techniques. Our aim here is to discuss some examples in which different analyses lead to apparently contradictory results and to resolve those contradictions, at least partially. Specifically, we discuss examples of Simpson's paradox, which can arise from treating nonrandom samples as random or from misinterpreting probabilities even though they have been computed correctly.

Simpson's paradox is named after Edward Hugh Simpson, a statistician who gave a careful discussion of it. The best way to gain an appreciation of the surprising results associated with Simpson's paradox is to see concrete examples. One example described by Simpson involves a rookie who is trying to break into a Major League Baseball lineup by replacing a veteran. For our first example, we also present such a scenario.

Example 1. A rookie is told by the manager that starters are chosen on the basis of hitting ability. Since the rookie is batting 0.380 and the veteran is batting 0.320, the rookie is excited in anticipation of the first game. However, the rookie is dismayed to learn that the veteran is designated to start the season opener. When he questions the manager about this, he is told that the opponents are using a right-handed pitcher for the first game. Since the veteran is batting 0.300 against right-handers and the rookie is only batting 0.200 against right-handers, the rookie accepts his spot on the bench for game one.

He knows that the opponents have scheduled a left-handed pitcher for the second game, and he bats 0.400 against left-handers. When, to his horror, the rookie learns that the veteran is also slated to start the second game, he immediately confronts the manager. Again, the manager defends his choice and explains that the veteran is batting 0.500 against left-handers and is, hence, a better choice than the rookie. The rookie is left to sit on the bench in disbelief.

Has the manager been deceiving the rookie? Can the numbers be correct? Is it really possible for the veteran to have a better batting average against right-handers and left-handers and yet have a worse batting average overall? After all, right-handed pitchers and left-handed pitchers are the only kinds of pitchers that there are. The fact that this apparent contradiction is possible is the reason that we have a paradox on our hands. The data in table 1 certainly confirms that the batting averages are correct.

Table 1: Baseball Batting Averages

Player	Batting Average		
	v. left-handers	v. right-handers	overall
Veteran	$\frac{50}{100} = 0.500$	$\frac{270}{900} = 0.300$	$\frac{320}{1000} = 0.320$
Rookie	$\frac{36}{90} = 0.400$	$\frac{2}{10} = 0.200$	$\frac{38}{100} = 0.380$

Careful analysis of table 1 reveals the source of the paradox. The basic idea is that most of the veteran's at-bats came against right-handers, while most of the rookie's at-bats came against left-handers. Since the rookie is better against left-handers than the veteran is against right-handers, it is possible for the rookie to have a better overall average. A more careful explanation is possible once we discuss the concept of a weighted average. However, we first present a few more examples of Simpson's paradox.

Example 2. The batting averages of the veteran and the rookie in table 1 are not that far apart. Will that always be the case in examples of Simpson's paradox? Is it possible to have examples with extreme differences in the overall batting average? The data in table 2 shows that even the greatest extremes are possible.

Table 2: Extreme Baseball Batting Averages

Player	Batting Average		
	v. left-handers	v. right-handers	overall
Veteran	$\frac{1}{1} \approx 1.000$	$\frac{1}{1999} \approx 0.001$	$\frac{2}{1000} = 0.001$
Rookie	$\frac{1998}{1999} = 0.999$	$\frac{0}{1} = 0$	$\frac{1998}{2000} = 0.999$

The numbers in table 1 might have suggested that the manager's means of deciding his starters is a good one. However, the example in table 2 certainly warrants a decision based on overall batting average and not batting averages against different kinds of pitchers. In general, neither method is foolproof. This is one reason why they do not simply hire statisticians as baseball managers.

Example 3. Two treatments A and B are being considered for a deadly disease. A doctor tells his patient that the success rate of treatment A is 36 percent and the success rate of treatment B is 45 percent. Which should the patient choose? What if the patient was given the results in table 3? There, the study is split according to gender, and treatment A has greater success for both men and women.

Table 3: A Comparison of Two Treatments

Treatment	Fraction Recovering		
	among females	among males	overall
A	$\frac{40}{50} = 0.80$	$\frac{50}{200} = 0.25$	$\frac{90}{250} = 0.36$
B	$\frac{70}{100} = 0.70$	$\frac{20}{100} = 0.20$	$\frac{90}{200} = 0.45$

Which treatment is better?

The round numbers in Examples 1, 2, and 3 make it pretty obvious that they are fictional. To eliminate any thoughts that Simpson's paradox never actually happens, our next two examples are real.

Example 4. Is smoking good for women? In 1972–1974, a one-in-six survey was carried out in Whickham, United Kingdom (Tunbridge et al. 1977). For the sake of this example, we are only concerned with the fact that 1,314 women were asked their age and whether or not they smoked. Twenty years later, a follow-up study was conducted (Vanderpump et al. 1995). Our interest here is in which of the women from the original study were still living. The results of the study are presented in table 4. Notice that the women are categorized according to their age at the time of the first survey.

Table 4: Smoking and Twenty-Year Survival Rates

Habit	Twenty-Year Survival Rate (from age at start of study)						
	<35	35-44	45-54	55-64	65-74	75+	Overall
Non-Smoker	$\frac{207}{213}=0.97$	$\frac{114}{121}=0.94$	$\frac{66}{78}=0.84$	$\frac{81}{121}=0.66$	$\frac{28}{129}=0.21$	$\frac{0}{61}=0$	$\frac{502}{732}=0.68$
Smoker	$\frac{169}{174}=0.97$	$\frac{95}{109}=0.87$	$\frac{103}{130}=0.79$	$\frac{64}{115}=0.55$	$\frac{7}{36}=0.19$	$\frac{0}{13}=0$	$\frac{443}{582}=0.76$

Although the overall survival rate was greater among smokers, the opposite is true in the individual age groups. Which part of the results do you suppose the tobacco industry would report if they did the study? Does the study tell us that smoking is not bad (or even good) for women?

Example 5. The death rates in 1910 from tuberculosis in New York and Richmond have been separated according to race (Cohen and Nagel 1934). Table 5 shows that the death rates for both whites and nonwhites were greater in New York. However, the overall death rate was greater in Richmond.

Table 5: Death Rates from Tuberculosis in 1910.

Death Rates from Tuberculosis			
City	among nonwhites	among whites	overall
New York	$\frac{513}{91,709} = 0.00560$	$\frac{8,365}{4,675,174} = 0.00179$	$\frac{8,878}{4,766,883} = 0.00186$
Richmond	$\frac{155}{91,709} = 0.00332$	$\frac{131}{80,895} = 0.00162$	$\frac{286}{127,623} = 0.00224$

Public health officials might have reached very different conclusions depending on which part of the table received their attention.

Weighted averages. In each of tables 1 through 5, the last column seems to contradict the previous columns. The key to understanding how this can occur is to realize that in each row, the last entry is a weighted average of the previous entries. Everyone knows how to take a simple average of two values x and y. It is given by $\frac{x+y}{2} = \frac{1}{2}x + \frac{1}{2}y$. The right-hand side of this equation emphasizes the fact that the average gets an equal contribution from each of x and y. More generally, given two nonnegative numbers α and β such that $\alpha+\beta = 1$, there is an associated weighted average of x and y given by $\alpha x + \beta y$. If $\alpha > \beta$, then x carries more weight in the average. The simple average is just the special case in which the weights are equal since $\alpha=\beta=1/2$ Of course, weighted averages can also be computed for any number of values (not just two) provided the appropriate number of coefficients (α, β, γ, etc.) that sum to 1 is given.

The formal concept of a weighted average may seem strange at first, but it is definitely not unfamiliar. Consider what happens if someone invests in the stock market. Suppose, for example, that investments are made in three companies, and table 6 shows their rate of growth after one year.

Table 6: Stock Growth

Company	% Increase
A	20
B	0
C	-10

The profit made on the investment is a weighted average based on the relative amounts invested in each company. If $1,000 is invested by putting $500 into company A, $200 into company B, and $300 into company C, then the percent increase obtained from the investment is $\frac{500}{1000}(0.20)+\frac{200}{1000}(0)+\frac{300}{1000}(-0.10)=0.07$—that is, a 7 percent profit or $70 is made on the investment of $1,000. Another familiar example of a weighted average is the computation of students' grades in a class. For example, if a syllabus states that tests comprise 50 percent of the grade, homework 20 percent, and the final exam 30 percent, then the overall grade is a weighted average.

With the concept of a weighted average in hand, the numbers in tables 1 through 5 can be better understood. For example, the batting averages in the last column of table 1,

$$\frac{100}{1000}(0.500)+\frac{900}{1000}(0.300)=0.320$$

$$\frac{90}{100}(0.400)+\frac{10}{100}(0.200)=0.380$$

are weighted averages. The key to understanding how Simpson's paradox occurs is realizing that the weights are different for the veteran V and the rookie R. For V, the weights are $\alpha_V=\frac{100}{1000}=0.10$ and $\beta_V=\frac{100}{1000}=0.90$. For R, the weights are $\alpha_R=\frac{90}{1000}=0.90$ and $\beta_R=\frac{10}{100}=0.10$. The veteran's overall average gets most of its weight from at-bats against right-handers, while the rookie's average gets most of its weight from at-bats against left-handers. Since the weights are completely different, comparing their overall averages is like comparing apples and oranges. It does seem reasonable to compare the veteran and the rookie versus left-handed pitchers since they have faced similar numbers. However, the rookie's ability against right-handers is certainly not yet clear. The manager would probably want to see several more at-bats before making a long-term decision. The point is that statistical analysis requires more than merely looking at the data. The underlying application is relevant to deciding how to interpret the results.

In table 2, it seems that the last column is the most important; whereas, in table 4, the columns representing each age group give the important results. One reason the overall survival rate in the last column of table 4 may seem counterintuitive can be seen by thinking about the weighted average computation. Few of the older women in the study were smokers, but many of them had died by the time of the follow-up. Perhaps there were not too many older smokers around to participate in the original study because smokers die at an earlier age. However, this possibility is not accounted for in the last column containing overall results.

Careful inspection of tables 1 through 5 shows the source of the paradox. In each case, the weights used for the weighted average in the first row are very different from those used in the second row. Hence, the summary statistics in the final column containing overall averages seems counterintuitive. Statistics don't lie, but they can mislead. It pays to have some understanding of statistical analysis and ability to scrutinize reported results. Another very readable explanation of Simpson's paradox and another real-world example can be found in Westbrook (1998).

Zeno's Paradox

Suppose that a person is given a hamburger and is asked to first divide it into two equal pieces and eat one piece. Then do the same with the other half—that is, divide it into two pieces, eat one, and repeat the process using the other one. Would this person ever finish the hamburger? How long would it take?

Funny version: A group of boys are lined up on one wall of a dance hall, and an equal number of girls are lined up on the opposite wall. They are instructed to advance toward each other by one-quarter the distance separating them every ten seconds. Here is the question: when do they meet at the center of the dance hall? And here are the answers. Mathematician: never because the series is infinite. Physicist: when time equals infinity. Engineer: they would get close enough for all practical purposes.

The great Greek philosopher Zeno of Elea (born sometime between 495 and 480 BC) proposed four paradoxes in an effort to challenge the accepted notions of space and time that he encountered in various philosophical circles. His paradoxes confounded mathematicians for centuries, and it was not until Cantor's development (in the 1860s and 1870s) of the theory of infinite sets that the paradoxes could be fully resolved.

Zeno's paradoxes focus on the relation of the discrete to the continuous, an issue that is at the very heart of mathematics. Here, we will present the first of his famous four paradoxes. Zeno's first paradox attacks the notion held by many philosophers of his day that space was infinitely divisible, and that motion was therefore continuous.

The St. Petersburg Paradox

Suppose that you are offered the following game knowing that you are a 50 percent free-throw shooter. You will try free throws and will receive 2^n pennies, where n is the number of shots made before the first miss. Of course, there is a price to play this game. How much would you pay to play this game?

Solution: The probability that the award will be 2^n pennies is equal to the probability that you will make n shots and then miss one, which is equal to $1/2^{n+1}$. Hence, the expected value of the award (in pennies) is equal to sum of the $2^n \times 1/2^{n+1}$ for n= 1, 2, 3, . . . The answer is than infinity—that is, the average amount of the award is infinite! Hence, according to the interpretation of expected value, it seems that you should be willing to pay an infinite amount.

Now, suppose that it is agreed that the award will be truncated at 2^{30} cents (which is just over \$10 million!)—that is, the award will be frozen once it exceeds 2^{30} cents. For this case, the award is equal to $2^{\min(30, n)}$ pennies, where n is as before. How much would you be willing to pay for this new award?

Surprisingly, the expected value of the new award (in cents) is equal to $30/2 + 1 = 16$—that is, truncating the award at just over \$10 million changes its expected value enormously from infinity to sixteen cents.

The Motionless Runner

A runner wants to run a certain distance—let us say 100 meters—in a finite time. But to reach the 100-meter mark, the runner must first reach the 50-meter mark, and to reach that, the runner must first run 25 meters. But to do that, he or she must first run 12.5 meters. Since space is infinitely divisible, we can repeat these "requirements" forever. Thus, the runner has to reach an infinite number of "midpoints" in a finite time. This is impossible, so the runner can never reach his goal. In general, anyone who wants to move from one point to another must meet these requirements, and so motion is impossible, and what we perceive as motion is merely an illusion.

Although, practically, no scholars today would agree with Zeno's conclusion, we cannot escape the paradox by jumping up from our seat and chasing down a tortoise, nor by saying Zeno should have constructed a new argument in which Achilles takes better aim and runs to some other target place ahead of where the tortoise is. Because Zeno was correct in saying Achilles needs to run at least to all those places where the tortoise once was; what is required is an analysis of Zeno's *own* argument.

Mathematics Movies

- *A Beautiful Mind,* winner of four Academy Awards, including best picture, is about a brilliant mathematician John Nash on the brink of international acclaim when he becomes entangled in a mysterious conspiracy.
- In *Pi,* again, a brilliant mathematician teeters on the brink of insanity as he searches for an elusive numerical code in this critically acclaimed schizophrenic thriller! He is on the verge of that most important discovery of his life. For the past ten years, he has been attempting to decode the numerical pattern beneath the ultimate system of ordered chaos—the stock market. As Max verges on a solution, chaos is swallowing the world around him. Pursued by an aggressive Wall Street firm set on financial domination and Kabbalah sect intent on unlocking the secrets behind their ancient holy texts, he races to crack the code, hoping to defy the madness that looms before him. Instead, he uncovers a secret for which everyone is willing to kill.
- *Proof* is the compelling story of an enigmatic young woman haunted by her father's past and the shadow of her own future, exploring the links between genius and madness, the tender relationships between fathers and daughters, and the nature of truth and family. On the eve of her twenty-seventh birthday, a young woman who has spent years caring for her brilliant but unstable father, a mathematical genius must deal not only with the arrival of her estranged sister but also with the attentions of a former student of her father who hopes to find valuable work in the 103 notebooks of his.
- In *Rain Main,* a man has just discovered he has an autistic brother and is now taking him on the ride of his life. Or is it the other way around? From his refusal to drive on major highways to a "four minutes to Wapner" meltdown at an Oklahoma farmhouse, he first pushes hotheaded brother to the limits of his patience . . . and then pulls him

completely out of his self-centered world! But what begins as an unsentimental journey for the Babbitt brothers becomes much more than the distance between two places—it's a connection between two vastly different people.

- In *Enigma,* an East German defector who has turned CIA agent is assigned to discover the names of five Soviet dissidents targeted by a ruthless KGB hit squad. To stop this diabolic plot, he must infiltrate Soviet intelligence and obtain information from Enigma, a Russian computer component. But when he arrives in Berlin, his cover is blown, and he quickly seeks shelter with his former lover. But as they get deeper in this deadly game of pursuit, he must outwit the assassins before they fall into the hands of the enemy.

- *Good Will Hunting,* nominated for nine Academy Awards, is about a headstrong working-class genius. After one too many run-ins with the law, Will's last chance is a psychology professor.

- *Numbers* is a drama about an FBI agent who recruits his mathematical-genius brother to help the bureau solve a wide range of challenging crimes in Los Angeles. The two brothers take on the most confounding criminal cases from a very distinctive perspective. Inspired by actual events, the series depicts how the confluence of police work and mathematics provides unexpected revelations and answers to the most perplexing criminal questions. A dedicated FBI agent couldn't be more different from his younger brother, a brilliant mathematician who, since he was little, yearned to impress his big brother. He is joined on his team by fellow agents, a behavioral specialist who brings psychological insight to their investigations.

After some initial reluctance, their team welcomes the innovative methods to crime-solving. Their father is happy to see his sons working together even though he doesn't understand the intricacies of what he does for a living. It is his coworkers who further refine his approach and help him stay focused. Physicist friend constantly challenges him to employ a broader point of view to his work with the FBI. His former grad student frequently helps him see cases in a new light. Despite their disparate approaches to life, brothers are able to combine their areas of expertise and solve some killer cases.

Love of Music

From early childhood, I felt something deeply mysterious about music partly because harmonies are relationships of numbers. Since ancient times, music has been understood as numbers made audible.

To me, it is something in science and mathematics in many ways, as well as in everything else. It affects me so powerfully. I found it something helpful, joyful, and encouraging. A subtle form of communication. It may also seem strange to us that numbers express qualities, yet music shows that this is so. We might say that music specifically expresses numbers as qualities.

I became interested in fundamental questions about the nature and value of music and my experience of it. I even became excited about philosophy of music as I moved from Middle East to Europe and then to the United States and experienced different kind of music. Music is perhaps the art that presents the most philosophical puzzles. Unlike painting, its works often have multiple instances, none of which can be identified with the work itself. Music brings to the sense of hearing the numerical order of nature. As pointed out by Joseph Milne, "In music we hear the qualities of the numbers, not simply their quantities, and it is these qualities that move the heart."

Music has been and is my sanctuary and addiction/drug; I consider music a universal language. For me, music could address physical, emotional, cognitive, and social needs of individuals. It can also provide avenues for communication that can be helpful to those who find it difficult to express themselves in words. It moves through time, evokes so much emotion, provides understanding and comfort, excitement and joy, and is anything from calming to exiting . . .

I think the world would be a better place if music were a religion rather than what we have today, since, at least, music evolves and adapts over time . . . It also can bring people in our world together in many ways. Whether it be through the same taste in music or the willingness to try something new or even performing music with others. Many people like the same genres or styles of music that are out in the world right now. In sum, without music, life would be boring and dull.

Wolfgang Amadeus Mozart (1756–1791), an Austrian musician and composer, was truly a child prodigy: he wrote his first symphony at age nine. His music endures more than two hundred years after his death as some of the most popular and beautiful classical music ever composed. He was one of the first composers to write in what became known as the "classical" style, along with Haydn and later Beethoven.

The classical style introduced shifts in musical phrasing away from the earlier Baroque style. Classical music uses short, articulated phrases, in contrast to the sweeping continuity of Baroque music. Classical phrases assume more of an independent character rather than simply leading into each other. As a result of this different structure, classical music reflects a sensitivity to symmetry and a varied rhythmic texture.

Mozart put an interesting spin on one of his works, the *Musikalisches Würfelspiel*, or *Musical Dice Game*, by introducing the laws of probability into its composition. For every sixteen bars of music in this piece, Mozart offers two choices for the eighth and sixteenth bars and eleven choices for every other bar. Any combination of choices results in a lovely minuet conforming to harmonic and compositional requirements for the Viennese minuets of his time. Mozart suggests the use of a pair of dice to make the choices: throw the dice and take the sum of the resulting numbers as the choice. More melodies can be made from this piece than there are people on Earth today!

Fibonacci Sequence

Throughout the ages, mathematicians have found connections between mathematics and music. One of the first to do so was the famous mathematician Pythagoras (sixth century BC), who first noted the fractional pitch relationships in the lengths of strings (i.e., if one halves a string and plucks it, the pitch is an octave higher). Other mathematicians have noticed particular patterns in music that have mathematical properties.

Around the year 1200 BC, the famous mathematician Leonardo of Pisa, who was called Fibonacci, wrote a book that promoted the use of the number system that is used today. He also discovered an interesting number pattern: 1, 1, 2, 3, 5, 8, 13, 21, 34, 55, 89, 144, 233, 377, 610, . . .

The pattern continues forever. It is generated by adding two preceding numbers to find the third. Through the centuries, mathematicians have noted that the Fibonacci pattern has many connections to science, art, literature, and music (*In Harmony with Education Program: Bose's Teacher Guide*).

Numbers and Songs

Students will create and perform songs based on math concepts and facts with this lesson plan. It is designed for students in fifth through seventh grades but can be customized to meet your classroom needs.

A novel and enjoyable way for students to highlight math concepts or facts is to write a song about them. Even if they don't have any particular musical talent, most students will plunge into this project with great enthusiasm.

Goal: Working in groups of two to four, students will compose songs that express an idea in mathematics. Upon completion of the project, students will be encouraged to perform or record their song for the class.

Suggested time: Two to three class periods.

State of Mathematics in the United States

When people find out that I am a mathematician, they either call me crazy or brilliant. They also often talk about their own inability to learn and deal with it. Both responses, however, point to a fundamental misunderstanding about what mathematics is and does, a problem American society and education face. What is even more problematic is the fact that people are not shy and even occasionally proud to announce that they are not good in mathematics as if this makes them normal and socially acceptable. It also seems that both in high schools and universities, most students learn to detest—or at best, endure—mathematics.

This is happening despite the fact that a large number of Americans believe that good mathematics skills are essential for the kids to keep up with new advances in science and technology. Research shows that the majority of Americans have a difficult time doing everyday mathematics such as figuring out tax on items. Some even expressed that they prefer cleaning the bathroom to solving a mathematics problem. Also, many proudly say they are not good in mathematics.

Although some young Americans graduate from high school with superb academic skills, a great many leave high school with little knowledge in crucial subjects such as reading, writing, basic mathematics, critical thinking, and reasoning. One of the key reasons why that is so is that many of their teachers are not very good themselves. Yes, they have their college degrees, but those degrees are easily acquired by some of the weakest students colleges admit.

When I meet new people, they often ask me what I do for a living. When they hear mathematician, they either tell me that I must be brilliant or talk about their own inability to understand it. Both responses, however, point to a fundamental misunderstanding about what mathematics is and does, a problem American education faces.

What is even more problematic is the fact that people are not shy and even occasionally proud that they are not good in math.

It seems that we are one of the few countries that teaches math by forcing students to memorize formulas and procedures rather than helping them to experience with mathematical puzzles, experiment and search for patterns, and find delight in their own discoveries. Most students learn to detest—or at best, endure—math, and this is why our students are falling behind their international peers.

- Most American eighth graders cannot do relatively simple tasks—such as add fractions—that students in other countries master by fourth grade.
- Just under half of U.S. eighth graders can correctly place three fractions—2/7, 1/2, 5/9—in ascending order.
- Only 39 percent of fourth graders, 34 percent of eighth graders, and 23 percent of twelfth graders score at or above the "proficient" level in mathematics.
- Sixty percent of community college students must take at least one so-called developmental course before enrolling in college-level classes. Many high school graduates arrive at college needing to learn not just basic algebra but also basic arithmetic.
- The most recent PISA results, from 2015, placed the United States at an unimpressive thirty-eighth out of seventy-one countries in math.
- Among the thirty-five members of the Organization for Economic Cooperation and Development, which sponsors the PISA initiative, the United States ranked thirtieth in math.
- The 2015 NAEP rated 40 percent of fourth graders, 33 percent of eighth graders, and 25 percent of twelfth-graders as "proficient" or "advanced" in math. While far fewer fourth and eighth graders now rate at "below basic," the lowest performance level (18 percent and 29 percent, respectively,

versus 50 percent and 48 percent in 1990), improvement in the top levels appears to have stalled out.

- Results from an international assessment of basic skills found that fifty-eight million American adults have low skills in working with and using mathematical information.

Where Does United States Stand?

In recent years, we have been witness to many amazing inventions and breakthroughs mostly by individuals who have received their education here in the United States. Proud of the performance and contributions of our higher education institutions, we may look ahead and wonder whether this trend will continue and will guarantee our competitiveness in the global world. Unfortunately, this might not be the case, as recent investigations point to a decline in basic knowledge of sciences and mathematics both in our schools and universities. In fact, a large percentage of positions requiring advance knowledge of sciences and mathematics are now filled by people who have recently moved to the United States from other countries. As a result, last year, for the first time ever, the U.S. Patent and Trademark Office issued more patents to foreigners than to Americans. This trend is likely to continue, since, presently, the two primary sources for graduate students in science and technology at American universities are China and South Korea.

To see where we stand, let us look at few more facts and statistics. Very recently, the biggest ever global school rankings have been published by the OECD (Organization for Economic Cooperation and Development). The study may have some drawbacks considering the facts that (1) some countries have a more homogeneous population and schools than others and (2) students in some countries are better test takers than others. Despite this, the study is considered significant, as it gives access to all countries to compare themselves against the world's education leaders, find out about their relative strengths and weaknesses, and see what long-term economic gains an improved quality in schooling could provide for them. As expected, the top-five positions were taken by Asian countries—Singapore, Hong Kong, South Korea, Taiwan, and Japan. The five lowest-ranked countries were Oman, Morocco, Honduras, South Africa, and Ghana. Out of seventy-six countries included, United States was ranked twenty-eighth. According to this report, the standard of education is a powerful predictor of the wealth that countries can generate in the long run. Also, last year, among the twenty-nine wealthiest countries, the United States ranked twenty-seventh in the proportion of college students with degrees in science and engineering. Additionally, among developed countries, the United States was ranked thirty-first in mathematics and twenty-third in science. This is despite the fact that our schools and universities have, by far, the best facilities in the world. So, considering these, it is logical to think that we may lose ground to foreign rivals unless we find a way to enhance the quality of our mathematics and science education and increase the number of students majoring in this disciplines. This, of course, needs long-term planning to find ways to motivate more students to major in science and mathematics as well as to learn more about other countries' approach and make changes or adjustments if necessary. Finally, here is the ranking at age fifteen:

<u>Singapore</u>
Hong Kong
South Korea
Japan
Taiwan
Finland
Estonia
Switzerland
Netherlands
Canada
Poland
Vietnam

Here are some reasons for the success of Asian countries: Teachers expect every student to succeed. There's a lot of rigor, focus, and coherence. Success in attracting the most talented teachers so that students have access to excellent teachers.

The Organization for Economic Cooperation and Development (OECD) is a unique forum where the governments of thirty-four democracies with market economies work with each other as well as with more than seventy nonmember economies to promote economic growth, prosperity, and sustainable development.

Mr. Inconsistent May Set a Record

A consistently inconsistent runner may give the coach what he wants.

Most coaches consider an athlete's consistency as a positive attribute, for obvious reasons. The performance of a consistent player exhibits less variability, and, therefore, such players are more predictable, something that coaches can utilize for their game plan. The following example illustrates how in some sports, variability could prove helpful for a different reason.

Think about the top two members of a college track team, A and B. Suppose that the coach has kept their statistics and has concluded that their times, in seconds, follow a bell curve (normal distribution) with means of 10.4 and 10.6 and standard deviation of 0.2 and 0.4, respectively (N (10.4, 0.2) and N (10.6, 0.4)). Suppose also that the college competition's best/present record is ten seconds. The coach thinks that on a good day, both runners have a chance to break the record, but he's not sure which runner has the better chance. His assistant thinks A, who has averaged 10.4, which is closer to 10.

The coach is wondering about another factor—namely, B's inconsistency—because he has heard that athletes with larger mean value and greater irregular performance (larger standard deviation) can have a higher chance to drop below a chosen threshold. So he uses the tables for normal distribution and calculates the probability of breaking the record for A and B. He finds that they are 0.0228 and 0.0668, respectively, which shows that the athlete with larger mean value and greater irregular performances (larger standard deviation) has a better chance of breaking the ten-second record.

To better appreciate the difference, suppose that both runners can participate in ten tournaments during a season. Based on the above calculations, the probabilities that for the season neither of them would run under ten seconds are respectively $0.794 = (1 - 0.0228)^{10}$ and $0.501 = (1 - 0.0668)^{10}$. This implies that their probabilities of breaking the record in one season are respectively 0.206 and 0.499.

We can analyze this situation differently by considering distribution of the minima of samples of sizes to ten from a standard normal or any other parent population. This can be done through what is known as order statistics, which is the data ordered in ascending order.

It's interesting that the athlete with larger mean time and frequent irregular performance (larger standard deviation) can have a better chance to run below ten seconds. In other words, the latter athlete, though more irregular and with smaller mean time, should be chosen for the competition if only one athlete must be chosen. This example illustrates how extreme values look and feel different from central value problems.

Comparing Athletes

People who follow sports compare their favorite athletes. This is not easy, especially if these athletes are from different sports or different years. However, this may be done using relative standing of the athletes by comparing them to other players in the same sport.

Z-Score

Suppose that x presents a performance measure in a specific sport. For a sample of size n, $\{x_1, x_2, ..., x_N\}$ of such measure, we can calculate the sample mean x_{avg} and the sample standard deviation s. Then for each measurement x_i, the sample z-score is calculated by subtracting the mean and dividing it by standard deviation—that is, $z_i = (x_i - x_{avg})/s$. By definition, we see that the z-score of a value tells how many standard deviations that value is away from (either above or below) the mean. Since standard deviation is positive, it follows that if the z-score is positive, the value is above the mean; if the z-score is negative, the value is below the mean; and if the z-score is equal to zero, then the value is exactly equal to the mean. Thus, z-score provides some information regarding the relative standing of a given measurement. Note that if we change the unit of measurements by adding or multiplying the data values by a fixed number, the face values of x will change, but not their z-scores. In other words, z-scores are scale-invariant. The mean of the z-scores is always zero, and the standard deviation of the z-scores is always one.

Example

Suppose we wish to compare two students taking the same course at the same university, one in the morning and the other in the afternoon class, to determine who is the better student. Suppose that the following information is available.

Description	Grade	Class Average	Class Standard Deviation	z-score
Student in the morning class	80	85	5	$z = (80 - 85)/5 = -1$
Student in the afternoon class	76	72	4	$z = (76 - 72)/4 = 1$

From this table, it is clear that although the first student has a higher grade, she has a lower z-score, which means that the second student has a much better relative standing. In other words, the grade of the second student is one standard deviation above the average grade in her class, whereas the first student's grade is one standard deviation below the average grade in her class. This means that the second student has done better than many more students in her class than the first student in her class.

Since we don't know what the grades represent, a reasonable approach is to compare students using their z-score, which indicates their relative standing. This eliminates factors such as hard tests, bad teachers, and poor textbooks. In other words, each student is judged based on his or her relative standing in their own class. We can apply the same procedure when comparing athletes from different sports or students from different majors or universities.

Mathematics, Metrics, and Sports

Think about a basketball or an indoor soccer team with five players in the field. Suppose that in a given situation (perhaps a set play), each player can pass the ball to only certain teammates. To represent this situation in matrix form, we could construct a five-by-five incidence matrix A, where the rows represent the passing player with the ball and columns represent the receiving player. We place a 1 in the ith row and jth column of this matrix if there is possibility of passing the ball to player j from i; otherwise, insert a -. We also place Os on the principal diagonal, because player i can not pass the ball to himself. With the situation represented in the mathematical form of a matrix, we can perform operations on this matrix to obtain information about the game plan.

Receiving Player

$$\text{Passing Player } A = \begin{bmatrix} 0 & 1 & 0 & 1 & 0 \\ 0 & 0 & 1 & 0 & 0 \\ 1 & 0 & 0 & 0 & 1 \\ 0 & 0 & 1 & 0 & 0 \\ 0 & 0 & 0 & 1 & 0 \end{bmatrix}$$

$$A^2 = \begin{bmatrix} 0 & 0 & 2 & 0 & 0 \\ 1 & 0 & 0 & 0 & 1 \\ 0 & 1 & 0 & 2 & 0 \\ 1 & 0 & 0 & 0 & 1 \\ 0 & 0 & 1 & 0 & 0 \end{bmatrix}, A^3 = \begin{bmatrix} 2 & 0 & 0 & 0 & 2 \\ 0 & 1 & 0 & 2 & 0 \\ 0 & 0 & 3 & 0 & 0 \\ 0 & 1 & 0 & 2 & 0 \\ 1 & 0 & 0 & 0 & 1 \end{bmatrix}, A+A^2+A^3+A^4 = \begin{bmatrix} 0 & 1 & 0 & 1 & 0 \\ 0 & 0 & 1 & 0 & 0 \\ 1 & 0 & 0 & 0 & 1 \\ 0 & 0 & 1 & 0 & 0 \\ 0 & 0 & 0 & 1 & 0 \end{bmatrix}$$

In A^2, the 1 in row two, column one indicates that there is one way that player 2 can get the ball to player 1 with one intermediate pass—namely, 2-to-3-to-1. The 2 in row one, column three indicates that there are two ways for player 1 to get the ball to player 3—namely, 1-to-2-to-3, and to 1-to-4-to-3.

In A^3, the 1 in row four, column two indicates that there is one way to get the ball from player 4 to player 2 with two intermediate passes. The 2 in row one, column five indicates that there are two ways for player 1 to get the ball to player 4 with two intermediate passes.

$A + A^2 + A^3 + A^4$ shows that it is possible to pass the ball from any player to any other player with at most three intermediate passes.

CHAPTER 3

The Fascinating World of Numbers

I do not know why integers that are not divisible by two look odd to me.
Numbers will confess to anything you like if tortured long enough.
Remember that you are number 1. Without you, 10, 100, 1,000,
1,000,000, 1,000,000,000 are all just zero, just nothing.

IT SEEMS INCONCEIVABLE how much we rely on numbers in our daily lives and how natural it feels. Our birth is announced by a set of numbers representing date, time, height, weight, etc. We become a functioning member of the society only after a social security number is assigned to us. Our health and fitness are evaluated using numbers representing our blood pressure, heart rate, body temperature, etc. From that point onward, every action performed and every life encountered becomes a part of the ongoing use of numbers. Some numbers born and die with us. Some even live long after we are gone. Numbers resonate with our personal and social lives in a greater way than we might imagine.

Numbers also have an interesting world parallel to ours. Like us, some numbers play a more significant role than others. Some represent complicated concepts. Some never get used, and among those we use, some get used more often than others. Again, like us, some numbers are rational and some irrational. A rational number is one that can be expressed as fractions such as 5/7, 3/15, 50/17. Irrational numbers cannot be presented as fractions. Although there are significantly more irrational than rational numbers, irrational numbers are hardly used in real life. Some numbers have a special significance in certain cultures and even religions. Some are loved by scientists, artists, and even the general public. Few of such numbers are presented and discussed later in this chapter.

Numbers Are Friends

Social security number, telephone number, credit card number, etc., play important parts in our lives. Our houses, cars, workplaces, etc., are full of numbers playing specific roles. They make the life easier and help us to better understand the world around us. Unlike names, numbers are unlimited and as such are a great tool for identification/labeling of people and items uniquely regardless of the population size. They are also fascinating and mystical.

Living with Numbers

Compared to names, numbers make the identification of individuals and objects much simpler. Names cannot be made too long. However, numbers could be made as long as we wish. With the letters of English language, we can make

$$26 + 26 \times 25 + 26 \times 25 \times 24 + \ldots + 26 \times 25 \times 24 \times \ldots \times 17$$

names with no longer than ten letters.

Like different civilizations or cultures, some numbers belong to a group with a given name. Examples include integers, prime, rational, irrational, real, complex, etc. Many numbers are given specific names because of their unique property. It seems inconceivable how much we rely on numbers in our daily lives and how natural it feels. Without a social security number, we are not even a functioning member of the society. Every action performed, every life encountered has become a part of the ongoing use of numbers. Some numbers born and die with us. Some even live long after we are gone.

Much like human being through the history, numbers have grown, changed, matured, and achieved a high level of sophistication. Advances in this field are apparent to anybody who shops at grocery stores and malls. Every item we buy, piece of mail we receive, ticket we purchase, etc., has an identification number. Like people, there are some extraordinary, famous, and beautiful numbers that amaze us. Some numbers are celebrated in a certain day of the year. For example, March 14 is the Pi Day, since 3.14 is the beginning of this special number. Lots can be said about pi. For example, it has a decimal representation that never stops or repeats, although modern computers have now computed it to over a trillion decimal places!

Numbers Wearing Different Hats

Numbers play variety of roles and as such may seem confusing. Here are some examples:

- As concepts: 24-7. Catch-22.
- As names: 007.
- As rank: third building
- As numeral and name: Fifth Avenue.
- As mystical or religious icon: 5, 7, 12, and 40.
- In conspiracy theories: Example, 11. Think about September 11 (9/11). Some have tried to give it a meaning through pattern-seeking approach.

- Numbers that play key roles: e, Pi, and famous relations between them. Also numbers attached to frequently appearing shapes (e.g., pi and circle).
- Iconic numbers in one system may not be iconic in other systems.
- Numbers and scaling. For instance, one hundred degrees Celsius is iconic but not if you use Fahrenheit scale.
- Real numbers can be divided into rational numbers, which include integers, both positive and negative, and fractions; and irrational numbers, which include the rest. Rational numbers are countable and can be put on one-to-one correspondence with integers, 1, 2, 3; whereas, irrational numbers cannot and as such are not countable. In a simple language, rational numbers are just a small subset of real numbers and are nothing compared to irrational numbers. But what is interesting is that we do not use irrational numbers for either numeration or identification unless we have no other choice. This is because an irrational number written as decimal has no end or pattern in digits of decimal representation. So we do what we know how to do and what we understand. In short, we only use a little subset of real numbers to deal with the problems we face. In other words, communication of numbers is only done using rational numbers.
- Prime, imaginary, and complex numbers.
- Numbers make the identification much simpler compared to names. We cannot make the names to long. However, numbers could be made as long as we wish.

Do We Comprehend Numbers?

Many believe that we humans have a hard time to comprehend the numbers. To see this, we may argue like this: think about the numbers between zero and one on an interval of lengths one. Suppose that from this line, we drop the numbers we wish. Let us first drop the infinite sequence of numbers 1/2,1/3,1/4,... Next, drop the infinite sequence 2/3, 3/4, 4/5, ..., and all other fractions such as midpoints between 1/2 and 1/3 (5/12), 1/3 and 1/4 (7/24), and all other fractions between 0 and 1. Now, what do you think will happen to that line? In other words, what will be left if any? Well, it turns out that the line will stay intact—that is, nothing will happen to that line, absolutely nothing. Why? Well, because fractions known as rational numbers, though infinite, are countable—that is, they are nothing compared to irrational numbers that are uncountable and cannot be represented as fractions. Also, their decimal representation is unlimited digits with no pattern. But what is surprising is that even though irrational numbers are, let us say, far, far more than rational numbers, very few people could write even two or three irrational numbers between 0 and 1. Why? Well, we simply do not understand irrational numbers. In our life, we constantly deal with what we understand and apply—that is, rational numbers. Here is another way of explaining the situation. Think about discrete and continuous variables. There is a gap between the values of the discrete variable. For continuous variables, this is not the case as they are compact.

As a result of all these differences, most of us have an incorrect perception about numbers, especially irrational numbers. To discuss the consequences, we note we always simplify the numbers to a level we consider acceptable or practical. In other words, we discretize almost all continuous variables measured. When we say I am thirty years old, we are not really thirty. We ignore the months, weeks, days, hours, minutes, seconds, one-tenth of seconds, . . . We may know two or few friends with the same height. But is such a thing even possible? The answer is no. In fact, the chance of finding two people with the same height is zero. Why? Because between any two numbers, never mind how close, there are infinitely many other numbers. Moreover, this infinite is uncountable. The difference between the two numbers could be in 1,000, 10,000, 1,000,000, . . . decimal place. Recall that irrational numbers written in decimal format not only have no limit but have no pattern too. Pi is a good example of this.

History of Numbers

Through history, numbers and numeration systems have played a key role in human development, and many nations have contributed to its advancement. They have gone through a dramatic change and have reached a high level of sophistication. Early application of numbers was restricted to counting or numeration. Today, numbers are extensively used as labeling and identification tools. A number is an abstract concept used to describe a quantity. A numeration system is a set of basic symbols, like 0, 1, 2, called numerals, and some rules, like addition, for making other symbols from them. The invention of a precise and "workable" numeration system is one of the greatest accomplishments of humanity. The numerical digits we mostly use today are based on the Hindu-Arabic numeral system developed over a thousand years ago. Other than decimal or base 10 system, which uses ten symbols, 0, 1, . . ., 9, the most popular system is binary or base 2 system, which uses only two symbols, 0 and 1. Because computers use a sequence of switches that can be on or off (bit), base 2 works very well there. In recent years, several other systems have been developed for identification of items, letters and packages, tickets, etc.

HISTORY: The primitive notion of numbers, requiring thousands of years to be extracted from repeated concrete situations, appears to have evolved from the many physical contrasts prevalent in nature—the difference between one tree and a forest, one sheep and a herd. Likewise, it was noticed that objects within a group could be placed in one-to-one correspondence with objects from other groups; the hands can be matched (placed in one-to-one correspondence) with the feet. Consequently, there was a recognition of an abstract property shared by some groups: that ability to be placed in a one-to-one correspondence with each other. This property is what we now call number. Those groups that could not be placed in one-to-one correspondence did not share this property and are thus said to be different in number. Out of this realization of "sameness" was born the concept of number in mathematics.

It is of great importance to recognize that the concept of number was born directly from observations of real physical phenomena exhibited in nature. The development of the concept of number was a consequence of observing both the likeness and contrasts found within nature: the likeness of groups that could be placed in one-to-one correspondence, the contrast of a single object against many objects of the same form. Indeed, the very notion of form and size was developed from observing the likeness and contrasts of objects. Our ability to perceive and interpret such likeness and contrast is what ultimately led to the concept of number. In the beginning, everything was void, and J. H. W. H Conway began to create numbers. Conway said, "Let there be two rules which bring forth all numbers large and small. This shall be the first rule: Every number corresponds to two sets of previously created numbers, such that no member of the left set is greater than or equal to any member of the right set. And the second rule shall be this: One number is less than or equal to another number if and only if no member of the first number's left set is greater than or equal to the second number, and no member of the second number's right set is less than or equal to the first number."

The word "number" is a general term that refers to a member of a given (possibly ordered). The meaning of "number" is often clear from context (it can refer to a complex number, integer, real number, etc.). Wherever possible, the word "number" is used to refer to quantities, which integers are, and constant is reserved for nonintegral numbers, which have a fixed value.

Because terms such as real numbers, Bernoulli numbers, and irrational numbers are commonly used to refer to nonintegral quantities, however it is not possible to be entirely consistent in nomenclature. The first numbers were created to answer the question, how many? These are the counting or natural numbers. The smallest numbers of this infinitely large set are 1, 2, 3, 4, 5, and 6. Counting numbers also name the sum or product of any two counting numbers. As humans become more sophisticated, they needed numbers to express more sophisticated ideas; they created more numbers, more sophisticated numbers. When none are present, the number zero, the smallest whole number, is needed. To subtract any two counting numbers, more numbers needed to be created. One can subtract 8 from 10 and name the difference with a counting number, but one cannot subtract 10 from 8 using only counting numbers. To do this, integers are needed. The integers include the whole numbers and their opposites. The opposites of 2, –(2), is negative two, –2. Once the idea of –2 exists, one can subtract 10 from 8. The reminder is –2.

To indicate a particular numerical label, the abbreviation "NR" is sometimes used (deriving from *numero,* the ablative case of the Latin *numerus*), as is the less common "no." The symbol # (known as the octothorpe) is commonly used to denote "number."

While some authors prefer to include "and" between various parts of a number name, in this work, "ands" are omitted. For example, the number 101 is called "one hundred one" rather than "one hundred and one." All Arabic numbers we use today are ideograms created by Abu Ja'far Muhammad ibn Musa al-Khwarizmi (c. 778–c. 850). Using the abacus notations, he developed the manuscript decimal system. Based on additives angles, he defined the numbers 1, 2, 3, and 4. Using his knowledge about the abacus manuscript notations, he defined the numbers 5, 6, 7, 8, 9, 0.

The development of language was essential to the rise of mathematical thinking, for language is the means of communicating thinking. Yet words expressing numerical ideas were slow in arising. Number signs proceeded number words; it is easier to cut notches in a stick than it is to establish a universally acknowledged phrase to identify a number. Had language been more advanced than number systems, complex numbers were being used by mathematicians long before they were first properly defined, so it's difficult to trace the exact origin. One of the earlier number creations was by Cardano, in 1545, in the course of investigating roots of polynomials. During this period, the notation "was used, but more in the sense of a convenient fiction to categorize the properties of some polynomials. It was seen how the notation could lead to fallacies such as that described in what was considered a useful piece of notation when putting polynomials into categories but was not seen as a real mathematical object. Later,

Euler, in 1777, eliminated some of the problems by introducing the notation i and -i for the two different square roots of -1. With him originated the notation a + bi for complex numbers. He also began to explore the extension of functions like the exponential function to the case of complex-valued arguments. However, the numbers i and -i were called "imaginary" (an unfortunate choice of terminology that has remained to this day), because their existence was still not clearly understood.

Wessel, in 1797, and Gauss, in 1799, used the geometric interpretation of complex numbers as points in a plane, which made them somewhat more concrete and less mysterious. Finally, Hamilton, in 1833, put complex numbers on a sound mathematical footing by showing that pairs of real numbers with an appropriately defined multiplication form a number system and that Euler's previously mysterious "i" can simply be interpreted as one of these pairs of numbers. That was the point at which the modern formulation of complex numbers can be considered to have begun.

More than Just a Number

Among many other applications, numbers are used for communication and representation. For example, as mentioned earlier, catch-22 represents a concept. The 007 represents a name. Numbers 5, 7, 12, 13, 40, and 666 represent mystical, unlucky, or religious icons. Pi indicates a geometric object.

Golden ratio: The golden ratio (approximately 1.618) is a number used to create aesthetically pleasing designs and artworks like Leonardo da Vinci's *Mona Lisa*.

Fibonacci numbers: Fibonacci numbers are 0, 1, 1, 2, 3, 5, 8, 13, 21, 34, 55, . . ., where each number in the sequence is the sum of the two preceding numbers. They are related to golden ratio and appear frequently in nature. For example, the spiral shapes of sunflowers follow a Fibonacci sequence. It also appears in the foundation of aspects of art, beauty, music, and life.

Google: The name of the popular search engine Google came from a misspelling of the word "googol," which is a very large number (one followed by one hundred zeroes).

Prime Numbers

A prime number (or a prime) is a natural number greater than 1 that has no positive divisors other than 1 and itself. Numbers that have more than two factors are called a composite number. For example, 5 is prime because 1 and 5 are its only positive integer factors; whereas, 6 is composite because it has the divisors 2 and 3 in addition to 1 and 6. The fundamental theorem of arithmetic establishes the central role of primes in number theory: any integer greater than 1 is either a prime itself or can be expressed as a product of primes that is unique up to ordering. For example, 6 = 2 x 3, where both 2 and 3 are prime. The uniqueness in this theorem requires excluding 1 as a prime because one can include arbitrarily many instances of 1 in any factorization (e.g., 1 x 3 = 1 x 1 x 3, etc., are all valid factorizations of 3). Here are the first few prime numbers: 2, 3, 5, 7, 11, 13, 17, 19, 23, 29, 31, 37, 41, 43, 47, 53, 59, 61, 67, 71, 73, 79, 83, 89, 97, displayed below.

1	2	3	4	5	6	7	8	9	10
11	12	13	14	15	16	17	18	19	20
21	22	23	24	25	26	27	28	29	30
31	32	33	34	35	36	37	38	39	40
41	42	43	44	45	46	47	48	49	50
51	52	53	54	55	56	57	58	59	60
61	62	63	64	65	66	67	68	69	70
71	72	73	74	75	76	77	78	79	80
81	82	83	84	85	86	87	88	89	90
91	92	93	94	95	96	97	98	99	100

The following is a list of all the prime numbers up to 1,000:

2	71	167	271	389	503	631	757	883
3	73	173	277	397	509	641	761	887
5	79	179	281	401	521	643	769	907
7	83	181	283	409	523	647	773	911
11	89	191	293	419	541	653	787	919
13	97	193	307	421	547	659	797	929
17	101	197	311	431	557	661	809	937
19	103	199	313	433	563	673	811	941
23	107	211	317	439	569	677	821	947
29	109	223	331	443	571	683	823	953
31	113	227	337	449	577	691	827	967
37	127	229	347	457	587	701	829	971
41	131	233	349	461	593	709	839	977
43	137	239	353	463	599	719	853	983
47	139	241	359	467	601	727	857	991
53	149	251	367	479	607	733	859	997
59	151	257	373	487	613	739	863	more....
61	157	263	379	491	617	743	877	
67	163	269	383	499	619	751	881	

The property of being prime is called primality. A simple but slow method of verifying the primality of a given number is known as trial division. Algorithms much more efficient than trial division have been devised to test the primality of large numbers. These include the Miller–Rabin primality test, which is fast but has a small probability of error; and the AKS primality test, which always produces the correct answer in polynomial time but is too slow to be practical.

Particularly fast methods are available for numbers of special forms, such as Mersenne numbers. As of January 2018, the largest known prime number has 23,249,425 decimal digits. It has one million more digits than the previous record holder. The discovery was initially made on December 26, 2017, by a computer volunteered by Jonathan Pace as part of the Great Internet Mersenne Prime Search (GIMPS). Pace, a fifty-one-year-old electrical engineer based in Germantown, Tennessee, has been hunting primes for over fourteen years, and his belated Christmas present is eligible for a $3,000 reward from GIMPS.

The new prime number, known as M77232917, is one million digits larger than the previous record. It is also a particularly rare type of prime called a Mersenne prime, meaning that it is one less than 2 to the power of a number minus 1, such as $3 = 2^2 - 1$.

There are infinitely many primes, as demonstrated by Euclid around 300 BC. There is no known simple formula that separates prime numbers from composite numbers. However, the distribution of primes—that is to say, the statistical behavior of primes in the large—can be modelled. The first result in that direction is the prime number theorem, proven at the end of the nineteenth century, which says that the probability that a given randomly chosen number is prime is inversely proportional to its number of digits, or to the logarithm of *that number*.

Many questions regarding prime numbers remain open, such as Goldbach's conjecture (that every even integer greater than 2 can be expressed as the sum of two primes) and the twin prime conjecture (that there are infinitely many pairs of primes whose difference is 2). Such questions spurred the development of various branches of number theory, focusing on analytic or algebraic aspects of numbers. Primes are used in several routines in information technology, such as public-key cryptography, which makes use of properties such as the difficulty of factoring large numbers into their prime factors. Prime numbers give rise to various generalizations in other mathematical domains, mainly algebra, such as prime elements and prime idea.

Here is an interesting result.

For every prime number p, there exists a prime number p' such that p' is greater than p. This mathematical proof, which was demonstrated in ancient times by the Greek mathematician Euclid, validates the concept that there is no "largest" prime number. As the set of natural numbers $N = \{1, 2, 3, ...\}$ proceeds, however, prime numbers generally become less frequent and are more difficult to find in a reasonable amount of time.

How Do I Know if a Number Is Prime?

A computer can be used to test extremely large numbers to see if they are prime. But because there is no limit to how large a natural number can be, there is always a point where testing in this manner becomes too great a task even for the most powerful supercomputers. Various algorithms have been formulated in an attempt to generate ever-larger prime numbers.

Prime Numbers and Cryptography

Cryptography or cryptology is the practice and study of techniques for secure communication in the presence of third parties called adversaries. More generally, cryptography is about constructing and analyzing protocols that prevent third parties or the public from reading private messages.[3]

Cryptography prior to the modern age was effectively synonymous with *encryption*, the conversion of information from a readable state to apparent nonsense. The originator of an encrypted message shared the decoding technique needed to recover the original information only with intended recipients, thereby precluding unwanted persons from doing the same. The cryptography literature often uses the name Alice (A) for the sender, Bob (B) for the intended recipient, and Eve (eavesdropper) for the adversary.[5] Since the development of rotor cipher machines in World War I and the advent of computers in

World War II, the methods used to carry out cryptology have become increasingly complex and its application more widespread.

Modern cryptography is heavily based on mathematical theory and computer science practice; cryptographic algorithms are designed around computational hardness assumptions, making such algorithms hard to break in practice by any adversary. It is theoretically possible to break such a system, but it is infeasible to do so by any known practical means. These schemes are therefore termed computationally secure; theoretical advances (e.g., improvements in integer factorization algorithms) and faster computing technology require these solutions to be continually adapted. There exist information theoretically secure schemes that provably cannot be broken even with unlimited computing power—an example is the one-time pad—but these schemes are more difficult to implement than the best theoretically breakable but computationally secure mechanisms.

Encryption always follows a fundamental rule: the algorithm (the actual procedure being used) doesn't need to be kept secret, but the key does. Even the most sophisticated hacker in the world will be unable to decrypt data as long as the key remains secret—and prime numbers are very useful for creating keys. For example, the strength of public/private key encryption lies in the fact that it's easy to calculate the product of two randomly chosen prime numbers, but it can be very difficult and time consuming to determine which two prime numbers were used to create an extremely large product when only the product is known.

RSA (Rivest-Shamir-Adleman)encryption is based on prime numbers—two prime numbers multiplied together. The original two prime numbers are known as your "private key." When you multiply them together, the product (a number that's only divisible by one, itself, and those two prime numbers) is called the "public key." In RSA public key cryptography, prime numbers are always supposed to be unique. The primes used by the Diffie-Hellman key exchange and the Digital Signature Standard (DSA) cryptography schemes, however, are frequently standardized and used by a large number of applications.

Real Numbers

Real numbers include rational and irrational numbers. Rational numbers include integers, both positive and negative, and fractions; and irrational numbers include the rest. A sequence of numbers that can be put on one-to-one correspondence with integers 1, 2, 3, . . . is called countable. Rational numbers are countable; whereas, irrational numbers are not countable. In a simple language, rational numbers are nothing compared to irrational numbers, a much bigger infinity. However, what is interesting is that we do not use irrational numbers neither for numeration nor for identification unless we have absolutely no choice. This is because irrational numbers written as decimals never end, nor exhibit any pattern. So we do what we know how to do or understand. In other words, communication of numbers is only done by using rational numbers.

Imaginary Numbers

When you look in the mirror, you see your picture. Your picture is not real but imaginary. However, it is very useful and can give you lots of information and insight.

We have always been told that we cannot take the square root of a negative number because the square of any number is always positive. So you could not very well square root a negative and expect to come up with anything sensible.

Now, however, you can take the square root of a negative number, but it involves using a new number to do it. This new number was invented (discovered?) around the time of the Reformation. At that time, nobody believed that any "real world" use would be found for this new number other than easing the computations involved in solving certain equations, so the new number was viewed as being a pretend number invented for convenience's sake. But then, when you think about it, aren't *all* numbers inventions? It's not like numbers grow on trees! They live in our heads. We made them *all* up! Why not invent a new one as long as it works OK with what we already have?

Anyway, this new number was called *i*, standing for "imaginary," because "everybody knew" that *i* wasn't "real." (That's why you couldn't take the square root of a negative number before: you only had "real" numbers—that is, numbers without the "*i*" in them.) The imaginary is defined to be $i = \sqrt{-1}$. Then: $i^2 = (\sqrt{-1})^2 = -1$.

Why Imaginary Numbers?

The answer is simple. The imaginary unit allows us to find solutions to many equations that do not have real number solutions. This may seem weird, but it is actually very common for equations to be unsolvable in one number system but solvable in another more general number system. Here are some examples with which you might be more familiar. First, as you continue to study mathematics, you will begin to see the importance of these numbers.

Imaginary numbers come up a lot in electricity and magnetism equations, as well as optics. Basically anything that moves as a wave, like the surface of a body of water or the EM field of an electron or photon, will often be written in terms of sine and cosine, which in turn can be written as imaginary powers of e. It also are used in design, simulation, analysis of normal and semiconductor circuits, acoustics and speakers, physics., mechanical system vibration, automotive exhaust note tuning, guitar pickups and boutique high power tube/ solid state amplifiers, chemical engineering linear/nonlinear flows, financial modeling, statistics and big data.

A Superstars

(π = 3.1415926535….)

A SLICE OF PI
****

3.14159265358979
1640628620899
23172535940
881097566
5432664
09171
036
5

Recall that numbers are used to count, measure, label, etc. Some numbers never get used, and among those we use, some get used more often than others. Although irrational numbers are hardly used, there is one known as pi (π) so famous that even a particular day is named after it. This is partly because it relates to circle, a cornerless geometric object or, as many mathematicians claim, a geometric object with an infinite number of corners. In fact, its ubiquity makes it one of the most widely known numbers both inside and outside the scientific community.

Who Cares?

Ancient civilizations needed the value of pi to be computed accurately for practical reasons. In the twentieth and twenty-first centuries, mathematicians and computer scientists have gone far more than that to extend the decimal representation of pi to, as of 2015, over 13.3 trillion (10^{13}) digits. Why? It is believed that the primary motivation for these computations is the human desire to break records and, in this case, find out if digits of pi follow a random distribution.

Pi Day

March 14 is Pi Day, as 3.14 is the beginning of this special number. Although number enthusiasts have been celebrating Pi Day for more than three hundred years, only recently, on March 11, 2009, the House of Representatives passed a resolution officially declaring March 14 as National Pi Day. Pi is the ratio of a circle's circumference to its diameter. It is also the ratio between a circle's area and its radius squared. In year 2015, in addition to its first three digits, its fourth and fifth digits also matched the common way of writing the calendar date (3/14/15). This happens only once in a century. Pi Day official celebration begins at 1:59 p.m. to make an appropriate 3.14159 when combined with the date.

Pi has a long history. The Egyptians and the Babylonians were the first cultures to discover it about four thousand years ago. The number is even mentioned in the Bible.

So let us celebrate Pi Day but remember:

1. Too much pi gives us a large circumference.
2. Our opinion without pi is just an onion.

Piem

Number enthusiasts enjoy memorizing the digits of pi. Many use mnemonic techniques, known as piphilology, to help them remember. Akira Haraguchi from Japan is able to recite the pi's digits from memory to 83,431 places and holds the world record. Some nonmathematicians also like to memorize digits of pi. To help them, few interesting tricks are developed. Here is an example called a piem (argh!). It is a little poem where the length of each word presents a digit of pi.

> How I need a drink, (3 1 4 1 5)
> alcoholic of course, (9 2 6)
> after the heavy lectures, (5 3 5 8)
> involving quantum mechanics. (9 7 9).

Pi and Pop Culture

The concept of Pi is deceivingly simple. It is the ratio between a circle's circumference and its diameter. Although never ending, the record for finding its consecutive digits from 3.14 onward is held by Fabrice Bellard who announced that he had calculated pi to 2.7 trillion digits. Attempting to recite, for example, only the first twelve billion digits of pi without stopping requires approximately 250 years. If typed in a normal-size font, it will cover the distance between Kansas and New York City.

People around the world celebrate the day by participating in variety of activities. Some gather and tell jokes about it. Some read poems with interesting connections to pi. Here are some:

> There once was a girl who loved Pi
> I never could quite fathom why
> To her it's a wonder
> To me just a number
> Its beauty revealed by and by.
> —Eve Anderson

> Three point one four one five nine
> Makes the lazy student whine,
> But give this ratio a try—
> You'll find that it's as easy as pi!
> —Fred Russcol

> Never talk to Pi, he'll go on forever
> Consider humble Pi. It is a number never ending.

It never repeats itself as its value keeps ascending.
Based upon a circle, many men have tried,
to calculate the ratio of its width to its outside.
It is called irrational because it cannot be made a fraction.
The challenge of its nature has been a call to action.
—Ken Johnson

The number Pi has played a role in every life on earth.
From physics to statistics, it's always proved its worth.
The tires that you ride on, the table where you dine,
little Pi was there throughout its concept and design.
—Ken Johnson

Humble Pi is constant—it's been a great addition.
and quietly serves us each day without recognition.
If you cannot appreciate why I hold this number high,
then shame! It is you that should be eating humble pie.
—Ken Johnson

It is a habit of mine,
A new value of Pi to assign
It's simple you see,
Just to use three
Instead of three point one four one five nine.
—Clifford Morse

Pi, an Icon

- Literary nerds invented a dialect known as Pilish in which the numbers of letters in successive words match the successive digits of Pi, such as these excerpts:
    "May I have a large container of coffee (3.1415926)."
    "How I want a drink, alcoholic of course, after the heavy lectures involving quantum mechanics."

- Mike Keith has written the book *Not a Wake* (Vinculum Press 2010) entirely in Pilish.
- The Band A Love Like Pi is an electronic rock band with two pi-related albums.
- A 1998 movie, *Pi*, directed by Darren Aronofsky, is about a lonely, paranoid mathematician named Max who believes that everything in life is governed by numbers.
- In a 1995 movie, Sandra Bullock is a computer software expert who stumbles across a nefarious online plot. She solves it by clicking on a pi symbol that appears on her computer screen.
- Pi plays a part in the season two *Star Trek* episode "Wolf in the Fold." Captain Kirk and Mr. Spock are able to conquer an evil being by using their computer to calculate all decimal places of pi.
- There is a music video about pi that has wizards, robots, and a graffiti artist in it.
- The song "Bye-Bye Miss American Pi" is a parody of Don Mclean's song "American Pie," rewritten as "American Pi."
- The pi-related song "Lose Yourself" is a parody of Eminem's rap song "Lose Yourself."
- The song "Pi" by the musician Kate Bush is composed of her singing the decimal places of pi with music in the background.
- Givenchy makes cologne for men called Pi. It smells like citrus and forests and costs about sixty dollars.
- The TV show that discusses what people search for on Google each day has a segment all about Pi Day.
- There is a website that allows one to download up to fifty million digits of pi. Its different sections include Pi Records, Pi People, Pi Literature, Pi News, and Pi Aesthetics.
- There is a website that includes all advanced mathematical properties of pi.
- The Pi Trivia Game allows testing one's knowledge about pi. It includes twenty-five questions ranging from pi history to its mathematical properties.

- There are several funny cartoons about pi and videos that use colored shapes and music to teach the basic mathematical properties of pi.
- Finally, there are many T-shirts made by the clearly nerdy company Pi-Dye T-Shirts.

Pi at a Glance

- March 14 is the Pi Day, as 3.14 is the beginning of this special number.
- Although number enthusiasts have been celebrating Pi Day for more than three hundred years, only very recently, March 11, 2009, the House of Representatives passed a resolution making March 14 officially the National Pi Day.
- Pi is the ratio of a circle's circumference to its diameter.
- It is also the ratio between a circle's area and its radius squared.

In honor of Pi Day, mathematical enthusiasts often amuse or enervate themselves and others by telling jokes involving pi.

Most pi jokes are a pun on the name of the dessert food pie.

Pi is an irrational number, and as such, it takes an infinite number of digits to give its exact value.

Pi has a long history. The Egyptians and the Babylonians were the first cultures to discover it about four thousand years ago.

The number is even mentioned in the Bible.

If you want to celebrate, make a circular pie with a diameter four, cut it into four equal pieces. Then you have four pieces of pie of size pi.

Recall that numbers have an interesting history and a world of their own. Like human beings, some numbers play a more significant role than others. Some represent certain concepts, some very complicated. Some get used more often than others. Some never get used. For example, although there are significantly more irrational than rational numbers; irrational numbers are hardly used. Some numbers have a special place in certain cultures and even religions. Some are loved by scientists, artists, and even the general public. But there is one and only one number so famous that a particular day is named after it.

Pi and Mysteries of the Universe

This is what is amazing about pi. It is an irrational number, and as such, it takes an infinite number of digits to give its exact value. This means that we can neither get to the end of it nor find the next digit using a pattern in its digits. In other words, it is an infinite, nonrepeating decimal.

Now, being infinite and nonrepeating, it is sensible to assume that every possible number combination exists somewhere in pi. This assumption implies that when converted into numbers (ASCII text), somewhere in that string of digits are answers to all the great questions of the universe, the DNA of every being, and everything that has ever been written or will be written. In short, everything trivial or amazing man could think of would be somewhere in the ratio of a circumference and a diameter of a circle.

In fact, right now, it is possible to check to see if somebody's birth date or social security number occurs as a string of digits in the first two hundred million digits of pi at the Pi Search Site, http://www. angio.net/pi/.

In sum, anything meaningless, trivial, important, or amazing could be found somewhere in the ratio of a circumference and a diameter of a circle. Maybe it is not a coincidence that Albert Einstein was born on Pi Day.

But could this be true? Well, nobody is really sure. How could we possibly find out if pi does or does not include all possible finite-length sequences of digits? There are some such sequences that include all possible finite sequences, and there are some that do not. So we can only say that this is a reasonable possibility.

Here are a few other surprising facts about pi:

- Pi is four times the infinite sum and difference of odd number's reciprocals
  Pi = 4(1/1 − 1/3 + 1/5 − 1/7 + 1/9 − 1/11 + 1/13 . . .)
- 2/Pi = 0.63662 is the probability that an inch-long needle will land on a line if it is dropped on a floor with parallel lines an inch apart.
- Pi appears in the complex equation representing the bell curve.
- Pi Day is Albert Einstein's birthday.

There's also a trick for memorizing pi called a piem (argh!). It is a little poem where the length of each word is a digit of pi.

How I need a drink, (3 1 4 1 5)
alcoholic of course, (9 2 6)
after the heavy lectures, (5 3 5 8)
involving quantum mechanics. (9 7 9).

So on Pi Day, take a minute to appreciate the day of the world's most famous number. To celebrate, make a circular pie with a diameter four, cut it into four equal pieces. Then you have four pieces of pie of size pi. Have a piece and think about the role of mathematics in your life for 3.1415926535897932 . . .

Euler's e, a Superstar in the World of Science

The most powerful force in the universe is compound interest.
—Albert Einstein

Although most people find the pi fascinating, most mathematicians find the irrational number e, often called Euler's number, after Leonhard Euler, even more fascinating. Pi is very easy to understand, as it is simply the ratio of circle's circumstance to its diameter. e is not so easy to describe or explain. It is the base of the natural logarithms invented by John Napier. Both numbers appear in the complex equation representing the bell curve or normal distribution. However, e appears in significantly more places.

For example, suppose that there are one hundred eligible applicants for a job in your company. An optimum strategy advises you to interview the first 1/e percentage—that is, the first thirty-seven—but do not hire any of them. Then hire the first after that who is better than all who came before.

Think about total annual snowfall in a place like New York and its historical record. As time passes, we have to wait longer to see a new record because a new records must be bigger than all past records. Mathematicians have shown that the ratio of the median wait times between two successive records tends to e.

The first few digits of e are 2.718281828459045. There are many ways of calculating the value of e, but none of them ever give an exact answer, because e is irrational (not the ratio of two integers). But it is known to over one trillion digits of accuracy! The simplest is to replace 1, 2, 3, . . . into $(1+ 1/n)^n$. As you increase n, the answer gets closer to e. For example, $(1 + 1/100)^{100} = 2.704829$. An alternative calculation is based on $1 + 1/(1) + 1/(1 \times 2) + 1/(1 \times 2 \times 3) + 1/(1 \times 2 \times 3 \times 4) + . . .$

Often, the number e appears in unexpected places. For example, consider the following fun game: let us say that we cut a number into equal parts and then multiply those parts together.

For example, cut twenty into four pieces and multiply them.

Each "piece" is 20/4 = 5 in size $5 \times 5 \times 5 \times 5 = 625$.

Now, how could we get the answer to be as big as possible? What size should each piece be?

Next, try five pieces. Each "piece" is 20/5 = 4 in size $4 \times 4 \times 4 \times 4 \times 4 = 4^5 = 1024$. The answer is bigger! But is there a best size? How could we get the answer to be as big as possible? What size should each piece be?

The answer: make the parts "e" (or as close to e as possible) in size.

For this example, 20/7 = 2.857 gives 1554, as 2.857 is the closest to e.

For ten: ten cut into three parts, we get 3.3 . . . and 3.3 . . . ×

3.3 ... × 3.3 ... = 37.037 ... Ten cut into four equal parts is 2.5; 2.5 × 2.5 × 2.5 × 2.5 = 2.5^4 = 39.0625. Ten cut into five equal parts is 2; 2 × 2 × 2 × 2 × 2 = 2^5 = 32. The winner is the number closest to "e"; in this case, 2.5.

Does the number e have any real physical meaning, or is it just a mathematical convenience? The answer is, yes, it has a physical meaning. It occurs naturally in any situation where a quantity increases at a rate proportional to its value, such as a bank account producing interest or a population increasing as its members reproduce.

Obviously, the quantity will increase more if the increase is based on the total current quantity (including previous increases) than if it is only based on the original quantity (with previous increases not counted). How much more? The number e answers this question.

To put it another way, the number e is related to how much more money you will earn under compound interest than you would under simple interest.

The more math and science you encounter, the more you run into the number e. Since its discovery, it has shown up in a variety of useful applications including (but definitely not limited to) solving for voltages, charge buildups, and currents in dynamic electrical circuits, spring/damping problems, growth and decay problems, Newton's laws of cooling and heating, plane waves, compound interest, etc.

History of e

e makes profound appearances in calculus. That said, it was first discovered with algebra in the seventeenth century. Mathematicians indirectly came close to many times without directly calculating or recognizing it as anything out of the ordinary. It was almost discovered when logarithms were invented in 1618 by John Napier. Logarithms immensely aided longhand computation by allowing multiplication and division to be carried out by addition and subtraction. These logarithms were not the same as our current conception of a logarithm. They were numbers that aided computation and were not yet thought of as functions that relate exponents to their bases. Mathematicians effectively computed tables of the natural logarithm but did not know that the number was anywhere behind them. Historians are not sure who exactly was the first person to calculate and recognize it as special. Most likely, it was not discovered by a mathematician but by someone with more worldly motivations. The number lies at the foundations of one of the most fundamental processes of finance: compound interest.

Finally, when estimating the ultimate records in sports, the three major contributing factors are the time between the present and penultimate record, time the last record has held to date, and Euler's number e.

> There once was a number named e
> Who took way too much LSD.
> She thought she was great.
> But that fact we must debate;
> We know she wasn't greater than 3.
> —*Mathfail*

> There once was a log named Lynn
> Whose life was devoted to sin.
> She came from a tree
> Whose base was shaped like an e.
> She's the most natural log I've seen.
> —*Mathfail*

Rare Events

Suppose that a special event is occurring with the rate (mean) of r times per unit time or per unit area (e.g., number of car accidents in a day in a given location, number of incoming telephone calls per minute, number of goals scored per game, number of trees per acre, and number of earthquakes in a given region per year). Let X denote the number of such events per unit time or unit area, etc. Then under certain conditions, X has what is known as Poisson distribution that has e in its probability mass function. Moreover, the distribution of time between occurrences of any two successive events, known as exponential distribution, also involves e. The Poisson distribution plays a very important role in science and engineering, since it represents an appropriate probabilistic model for a large number of observational phenomena. It also arises in discussion of an extremely useful process known as Poisson.

For example, suppose a new record for the four-hundred-meter run has just been set. Let X denote the number of times this record gets broken in the next ten years. Then X has a Poisson distribution. It has been shown that the number of goals scored in a soccer game during a fixed time interval can be modeled by Poisson distribution.

The exponential distribution is an appropriate model for the lifetime of most electronics such as transistors. Exponential distributions have an interesting property known as lack of memory. Let me demonstrate another property of exponential distribution through an example.

The lifetime of a bulb has an exponential distribution with a mean of three hundred hours. Find the probability that one such a bulb lasts more than three hundred hours. What percentage of the bulbs will last more than three hundred hours? The answer to this is surprisingly 1/e.

We end this part by noting that Poisson distribution has also been applied to sports such as hockey with success. Research indicates that the goals scored by and against teams in the National Hockey League are surprisingly well described by Poisson distributions, and even more surprisingly, the goals for and against a team seem to be independent.

Calculation of e

The value of e is also equal to $1/0! + 1/1! + 1/2! + 1/3! + 1/4! + 1/5! + 1/6! + 1/7!+ ...$, where ! means factorial. For example, 4! $= 1 \times 2 \times 3 \times 4$. The first few terms add up to $1 + 1 + 1/2 + 1/6 + 1/24 + 1/120 = 2.718055556$. To remember the value of e (to ten places), just remember this saying (count the letters!):

"To express e remember to memorize a sentence to simplify this."

Or you can remember the curious pattern that after the 2.7, the number 1828 appears twice—that is, 2.7 1828 1828. And following that are the angles 45°, 90°, 45° in a right-angled isosceles (two equal angles) triangle: 2.7 1828 1828 45 90 45 (an instant way to seem really smart!).

Since its discovery, it has shown up in a variety of useful applications including (but definitely not limited to)

- solving for voltages, charge buildups, and currents in dynamic electrical circuits;
- compound interest;
- spring/damping problems;
- growth and decay problems;
- Newton's laws of cooling and heating; and
- plane waves.

As pointed out earlier, e makes profound appearances in calculus. That said, it was first discovered with algebra in the seventeenth century. Mathematicians indirectly came close to many times without directly calculating or recognizing it as anything out of the ordinary. It was almost discovered when logarithms were invented in 1618 by John Napier. Logarithms immensely aided longhand computation by allowing multiplication and division to be carried out by addition and subtraction. These logarithms were not the same as our current conception of a logarithm. They were numbers that aided computation and were not yet thought of as functions that relate exponents to their bases. Mathematicians effectively computed tables of the natural logarithm but did not know that the number was anywhere behind them. Historians are not sure who exactly was the first person to calculate and recognize it as special. Most likely, it was not discovered by a mathematician but by someone with more worldly motivations. The number lies at the foundations of one of the most fundamental processes of finance: compound interest.

Compound Interest

The seventeenth century was a time of rapid change. It was the era of the scientific revolution, the proliferation of colonialism, the emergence of mass literacy, and an explosion of international trade. The European age of exploration (and exploitation) brought disparate cultures of the world in contact, conflict, and business with each other to a degree that none of the large empires of old ever approached. Increasing the scale of commerce increased the demands for capital. Money lending began to play an increasingly large role in the prosperity of individuals, business ventures, and nations.

Given the growing presence of finance in the seventeenth century, historians believe that the first person to calculate was most likely a banker or trader exploring the properties of compound interest.

Interest is a fee charged by a lender on a borrower for the service of providing a loan. Interest fees offset the opportunity cost to the lender of not being able to do anything else with their assets while they are controlled by the borrower. These interest fees accumulate over time, depending on the interest rate. Borrowers think of interest as the cost of having debt, while lenders think of interest as the return on their loan as an investment.

Simple interest is where the interest accumulated is always the same proportion of the initial amount borrowed or invested (this initial amount is called the principal). For instance, if you borrow $1.00 at 5 percent interest per year, after one year, you will owe $1.05. After two years, you will owe $1.10, etc. In effect, simple interest creates an arithmetic progression, where a debt or investment grows at the same rate over all periods. Simple interest is rare and usually appears only in short-term loans.

Compound interest is where interest accumulates on both the principal and the prior interest. Each time the interest rate is applied to the total accumulated debt or investment, it is referred to as compounding. If you invest $1.00 at 5 percent annual compounding interest, after the first year, you would have $1.05. Unlike simple interest, in the second year, it would increase by 5 percent of $1.05, yielding $1.1025.

Under compound interest, debts and investments grow by a geometric progression. Interest can be compounded multiple times within a given year or interest rate period. For instance, if you put one dollar into a bank account that returns 5 percent interest per year compounded biannually, then your account will grow by 2.5 percent twice in a given year. Compounding more times in a given period causes your debt or investment to grow more often but at a smaller rate each time it is compounded. As you can see, it would not take much imagination for an ambitious banker to wonder how much money could be made if the interest rate was really high and it was compounded as much as possible (daily, hourly, infinitely, etc.). If more compounding intervals make investments grow more often, would more compounding result in investments growing faster?

Historians believe that a businessman or banker likely beats the mathematicians to the thought experiment: if one dollar is invested at 100 percent interest over one year, how much more money will be made if it is compounded often as opposed to only compounding it a few times in a year? The equation for compound interest in this special case would simplify to a form involving e.

It turns out that compounding weekly barely yields any more money than compounding monthly, and at higher values of, it gets closer and closer to what we recognize as the number. As can be seen, compounding more often does yield more money up to a point but rapidly reaches an upper bound where growing more often does not yield faster growth. Mathematicians (led by Jacob Bernoulli, in 1683) would later go on to define as the limit.

With going to infinity, growth occurs continuously at every possible instant no matter how small the time interval. Simple interest reflects an arithmetic growth progression, compound interest reflects a discrete geometric growth progression, and infinitely compounding interest reflects a continuous growth progression (exponential growth). The number is thought of as the base that represents the growth of processes or quantities that grow continuously in proportion to their current quantity. This is why it appears so often in modeling the exponential growth or decay of everything from bacteria to radioactivity.

Here is a fun problem to try. Based on the rule of seventy-two, approximately how many years will it take to double the principal in an account that earns 3 percent of interest per year?

Bernoulli's Observation

Often the number e appears in unexpected places—for example, wherever the rate of change is proportional to the present value of whatever is measured.

e is used in continuous compounding for loans and investments. Jacob Bernoulli discovered this constant in 1683 by studying a question about compound interest.

An account starts with $1.00 and pays 100 percent interest per year. If the interest is credited once, at the end of the year, the value of the account at year-end doubles to $2.00. What happens if the interest is computed and credited more frequently during the year? If the interest is credited twice in the year, the interest rate for each six months will be 50 percent; so after the first six months of the year, the initial $1.00 is multiplied by 1.5 to yield $1.50. Reinvesting this by the end of the year, it becomes $1.50 x $1.50, yielding $1.00 × $1.5^2 = $2.25 at the end of the year. Compounding quarterly yields $1.00 × 1.25^4 = $2.4414 . . . and compounding monthly yields $1.00 × $(1+1/12)^{12} = $2.613035 . . . If there are n compounding intervals, the interest for each interval will be 100%/n, and the value at the end of the year will be $1.00 \times (1 + 1/n)^n$.

Bernoulli noticed that this sequence approaches a limit (the force of interest) with larger n and, thus, smaller compounding intervals. Compounding weekly (n = 52) yields $2.692597 . . ., while compounding daily (n = 365) yields $2.714567 . . ., just two cents more. The limit as n grows large is the number that came to be known as e; with continuous compounding, the account value will reach $2.7182818.

Zero

What is zero? Is zero nothing or is it something? Most of us would like to have a lot of them behind some numbers in our bank account or one of them where it says "balance due" on a bill. Most of us think the number zero is just like any other number. There's nothing special about it. Well, I found out that nothing could be further from the truth.

The idea of numbers like one, two, and three developed a long time before the concept of zero. This was mostly because the number zero was not a requirement in the calculations that early man was trying to perform. However, zero appeared, disappeared, and reappeared again and again as if the mathematicians searching for it did not recognize it even when they saw it.

There are two uses of zero, and both uses are very important but are somewhat different. One use for zero is as a placeholder in our place-value number system. The second use of zero is to indicate a quantity of nothing. The name "zero" is derived from the Arabic *sifr*, which also gives us the word "cipher." Neither of the uses has an easily described history. Someone did not just invent the ideas and then everybody started to use them.

The Babylonians had a place-value system without the zero for over one thousand years. The Babylonians wrote on tablets of unbaked clay using cuneiform writing. Tablets from around 1700

BC exist to show they did not distinguish between 3208 and 328. It is shown, around 400 BC, that the Babylonians used two wedge-shaped symbols in the place where we would put a zero. So 3208 would be written 32 "8, but they did not use it at the ends of numbers. So 328 and 3280 would still look the same.

A tablet found in what is today south-central Iraq, thought to date around 700 BC, uses three hooks as a placeholder. The early use of zero as a placeholder is just that; it is not really being used as the number.

The ancient Greeks began contributing to mathematics around the time that zero as a placeholder was beginning to be used in Babylonian mathematics. The Greeks, however, did not use a place-value number system because Greek mathematics was based on geometry with the exception of the mathematicians who recorded astronomical data. The first use of the symbol O began with the Greek astronomers.

Some historians believe it stands for the Greek word for nothing, *ouden,* and others believe it to be for obol, a coin of almost no value. The fact that counters were used for counting on a sand board and a counter was removed to leave an empty column, it left a depression in the sand that looked like 0.

Around AD 130, in *Almagest,* written by Ptolemy, the zero as a placeholder is used together with the Babylonian sexagesimal system. Ptolemy uses the symbol both between digits and at the end of numbers. Only a few astronomers used the O, and it was not used again until it next appears in Indian mathematics. Some historians believe that the Indian use of zero evolved from its use by the Greek astronomers.

Around AD 650, the use of zero as a number came into Indian mathematics. The Indians also used a place-value system, and zero was used as a placeholder. The first record of the Indian use of zero, which is dated and agreed on to be genuine, was written in AD 876. It is the date written on a stone tablet that contains two numbers, 270 and 50, written almost as they are today, except the O is smaller and slightly raised.

Now, zero is making its first appearance as a number. The work of the Indian mathematicians was transmitted to the Islamic and Arabic mathematicians further west and also spread east to China.

There was another civilization that developed a place-value number system with a zero. This was the Mayans who lived in Central America. By 665, they used a place-value number system to base 20 with a symbol for zero. Their use of zero goes back further than this and was in use before they introduced the place-value number system.

However, it did not influence other mathematicians.

Fibonacci was one of the main people to introduce the notion of zero and the nine Indian symbols to Europe at around 1200, but it was not widely used for a long time after that. It was not until the 1600s that zero began to come into widespread use but still only after encountering a lot of resistance.

When considering 0 as a number, the problem of how it interacts in regard to the operations of arithmetic arises. Three Indian mathematicians, Brahmagupta, Mahavira, and Bhaskara, tried to answer this in three important books. All three mathematicians struggled to explain division by zero. Today, we simply say it is undefined; how can you divide by nothing?

Even in the year 2000, the problems caused by 0 were evident. People celebrated the new millennium on January 1, but most people could not understand that the third millennium and the twenty-first century did not begin until January 1, 2001. When the calendar was set up, there was no year zero, so on January 1, 2000, only 1,999 years had passed. Zero is still causing problems and not fully understood!

Numbers and Communication

Numbers resonate with the mind in a greater way than you might imagine. In the movie *What the Bleep Do We Know,* they refer to the processing capacity of the brain. It processes over four billion bits of information per second—yes, that is *four billion per second.*

You are *aware* of only two thousand. Which do you want working for you consciously? What will be the effect on you if you were aware of significantly more than two thousand? Could you train your subconscious to process even more and make you aware of the result?

In your résumé, I recommend you change the name to background, put in more numbers. The amazing part is that it does not even make any difference what kind of numbers they are. It does not have to be *great* accomplishments. I supervised ten people supporting a $6 million budget.

In listening, begin to be aware of this and choose which numbers to question. The president recently said that oil production had gone up on his watch. He used numbers to illustrate the point. He was telling the *truth* and *lying* at the same time. Oil production on federally controlled acreage has gone *down;* however, oil production on acreage not controlled by his administration has gone *up.* The net total of *up* and *down* was positive and was the overall premise he wanted to take credit for, in my opinion.

Shortly after that, he said his administration could do nothing about gas prices because it was already drilling on 75 percent of the available land. Again, telling the *truth* and *lying* at the same time. He did not define *land* or *available.* His numbers did not include anything on or in *water,* nor did his numbers include any of the *land* his administration has made *unavailable.* The normal mind is drawn to the numbers.

So my guess is that 90 percent of the people who read this will get it immediately, and less than 5 percent will look on http://www.thegreatermind.com; and of those who do look intently, 51 percent of them will find something they feel compelled to purchase and 100 percent of them will be delighted.

To emphasize the point, consider a situation that involves treating a certain disease discussed earlier. Say there are two methods, A and B, of treatment. Suppose 390 patients with this disease were given one of the two treatments: 160 were given treatment A, and 230 were given treatment B. The recovery rates were $60/160 = 0.38$ using treatment A and $65/230 = 0.28$ using treatment B. While chances of recovering from the T disease are extremely poor and hopefully much more research will be done on it, it seems from these data that it is more favorable to use treatment A than B. However, by using one classifier—namely, sex—the total picture changes as follows.

Fraction Recovering

Sex	Treatment A	Treatment B
Male	20/100 = 0.20	50/210 = 0.24
Female	40/60 = 0.67	15/20 = 0.75
Total	60/160 = 0.38	65/230 = 0.28

Now, it is clear from this summary that both men and women would prefer—treatment B, which is different from our earlier conclusion. The lesson here is that you must be extremely careful in reporting averages unless the groups under consideration are fairly homogeneous. Note, in this example, the experiences of men and women are quite different.

My Experience with Numbers

Here is a situation that still makes me wonder. We are all brought up with decimal system or base 10 numbering system. To my surprise, I had a student who had a much easier time dealing with numbers in binary system than in decimal system. Is it possible that this student has a brain that processes numbers differently from mine? Could I have been the same if I was brought up using binary system in the first place? Here is one line of discussion. Which system is more efficient—decimal system or binary system? Here, for example, efficiency may relate to how much time and effort it takes to say or write a number; how many symbols one needs to represent numbers in a system. Decimal system needs ten; whereas, binary system needs only two. No wonder this system plays a key role in new technologies and speed they possess.

I had many students who had problems with fractions or with irrational numbers. Many were confused about different roles a number can play. Some were confused when mathematics is expressed using symbolic language and had a hard time with learning the rules of that language. I noticed that those who had a good command of English or any other language were better able to understand symbolic language and its rules and role. To help the discussion, it is useful to summarize few things about numbers and their relation to the learning process.

We know that throughout history, numbers have played a key role in human development, and many civilizations have contributed to their advancement. Like human beings, numbers have grown, changed, matured, and achieved a high level of sophistication. Advances in this field are apparent to anybody who shops at grocery stores and malls. Every item we buy, piece of mail we receive, ticket we purchase, etc., has an identification number. Like people, there are some extraordinary, famous, and beautiful numbers that amaze us. Some numbers are celebrated in a certain day of the year. For example, March 14 is the Pi Day, since 3.14 is the beginning of this special number. Lots can be said about pi. For example, it has a decimal representation that never stops or repeats, although modern computers have now computed it to over several trillion decimal places! I did mention this to emphasize the elements of surprise and the complexity. Another famous and irrational number is e, which appears in a great deal of mathematical and physical theories. Lots are written about these numbers. In a book titled *The Great Pi/e Debate*, Colin Adams and Thomas Garrity settle once and for all the burning question that has plagued humankind from time immemorial: "Which is the better number, e or pi?" in what could be the most important debate of the millennium.

Like different civilizations or cultures, some numbers belong to a group with a given name. Examples include integers, rational, irrational, real, complex, etc. Many numbers are given specific names because of their unique property. It seems inconceivable how much we rely on numbers in our daily lives and how natural it feels. Without a social security number, we cannot even become a functioning member of the society. Every action performed, every life encountered has become a part of the ongoing use of numbers. Some numbers born and die with us. Some even live long after we are gone. For a person suffering from a common disease such as diabetes, it is important to keep track of the numbers, most notably cholesterol, blood pressure, and blood sugar. For some, such numbers can become a source of stress.

Extended Numbers

Nearly thirty years ago, John Horton Conway introduced a new way to construct numbers. Donald E. Knuth, in appreciation of this revolutionary system, took a week off from work on *The Art of Computer Programming* to write an introduction to Conway's method. Never content with the ordinary, Knuth wrote this introduction as a work of fiction—a novelette. If not a steamy romance, the book nonetheless shows how a young couple turned on to pure mathematics and found total happiness. The book's primary aim, Knuth explains in a postscript, is not so much to teach Conway's theory as "to teach how one might go about developing such a theory." He continues, "Therefore, as the two characters in this book gradually explore and build up Conway's number system, I have recorded their false starts and frustrations as well as their good ideas. I wanted to give a reasonably faithful portrayal of the important principles, techniques, joys, passions, and philosophy of mathematics, so I wrote the story as I was actually doing the research myself." It is an astonishing feat of legerdemain. An empty hat rests on a table made of a few axioms of standard set theory. Conway waves two simple rules in the air and then reaches into almost nothing and pulls out an infinitely rich tapestry of numbers that form a real and closed field. Every real number is surrounded by a host of new numbers that lie closer to it than any other "real" value does. The system is truly "surreal," quoted from Martin Gardner, *Mathematical Magic Show* (pp. 16–19). *Surreal*

Numbers, now in its thirteenth printing, will appeal to anyone who might enjoy an engaging dialogue on abstract mathematical ideas and who might wish to experience how new mathematics is created.

Computer Number System

Binary or base 2. There are only two numbers in binary, 0 and 1. Because computers use a sequence of switches that can be on or off (also called a bit), base 2 works very well for them. Math in base 2 is pathetically simple but incredibly time consuming.

Octal or base 8. Uses the numbers 0–7. There are eight bits in a byte, which is used very often in the computer field. (A bit is great, but it's too small to hold any useful data, thus the byte is used.) Math in octal is more complicated than decimal.

Decimal or base 10. Uses the numbers 0–9. Computers only display numbers in decimal; they actually do all their work in binary. Math is quite simple with this number system, although some may argue.

Hexadecimal or base 16. Uses the numbers 0–F. Because there are sixteen values per placeholder, five new numbers had to be created. Those numbers are A, B, C, D, E, and F (original, isn't it?). A has a value of 10, B is 11, and so on. This system in the computer field is used as a means of viewing lots of data much faster. Math in hexadecimal is not very simple compared to decimal. You may wonder why I only counted up to 256 (don't forget 0). Well, 256 is a very common number in the computer field because it is a natural multiple of 2 (the base computers use); 256 is 2 to the eighth power (eight bits in a byte). Got it? The term "computer numbering formats" refers to the schemes implemented in digital computer and calculator hardware and software to represent numbers. A common mistake made by non-specialist computer users is a certain misplaced faith in the infallibility of numerical computations. For example, if one multiplies, one might perhaps expect to get a result of exactly 1, which is the correct answer when applying an exact rational number or algebraic model. In practice, however, the result on a digital computer or calculator may prove to be something such as 0.9999999999999999 (as one might find when doing the calculation on paper) or, in certain cases, perhaps 0.99999999923475. The latter result seems to indicate a bug, but it is actually an unavoidable consequence of the use of a binary floating-point approximation. Decimal floating point, computer algebra systems, and certain big numbering systems would give either the answer of 1 or 0.9999999999999999 . . .

Bits, Bytes, Nibbles, and Unsigned Integers Bits

The concept of a bit can be understood as a value of either 1 or 0, on or off, yes or no, or encoded by a switch or toggle of some kind. A single bit must represent one of two states. A one-digit binary value and a decimal value 0 01 1 are two distinct values.

Selecting a Partner

Suppose that there are one hundred eligible partners in your village, tribe, or social network; this strategy advises you to sample the population by dating the first thirty-seven, and then choose the first after that who's better than all who came before. Of course, you might be unlucky in a number of ways. For example, the perfect mate might be in the first thirty-seven and get passed over during the sampling phase. In this case, you continue dating the rest but find no one suitable and grow old alone, dreaming of what might have been. Another way you might be unlucky is if you have a run of really weak candidates in the first thirty-seven. If the next few are also weak but there's one who's better than the first thirty-seven, you commit to that one and find yourself in a suboptimal marriage. But the mathematics shows that even though things can go wrong in these ways, the strategy outlined here is still the best you can do. The news gets worse, though: even if you stringently follow this best strategy, you still only have a bit better than a one-in-three chance of finding your best mate.

Records

Records occur everywhere from sports to the stock market. *The Guinness Book of World Records* is popular reading around the world. We plan to discuss the mathematics behind the theory of records and present some applications.

Theory of records deals with

1. number of records in a sequence of n observations;
2. record times (times records occur);
3. waiting time between the records; and
4. record values.

Example: Think about total annual snowfall in New York in the next ninety or one hundred years.

Q: What is the chance that a baby born this year would experience, for example, six personal record snowfalls? ten personal record snowfalls?
Q: How many personal record snowfalls should she expect to experience?
Q: How many record snowfalls is the most likely case?
The first year is a record. No history. What about the second (next) year? The chance that the second year is a record is 50 percent or 1/2. What about the third year? The chance that the third year is a record is 1/3.
R_n: Number of records in n years

Term Years	Expected Number of Records
1	1
2	1/2

3	1/3
4	1/4
·	
·	
·	
n	1/n

$$E(R_n)=1+\tfrac{1}{2}+\tfrac{1}{3}+\tfrac{1}{4}+\cdots+\tfrac{1}{n}$$

For $n = 90$

$$E(R_n)=1+\tfrac{1}{2}+\tfrac{1}{3}+\tfrac{1}{4}+\cdots+\tfrac{1}{90}-5.08 \approx 5$$

Note: 5 is also the most likely scenario (probability = 0.21). What is the value of this series for $n = 1000$? $n = 1000,000$? $n = 1000,0000,0000$?

As n tends to infinity, what does $E(R_n)$ tend to? Well, using the approximation $\int(1/x)\,dx$, it tends to $Ln(n) + \gamma(=0.5772)$, where γ is also an Euler's number.

Note that $Ln(n)$ tends to infinity as n tends to infinity. It is known among people who follow records that every record will eventually be beaten. In fact, surprisingly, the intuitive idea that every record will be beaten also leads to mathematical proof that the harmonic sum $1+1/2+1/3+\cdots$ grows without bound, becoming bigger than any finite number.

W_r: waiting time between the $(r-1)^{th}$ and r^{th} records

The expected value of W_r is infinite even for $r = 2$.

Record Number r	2	3	4	5	6	7	8
Median W_r	4	10	26	69	183	490	1316
Med W_r /Med W_{r-1}		2.50	2.60	2.65	2.65	2.68	2.69

Med W_r /Med W_{r-1} tends to e. The median number of attempts required to arrive at a new record is $e=(2.718\ldots)$ times the median number of attempts that was required to arrive at the previous record.

This suggests a geometric increase with rate $e = 2.718\ldots$ If we assume that one unit of attempts was needed to arrive at the second record, the total number of attempts to arrive at record number twenty may be calculated as

$$1 + e + e^2 + \cdots + e^{18} = 103{,}872{,}541.$$

This is a slight overestimation, as, for early records, the ratios are less than e. This leads to probability estimates for a new record for one-hundred-meter dash as

$$0.152461 \text{ for the next year } 0.562681 \text{ for the next}$$
$$5 \text{ years } 0.808753 \text{ for the next } 10 \text{ years}$$

Recall that after seeing the second record, the median wait time to the third record is ten observations (attempts). Other results regarding W_r include a law of large numbers, $\log W_r / r \to 1$, and a result indicating that $\log W_r$ is approximately equivalent to the arrival time sequence of a Poisson process. Since sports records are more frequent than records generated by independent and identically distributed sequences, it is possible to model $\log W_r$ as a nonhomogeneous Poisson process.

Let us look at the data for long jump. For long jump, the fifth record was set in 1991. Using the theory of records, seventy-three attempts are needed to produce five records, and these should have occurred during the period 1962–1991 (thirty years). This leads to geometric increase with rate $i = 1.055$. Noting that the waiting time to the sixth record is 183 attempts, it takes (in median sense) forty-nine years for a new record to be set. This means waiting till the year 2040. The return period of the present record (8.95) was found to be 64.5 years based on the tail model I obtained.

We end this part by noting that, rather than records and waiting times between them, one could consider improvements of equal size and analyze the corresponding waiting times. This seems a reasonable approach, since, as records improve, increase in number of attempts could offset the decrease in number of record-breaking performances. For example, consider the rise in pole vault records and their waiting times shown in the following table.

Data for Pole Vault

Improvement (feet)	Number of Years
14 to 15	13
15 to 16	22
16 to 17	1.5
17 to 18	7
18 to 19	10
19 to 20	10

Here, one can consider smaller improvements and apply some of the classical statistical methods. In the case of pole vault, for example, the goal of such analysis should be to predict the number of years it would take to go from twenty to twenty-one.

Lover's Question

Consider the question of how many people you should date before you commit to a more permanent relationship such as marriage. Marrying the first person you date is, as a general strategy, a bad idea. After all, there's very likely to be someone better out there; but by marrying too early, you're cutting off such opportunities. But at the other extreme, always leaving your options open by endlessly dating and continually looking for someone better is not a good strategy either. It would seem that somewhere between marrying your first high-school crush and dating forever lies the ideal strategy. Finding this ideal strategy is an optimization problem and, believe it or not, is particularly amenable to mathematical treatment. In fact, if we add a couple of constraints to the problem, we have the classic mathematical problem known as the *secretary problem*. The mathematical version of the problem is presented as one of finding the best secretary (which is just a thin disguise for finding the best mate) by interviewing (i.e., dating) a number of applicants. In the standard formulation, you have a finite and known number of applicants, and you must interview these n candidates sequentially. Most importantly, you must decide whether to accept or reject each applicant immediately after interviewing him or her; you cannot call back a previously interviewed applicant. This makes little sense in the job-search context but is very natural in the dating context: typically, boyfriends and girlfriends do not take kindly to being passed over for someone else and are not usually open to the possibility of a recall. The question, then, is, how many of the n possible candidates should you interview before making an appointment? Or in the dating version of the problem, the question is, how many people should you date before you marry? It can be shown mathematically that the optimal strategy for a large applicant pool (i.e., when n is large) is to pass over the first ne (where e is the transcendental number from elementary calculus—the base of the natural logarithm, approximately 2.718) applicants and accept the next applicant who's better than all those previously seen. This gives a probability of finding the best secretary (mate) at e or approximately 0.37. For example, suppose that there are one hundred eligible partners in your village, tribe, or social network; this strategy advises you to sample the population by dating the first thirty-seven, and then choose the first after that who's better than all who came before. Of course, you might be unlucky in a number of ways. For example, the perfect mate might be in the first thirty-seven and get passed over during the sampling phase. In this case, you continue dating the rest but find no one suitable and grow old alone, dreaming of what might have been. Another way you might be unlucky is if you have a run of really weak candidates in the first thirty-seven. If the next few are also weak but there's one who's better than the first thirty-seven, you commit to that one and find yourself in a suboptimal marriage. But the mathematics shows that even though things can go wrong in these ways, the strategy outlined here is still the best you can do. The news gets worse, though: even if you stringently following this best strategy, you still only have a bit better than a one-in-three chance of finding your best mate.

This problem and its mathematical treatment are instructive in a number of ways. Here, I want to draw attention to the various idealizations and assumptions of this way of setting things up. Notice that we started with a more general problem of how many people you should date before you marry, but in the mathematical treatment, we stipulate that the population of eligible partners is fixed and known. It's interesting that the size of this population does not change the strategy or your chances of finding your perfect partner—the strategy is as I just described, and so long as the population is large, the probability of success remains at 0.37. The size of the population just affects the number of people in the initial sample. But, still, stipulating that the population is fixed is an idealization. Most pools of eligible partners are not fixed in this way—we meet new people, and others who were previously in relationships later become available, while others who were previously available enter new relationships and become unavailable. In reality, the population of eligible candidates is not fixed but is open-ended and in flux. The mathematical treatment also assumes that the aim is to marry the best candidate. This, in turn, has two further assumptions. First, it assumes that it is in fact possible to rank candidates in the required way and that you will be able to arrive at this ranking via one date with each. We can have ties between candidates, but we are not permitted to have cases where we cannot compare candidates. The mathematical treatment also assumes that we're after the *best* candidate, and anything less than this is a failure. For instance, if you have more modest goals and are only interested in finding someone who'll meet a minimum standard, you need to set things up in a completely different way—it then becomes a satisficing problem and is approached quite differently.

Another idealization of the mathematical treatment—and this is the one I am most interested in—is that finding a partner is assumed to be one-sided. The treatment we're considering here assumes that it is an employer's market. It assumes, in effect, that when you decide that you want to date someone, he or she will agree, and that when you decide to enter a relationship with someone, again they will agree. This mathematical equivalent of wishful thinking makes the problem more tractable but is, as we all know, very unrealistic.

A natural way to get around this last idealization is to stop thinking about your candidate pool as a row of wallflowers at a debutante's ball and instead think of your potential partners as active agents engaged in their own search for the perfect partner. The problem, thus construed, becomes much more dynamic and much more interesting. It becomes one of coordinating strategies. There is no use setting your sights on a partner who will not reciprocate. For everyone to find someone to reciprocate their interest, a certain amount of coordination between parties is required. This brings us the game theory.

Other Interesting Numbers

$$1 \times 142,857 = 142,857$$
$$2 \times 142,857 = 285,714$$
$$3 \times 142,857 = 428,571$$
$$4 \times 142,857 = 571,428$$
$$5 \times 142,857 = 714,285$$
$$6 \times 142,857 = 857,142$$
$$7 \times 142,857 = 999,999 \ (= 142857 + 857142)$$

If you multiply by an integer greater than 7, there is a simple process to get to a cyclic permutation of 142857. By adding the rightmost six digits (ones through hundred thousands) to the remaining digits and repeating this process until you have only the six digits left, it will result in a cyclic permutation of 142857:

$$142857 \times 8 = 1142856$$
$$1 + 142856 = 142857$$
$$142857 \times 815 = 116428455$$
$$116 + 428455 = 428571$$

$$142857^2 = 142857 \times 142857 = 20408122449$$
$$20408 + 122449 = 142857$$

Multiplying by a multiple of 7 will result in 999999 through this process:

$$142857 \times 7^4 = 342999657$$
$$342 + 999657 = 999999$$

If you square the last three digits and subtract the square of the first three digits, you also get back a cyclic permutation of the number.

$$857^2 = 734449$$
$$142^2 = 20164$$
$$734449 - 20164 = 714285$$

It is the repeating part in the decimal expansion of the rational number $1/7 = 0.142857$. Thus, multiples of $1/7$ are simply repeated copies of the corresponding multiples of 142857:

$$1 \div 7 = 0.142857$$
$$2 \div 7 = 0.285714$$
$$3 \div 7 = 0.428571$$
$$4 \div 7 = 0.571428$$
$$5 \div 7 = 0.714285$$
$$6 \div 7 = 0.857142$$
$$7 \div 7 = 0.999999 = 1$$
$$8 \div 7 = 1.142857$$
$$9 \div 7 = 1.285714$$

. . .

1/7 as an infinite sum

There is an interesting pattern of doubling, shifting, and addition that gives 1/7.

$1/7 = 0.142857142857142857\ldots$
$= 0.14+0.0028+0.000056+0.00000112+0.0000000224+0.0000$
$00000448+0.00000000000896$
$+\cdots$

$$=\frac{14}{100}+\frac{28}{100^2}+\frac{56}{100^3}+\frac{112}{100^4}+\frac{224}{100^5}+\cdots+\frac{7\times2^n}{100^n}+\cdots$$

$$=(\frac{7}{50}+\frac{7}{50^2}+\frac{7}{50^3}+\frac{7}{50^4}+\frac{7}{50^5}+\cdots\frac{7}{50^n}+\cdots)=\sum_{k=1}^{\infty}\frac{7}{50^k}$$

Each term is double the prior term shifted two places to the right. Another infinite sum is

$$1/7 = 0.1+ 0.03+0.009+ 0.0027\ldots..= 1/10+$$
$$3/100+ 9/1000+ 27/10000+\ldots\ldots$$

Connection to the Enneagram

The 142857 nm figure, a symbol of the Gurdjieff Work used to explain and visualize the dynamics of the interaction between the two great laws of the universe (according to G. I. Gurdjieff), the Law of Three and the Law of Seven. The movement of the numbers of 142857 divided by 1/7, 2/7. etc., and the subsequent movement of the enneagram, are portrayed in Gurdjieff's sacred dances known as the movements.

Patterns

We end this part by looking at a couple of number patterns that I sometimes use in the class. Many students consider these as coincidences. Some find these more surprising than some of the well-known coincidences. I usually remind students that noticing coincidences has played an important role in scientific discoveries. The ability to see obscure connections (those that most people miss) is believed by some to be among the more important abilities a scientist can have. A good example of this is stories about Isaac Newton and his discoveries.

One Pattern

- 1 times 1 = 1
- 11 times 11 = 121
- 111 times 111 = 12321
- 1111 times 1111 = 1234321
- 11111 times 11111 = 123454321
- 111111 times 111111 = 12345654321
- 1111111 times 1111111 = 1234567654321
- 11111111 times 11111111 = 123456787654321
- 111111111 times 111111111 = 12345678987654321

Permutation Pattern

- 1 times 142857 = 142857
- 3 times 142857 = 428571
- 2 times 142857 = 285714
- 6 times 142857 = 857142
- 4 times 142857 = 571428
- 5 times 142857 = 714285
- 7 times 142857 = 999999

Seventeen and Twenty-Three

For an unknown reason, there are many coincidences attributed to the number seventeen in important events in history. For example, many events have happened on the seventeenth day of the month. Here are some examples. The first impeachment in the United States happened on the seventeenth. This was William Blount, and the impeachment date was April 17, 1704. The first child born in the White House was President Jefferson's grandson, who was born on January 17, 1806. The Statue of Liberty arrived in the United States on June 17, 1885. The first plane flew on December 17, 1903, by the Wright brothers. The first president of Israel was elected on February 17, 1949. The Bay of Pigs invasion took place on April 17, 1961. The Watergate break-in was on June 17, 1971. Watergate prosecutor Archibald Cox was born on May 17, 1912. Nixon's vice president, Spiro Agnew, died on September 17, 1996. Three major earthquakes have occurred on the seventeenth: the San Francisco earthquake on October 17, 1989; the Northridge earthquake in Los Angles on January 17, 1994; and the Kobe, Japan, earthquake on January 17, 1995. The youngest person to win the French Open tennis tournament was Michael Chang, who was seventeen years old. David Koresh, leader of the Branch Davidian cult in Waco, Texas, in 1993, was born on August 17, 1959. O. J. Simpson was arrested for the murder of his ex-wife Nicole on June 17, 1994. There were seventeen fingerprints found at the murder scene. O. J. and Nicole knew each other for seventeen years, and their first child, Sydney, was born on October 17, 1985. Pres. Bill Clinton was investigated by the seventeenth independent counsel, Ken Starr, which led to his impeachment. The last president who was impeached, Pres. Andrew Johnson, was the seventeenth president of the United States. So far, there have been two space shuttles that exploded in the air, the *Challenger* in 1986 and the *Columbia* in 2003, which occurred seventeen years apart. Finally, many analysts have pointed out that United States presidential elections are decided by the results in seventeen states.

The Number Twenty-Three

Many people think the number twenty-three has mystical properties . . . and even changed the course of history. First, Saddam Hussein was the president of Iraq for twenty-three years. Second, the controversial book *Satanic Verses,* by Salman Rushdie, is about twenty-three years of Prophet Mohammad's active life. Consider the events related to the death of Maynard Jackson Jr., the first black mayor in Atlanta. Maynard was born on March 23, 1938, and died on June 23, 2003. He died on the same day of the month he was born on. This is just one example of this type. Basketball legend Michael Jordan wore 23 for the Chicago Bulls. His dad was also murdered on July 23, 1993, during a botched robbery. William Shakespeare was born and died on April 23. His first sheet of plays came in 1623, and his wife Anne died in 1623. Car giant Nissan is touched by a numerical coincidence. In Japanese, *ni* is 2 and *san* is 3. So Nissan would be 23. William Burroughs's final TV appearance was in U2's "Last Night on Earth" video. The letter *U* is the twenty-first in the alphabet, and adding the two gives twenty-three. Sesame Street's Bert is a member of the national association of W lovers, the twenty-third letter of the alphabet. The Latin alphabet has twenty-three letters. The German movie *23* explored an obsession with the number, based on a real-life story.

When twenty-three or more people are in the same room, there is a better than even chance that at least two share the same birthday. Class sizes in schools in Britain are larger than twenty-three, so more than half classes in Britain have two pupils with the same birthdays. On average, every twenty-third wave crashing to shore is twice as large as normal. It takes twenty-three seconds for blood to circulate through the body. The average smoker gets through twenty-three cigarettes a day. Parents each contribute twenty-three chromosomes. Rock star Kurt Cobain was born in 1967 and died in 1994. Both years bizarrely add up to twenty-three if counted as individual digits: $1 + 9 + 6 + 7 = 23$ and $1 + 9 + 9 + 4 = 23$. Roman emperor Julius Caesar was stabbed twenty-three times when he was assassinated.

The human biorhythm is generally twenty-three days. The music group Psychic TV was obsessed with twenty-three. They released twenty-three live albums, each on the twenty-third day of twenty-three months running. In the final assault on the Death Star in *Star Wars*, Luke Skywalker is in Red 5. Red 2 and Red 3 start bomb runs at twenty-three degrees. The cellblock holding Princess Leia was AA-23. U.S. Cavalry legend General Custer was promoted to the senior military rank at the age of twenty-three. The soldier was the youngest general in the United States Army at the time.

Movies about Math/Numbers

- *A Beautiful Mind,* the winner of four Academy Awards, including best picture, is about a brilliant mathematician John Nash, on the brink of international acclaim when he becomes entangled in a mysterious conspiracy.
- In *Pi,* again, a brilliant mathematician teeters on the brink of insanity as he searches for an elusive numerical code in this critically acclaimed schizophrenic thriller! He is on the verge of that most important discovery of his life. For the past ten years, he has been attempting to decode the numerical pattern beneath the ultimate system of ordered chaos—the stock market. As Max verges on a solution, chaos is swallowing the world around him. Pursued by an aggressive Wall Street firm set on financial domination and Kabbalah sect intent on unlocking the secrets behind their ancient holy texts, he races to crack the code, hoping to defy the madness that looms before him. Instead, he uncovers a secret for which everyone is willing to kill.
- *Proof* is the compelling story of an enigmatic young woman haunted by her father's past and the shadow of her own future, exploring the links between genius and madness, the tender relationships between fathers and daughters, and the nature of truth and family. On the eve of her twenty-seventh

birthday, a young woman who has spent years caring for her brilliant but unstable father, a mathematical genius must deal not only with the arrival of her estranged sister but also with the attentions of a former student of her father who hopes to find valuable work in the 103 notebooks of his.

- In *Rain Main*, a man has just discovered he has an autistic brother and is now taking him on the ride of his life. Or is it the other way around? From his refusal to drive on major highways to a "four minutes to Wapner" meltdown at an Oklahoma farmhouse, he first pushes hotheaded brother to the limits of his patience . . . and then pulls him completely out of his self-centered world! But what begins as an unsentimental journey for the Babbitt brothers becomes much more than the distance between two places—it's a connection between two vastly different people.

- In *Enigma,* an East German defector who has turned CIA agent is assigned to discover the names of five Soviet dissidents targeted by a ruthless KGB hit squad. To stop this diabolic plot, he must infiltrate Soviet intelligence and obtain information from Enigma, a Russian computer component. But when he arrives in Berlin, his cover is blown, and he quickly seeks shelter with his former lover. But as they get deeper in this deadly game of pursuit, he must outwit the assassins before they fall into the hands of the enemy.

- *Good Will Hunting,* nominated for nine Academy Awards, is about a headstrong working -class genius. After one too many run-ins with the law, Will's last chance is a psychology professor.

- *Numbers* is a drama about an FBI agent who recruits his mathematical-genius brother to help the bureau solve a wide range of challenging crimes in Los Angeles. The two brothers take on the most confounding criminal cases from a very distinctive perspective. Inspired by actual events, the series depicts how the confluence of police work and mathematics provides unexpected revelations and answers to the most perplexing criminal questions. A dedicated FBI agent couldn't be more different from his younger brother, a brilliant mathematician who, since he was little, yearned to impress his big brother. He is joined on his team by fellow agents, a behavioral specialist who brings psychological insight to their investigations.

After some initial reluctance, their team welcomes the innovative methods to crime-solving. Their father is happy to see his sons working together even though he doesn't understand the intricacies of what he does for a living. It is his coworkers who further refine his approach and help him stay focused. Physicist friend constantly challenges him to employ a broader point of view to his work with the FBI. His former grad student frequently helps him see cases in a new light. Despite their disparate approaches to life, brothers are able to combine their areas of expertise and solve some killer cases.

Numbers and Music

Music has always been understood as number made audible, a sense of numerical order of nature. Harmonies are order of nature. Numbers show how harmonic series arise and how these ratios can be applied to geometry.

The basis of music is the scale. Musical scales are based on Fibonacci numbers. They are produced through simple ratios. The Fibonacci series appears in the foundation of aspects of art, beauty, and life. Even music has a foundation in the series: There are thirteen notes in the span of any note through its octave. A scale is composed of eight notes, of which the fifth and third notes create the basic foundation of all chords, and are based on whole tone, which is two steps from the root tone, which is the first note of the scale.

Numeration Systems

The primitive notion of numbers, requiring thousands of years to be extracted from repeated concrete situations, appears to have evolved from the many physical contrasts prevalent in nature—the difference between one tree and a forest, one sheep and a herd. Likewise, it was noticed that objects within a group could be placed in one-to-one correspondence with objects from other groups; the hands can be matched (placed in one-to-one correspondence) with the feet. Consequently, there was recognition of an abstract property shared by some groups: that ability to be placed in a one-to-one correspondence with each other. This property is what we now call number. Those groups that could be placed in one-to-one correspondence did not share this property and are thus said to be different in numbers. Out of this realization of "sameness" was born the concept of number in mathematics.

Today, identification of objects and people has become much easier and more sophisticated. Thanks to numbers and codes to make this logically possible. Advances in this field are apparent to anybody who shops at grocery stores and shopping malls. Every piece of mail, ticket to a game, etc., has its own identification and even a check digit or number to make sure that it is not mixed up with any other item. This article reviews how the use of numbers and numeration systems have reached their current sophistication level. We start by defining a system.

A numeration system is a set of basic symbols and some rules for making other symbols from them to represent quantities or numbers. The symbols for a numeration system are called numerals. A number is an abstract concept used to describe quantity. The invention of a precise and "workable" system is one of the greatest inventions of humanity. Many nations have had contributions to this invention. They have approached the problem using different paths but have arrived at the same point. The table below demonstrates presentation of the number three.

Roman Numerals

The system of Roman numerals is a numeral system originating in ancient Rome and was adapted from Etruscan numerals. The system used in antiquity was slightly modified in the Middle Ages to produce the system we use today.

Roman Numeral	Hindu-Arab Face Value
I	1
V	5
X	10
L	50
C	100
D	500
M	1000

For example, XL = 50 – 10 = 40. That is why when you reach your fortieth birthday, you need extra-large clothes.

Arabic Numerals

Arabic numerals, also known as Hindu numerals or Indian numerals, are the most common set of symbols used to represent numbers. They are considered one of the most significant developments in mathematics. It is interesting to know that these numerals were neither invented nor widely used by the Arabs. In fact, they were developed in India by the Hindus at around 400 BC. However, Arabs transmitted this system to the West after the Hindu numerical system found its way to Persia. Arabs themselves call the numerals they use Indian numerals. Hindu numerals in the first century AD is shown below.

Maya Numerals

The pre-Columbian Maya civilization used a vigesimal (base 20) numeral system. Maya numbers as shown in Maya codices. The numerals are made up of three symbols: zero (shell shape), one (a dot), and five (a bar). For example, nineteen is written as four dots in a horizontal row above three horizontal lines stacked upon each other.

Babylonian Numerals

Babylonian numerals were written in cuneiform, using a wedge-tipped reed stylus to make a mark on a soft clay tablet, which would be exposed in the sun to harden to create a permanent record.

Cyrillic Numerals

Cyrillic numerals was a numbering system derived from the Cyrillic alphabet, used by South and East Slavic peoples. The system was used in Russia as late as the 1700s when Peter the Great replaced it with the Arabic system.

Methods of Numeration

In this section, three methods of numeration are discussed. These represent the historical development of the systems used in different parts of the world.

1.　Simple Grouping

A good demonstrating example for simple grouping is United States' coins. The general concept is as follows: simple grouping system uses one stroke for each item (tally method). For example:

$$9 = |\,|\,|\,|\,|\,|\,|\,|\,|$$

This is, of course, lengthy. To make this easier, let us, for example, introduce a new symbol /\ to replace five strokes, and then nine strokes can be written as /\ | | | |, one 5 and four 1s. To understand this, think of one nickel and four pennies. Here are more examples:

$$17 = /\ /\ /\ | | = 3 \text{ nickels, 2 pennies}$$
$$21 = /\ /\ /\ /\| = 4 \text{ nickels. 1 pennies}$$

2.   Multiplicative Grouping

In example (b) above, the four /\ symbols take up a lot of space. To save space, let us use a multiplier, along with only one /\ symbol to show how many /\s are there.

Let °, ° °, ° ° °, . . . represent one, two, three, . . . of any symbol. Then we can write the numbers seventeen and twenty-one as

$$° ° ° /\ ° ° | = 17 \qquad ° ° ° ° /\| = 20$$

3.   Positional Method

You may have discovered by now that there is a rather natural way to change the multiplicative grouping method and save even more space. To see this, consider the first example. The sequence of symbols, from right to left, means the following: a pair of symbols showing the number of 1s, and then a pair showing the number of 5s, and then a pair for 25s, and so on, and that this sequence always occurs in that order. Thus, it is possible to use the multipliers alone and save both space and symbols.

$$° ° ° \quad ° °$$
$$(3 \times 5) + (2 \times 1)$$

As we know, base 2, with only two symbols 0 and 1, is used in computers. Let us consider some examples.

Example: Convert base 2 numeral 10111 to base 10.

4. $(10111)_{two} = (1 \times 2^4) + (0 \times 2^3) + (1 \times 2^2) + (1 \times 2^1) + (1 \times 2^0) = 23,$

Also $27 = (11011)_{two}$ Thus $29 = (11101)_{two}$. Also, $99 = (1100011)_{two}$

Communicating Data

This section discusses some application of the material presented earlier. We start by noting that the sequence $a_k a_{k+1} \ldots a_1 a_0$ in the base 10 represents

$$a_k (10^k) + a_{k-1}(10^{k-1}) + \ldots + a_1(10^1) + a_0(1),$$

where $a_k, a_{k-1}, \ldots, a_1, a_0$ are digits in the range 0 to 9. Similarly, the binary sequence $b_k b_{k-1} \ldots b_1 b_0$ in the base 2 represents

$$b_k(2^k) + b_{k-1}(2^{k-1}) + \ldots + b_1(2^1) + b_0(1),$$

where $b_k, b_{k-1}, \ldots, b_1, b_0$ are either 0 or 1.

Computers use a special code such as EXTENDED ASCII in which each symbol is represented by a binary sequence of eight symbols, called a byte. For example, a and A are represented respectively by 01100001 and 01000001.

This allows conversions of names, sentences, etc., to string of 0s and 1s. As pointed out earlier, numbers are used as codes and identification tolls. For example, many industries use base 10 numbers with certain characteristics for identification. They even add a check digit to avoid errors and misidentification. For instance, airline tickets, Federal Express, UPS packages, Avis, National Rental Cars use remainder upon division by seven of the number itself as the check digit. For example, if a package is given a number such as 540047, then $540047 = 7 \times 7149 + 4$; a digit (in this case 4) is added to the end of the number as a check digit, making the number 5400474.

Note that this method will not detect the substitution of 0 for 7,

1 for 8, 2 for 9, and vice versa. However, unlike method used by the Postal Service Method, it will detect transpositions of adjacent digits with the exceptions of the pairs (0,7), (1.8), and (2,9).

Grocery products use a smart method known as Universal Product Code (UPC). Here is an example:

$$0 \qquad 16000 \qquad 65620 \qquad 8$$

broad category of goods   manufacturer   product   check digit

To check, they multiply every other digit by 3 and add the results. For the above example,

$$3(0) + 1 + 3(6) + 0 + 3(0) + 0 + 3(6) + 5 + 3(6) + 2 + 3(0) + 8 = 70.$$

If the result doesn't end with a 0, computer knows the entered number is incorrect.

This simple scheme detects all single-position errors and about 89 percent of all other kinds of errors. Banks use different weights. For example, the First Chicago Bank has the number 071000013. Here, the last digit is the check digit. The check digit is the last digit of the following:

$$7(0) + 3(7) + 9(1) + 7(0) + 3(0) + 9(0) + 7(0) + 3(1) = 33.$$

The digits 7, 3, and 9 are called weights. Next, consider the ZIP code.

ZIP code: Every city has a ZIP code. For example, where we live has a code 17815. Each digit is used to identify something. For 17815, 1 represents the geographic area. There are ten possibilities; 0 (east) to 9 (west) of state. Other numbers represent:

78: Central mail-distribution or Sectional Center Facility (SCF). Some states have more than one.

15: Town or local post office (alphabetical within delivery area).

ZIP+4 code uses extra digits. Again, where we live has a code. 17815 – 8879

88: Delivery sector (several blocks a group of streets, several office buildings, . . .)

79: Group of post office boxes, a department in a university, etc.

Next, consider barcodes used by the post office.

A barcode is a series of dark and light spaces that represent characters.

$$\text{Dark} \rightarrow 0 \qquad \text{Light} \rightarrow 1$$

Any system using only two symbols is a binary code. Another example is ZIP code barcode.

The simplest barcode is the Postnet Code used by postal service. For ZIP+4 code, there are fifty-two vertical bars of two possible lengths (long and short).

The long bars in the beginning and end are guard bars. The tenth digit of a Postnet code number is a check digit so that

$$a_1 + a_2 + \ldots + a_9 + c \text{ ends with } 0.$$

For example, the ZIP+4 code 17815 – 8879 has a check digit 6 because

$$1 + 7 + 8 + 1 + 5 + 8 + 8 + 7 + 9 + 6 = 60 \text{ end with } 0.$$

In a more advanced application, each digit is represented by exactly two long and three short bars. Any error of replacing one with the other is detected, since we will have only one or three long bars. Note that there are exactly ten arrangement $\frac{5!}{3!2!} = 10$. A misreading of a single bar in such a block is therefore recognizable.

Postnet is a barcode symbology that is used by the United States Postal Service to assist in directing mail. The ZIP code or ZIP+4 code is encoded in this unique symbology that encodes data in half- and full-height bars. Most often, the delivery point is added, usually being the last two digits of the address or PO box number.

The barcode starts and ends with a full bar (often called a guard rail) and has a check digit after the ZIP or ZIP+4. The check digit is calculated as follows:

Add up all the data being encoded. If you are sending a letter to somewhere in <u>Young America, Minnesota</u>, you might be sending to 55555-1234, which would have the sum of 35.

Find the number that would need to be added to this number to make it evenly divisible by 10, in this case **5**, which is your check digit.

The encoding table is shown on the right. | denotes a full bar and ₁ denotes a half bar.

Each individual digit is represented by a set of five bars, two of which are full bars. The full bars represent "on" bits in a pseudo-binary code in which the places represent, from left to right, 7, 4, 2, 1, 0. (Though in this scheme, zero is encoded as 11, or "binary" 11000.). The above-mentioned example of 55555-1234 yields

Value	Encoding
1	ııı‖
2	ıı|ı|
3	ıı‖ı
4	ı|ıı|
5	ı|ı|ı
6	ı‖ıı
7	|ııı|
8	|ıı|ı
9	|ı|ıı
0	‖ııı

1. Sum is 35.
2. The check digit is therefore 5.
3. The data encoded will be 5555512345.

Together with the initial and terminal guard bars, this would be represented as

|ı|ıı|ı|ıı|ı|ıı|ı|ı|ı|ıı|ıııı‖ıı|ı|ı|ıı‖ıı|ıı|ı|ı|ı|

Note that the delivery point is often added after the ZIP+4 and before the check digit.

There have been four formats of Postnet barcodes used by the postal service:

A five-digit barcode, containing the basic ZIP code only, is referred to as the A code.

A six-digit barcode, containing the last two digits of the ZIP code and the four digits of the ZIP+4 code, is referred to as a B code.

In the early stages of postal automated mail processing, the B code was used to "upgrade" mail that had been coded only with a five-digit A code. This barcode was only found on mail that received a five-digit barcode on the initial coding by an OCR. Now obsolete.

A nine-digit barcode, containing the ZIP code and ZIP+4 code, is referred to as the C code. The nine-digit barcode enabled the sorting of mail to the individual delivery carrier, and in some cases into a semblance of delivery sequence.

An eleven-digit barcode, containing the ZIP code, ZIP+4 code, and the delivery point code, is usually referred to as the DPBC, or delivery point barcode. This is the predominant barcode in use currently (2005), and it enables the postal service to sort mail into delivery point (address) sequence.

Barcodes

If you go look in your refrigerator or pantry right now, you will find that just about every package you see has a UPC barcode printed on it. In fact, nearly every item that you purchase from a grocery store, department store, and mass merchandiser has a UPC barcode on it somewhere.

Have you ever wondered where these codes come from and what they mean? In this article, we will solve this mystery so that you can decode any UPC code you come across.

UPC stands for Universal Product Code. UPC barcodes were originally created to help grocery stores speed up the checkout process and keep better track of inventory, but the system quickly spread to all other retail products because it was so successful.

UPCs originate with a company called the Uniform Code Council (UCC). A manufacturer applies to the UCC for permission to enter the UPC system. The manufacturer pays an annual fee for the privilege. In return, the UCC issues the manufacturer a six-digit manufacturer identification number and provides guidelines on how to use it. You can see the manufacturer identification number in any standard twelve-digit UPC code. The UPC symbol has two parts:

- the machine-readable barcode
- the human-readable twelve-digit UPC number

The manufacturer identification number is the first six digits of the UPC number—639382 in the image above. The next five digits—00039—are the item number. A person employed by the manufacturer, called the UPC coordinator, is responsible for assigning item numbers to products, making sure the same code is not used on more than one product, retiring codes as products are removed from the product line, etc.

In general, every item the manufacturer sells, as well as every size package and every repackaging of the item, needs a different item code. So a twelve-ounce can of Coke needs a different item number than a sixteen-ounce bottle of Coke, as does a six-pack of twelve-ounce cans, a twelve-pack, a twenty-four-can case, and so on. It is the job of the UPC coordinator to keep all of these numbers straight!

The last digit of the UPC code is called a check digit. This digit lets the scanner determine if it scanned the number correctly or not. Here is how the check digit is calculated for the other eleven digits, using the code 63938200039 from *The Teenager's Guide to the Real World* example shown above:

1. Add together the value of the digits in odd positions (digits 1, 3, 5, 7, 9, and 11). $6 + 9 + 8 + 0 + 0 + 9 = 32$
2. Multiply the resulting number by 3. We get $32 \times 3 = 96$.
3. Add together the value of all the digits in even positions (digits 2, 4, 6, 8, and 10). $3 + 3 + 2 + 0 + 3 = 11$
4. Add this sum to the value in step 2, we get $96 + 11 = 107$.
5. Take the number in step 4. To create the check digit, determine the number that, when added to the number in step 4, is a multiple of 10—that is, $107 + 3 = 110$. The check digit is therefore 3.

Each time the scanner scans an item, it performs this calculation. If the check digit it calculates is different from the check digit it reads, the scanner knows that something went wrong and the item needs to be rescanned.

Using price lookup codes (PLUs), the nutshell:

- PLU codes are four-digit numbers that identify different types of produce. For example, #4011 is the code for a standard yellow banana.
- The number 9 prefix added to a PLU signifies that an item is organic. For example, #94011 is the code for an organic yellow banana.
- A number 8 prefix added to a PLU signifies that an item is genetically engineered (GE). For example, #84011 is the code for a genetically engineered yellow banana.
- PLU codes and their organic prefixes are in wide use, but GE codes are rare at best.
- The Universal Product Code (UPC) is a barcode symbology that is widely used in the United States, Canada, United Kingdom, Australia, New Zealand, in Europe, and other countries for tracking trade items in stores.

UPC (technically refers to UPC-A) consists of twelve numeric digits that are uniquely assigned to each trade item. Along with the related EAN barcode, the UPC is the barcode mainly used for scanning of trade items at the point of sale, per GS1 specifications. [1] UPC data structures are a component of GTINs and follow the global GS1 specification, which is based on international standards. But some retailers (clothing, furniture) do not use the GS1 system (rather other barcode symbologies or article number systems). On the other hand, some retailers use the EAN/UPC barcode symbology, but without using a GTIN (for products, brands, sold at such retailers only).

Wallace Flint proposed an automated checkout system in 1932 using punched cards. Bernard Silver and Norman Joseph Woodland, a graduate student from Drexel Institute of Technology (now Drexel University), developed a bull's-eye style code and applied for the patent in 1949.

In the 1960s, railroads experimented with a multicolor barcode for tracking railcars, but they eventually abandoned it.

In mid-1971, William "Bill" Crouse invented a new barcode called Delta C.

David Savir, a mathematician, was given the task of proving the symbol could be printed and would meet the reliability requirements, and was most likely unaware of Baumeister's equations. He and Laurer added two more digits to the ten for error correction and detection.

Each UPC-A barcode consists of a scannable strip of black bars and white spaces above a sequence of twelve numerical digits. No letters, characters, or other content of any kind may appear on a UPC-A barcode. There is a one-to-one correspondence between twelve-digit number and strip of black bars and white spaces—that is, there is only one way to represent each twelve-digit number visually, and there is only one way to represent each strip of black bars and white spaces numerically.

UPC-A barcodes can be printed at various densities to accommodate a variety of printing and scanning processes. The significant dimensional parameter is called x-dimension (width of single module element). The width of each bar (space) is determined by multiplying the x-dimension and the module width (1, 2, 3, or 4 units) of each bar (space). Since the guard patterns each include two bars, and each of the twelve digits of the UPC-A barcode consists of two bars and two spaces, all UPC-A barcodes consist of exactly $(3 \times 2) + (12 \times 2) = 30$ bars, of which six represent guard patterns.

UPC Barcode

First used in1973 for grocery items, the universal product code (UPC) is a twelve-digit number and associated machine-readable barcode used to identify products being purchased in grocery stores. UPCs encode an individual product, but not its price (this part is done by a store's computer after reading the product identifier). The UPC is maintained by the Uniform Code Council of Dayton, Ohio. The first and last digits are separated from the others and written in a smaller font size.

Measurement Theory

Measurement of some attribute of a set of things is the process of assigning numbers or other symbols to the things in such a way that relationships of the numbers or symbols reflect relationships of the attribute being measured. A particular way of assigning numbers or symbols to a measure something is called a scale of measurement.

Measurement theory is a branch of applied mathematics that is useful in situations that involve measurement and data analysis. The fundamental idea of measurement theory is that measurements are not the same as the attribute being measured. Hence, if one wants to draw conclusions about the attribute, one must take into account the nature of correspondence between the attribute and the measurement.

The mathematical theory of measurement is elaborated in many publications. The theory was popularized in psychology by people originated the idea of levels of measurement.

Measurement theory helps one to avoid making meaningless statements. A typical example of such a meaningless statement is the claim by the weatherman on the local TV station that it was twice as warm today as yesterday because it was forty degrees it today but only twenty degrees yesterday. This statement is meaningless because one measurement (forty) is twice the other measurement (twenty) only in certain arbitrary scales of measurement, such as Fahrenheit. The relationship "twice as" applies only to the numbers, not to the attribute being measured (temperature).

When we measure something, the resulting numbers are usually, to some degree, arbitrary. We *choose* to use a 1 to 5 rating scale instead of a -2 to 2 scale. We choose to use Fahrenheit instead of Celsius. We choose to use miles per gallon instead of gallons per mile. The conclusions of our analysis should not depend on these arbitrary decisions, because we could have made the decisions differently. We want the analysis to say something about reality, not simply about our whims regarding meters or feet.

Levels of Measurement

There are different levels of measurement that involve different properties (relations and operations) of the numbers or symbols that constitute the measurements. Associated with each level of measurement is a set of permissible transformations. The most commonly discussed levels of measurement are as follows:

Nominal—two things are assigned the same symbol if they have the same value of the attribute. Examples: numbering of football players, numbers assigned to religions in alphabetical order (e.g., atheist = 1, Buddhist = 2. Christian = 3, etc.).

Ordinal—things are assigned numbers such that the order of the numbers reflects an order relation defined on the attribute. Examples include grades for academic performance (A, B, C).

Interval—things are assigned numbers such that differences between the numbers reflect differences of the attribute. Examples include temperature in degrees Fahrenheit or Celsius; calendar date.

Log-interval—things are assigned numbers such that ratios between the numbers reflect ratios of the attribute. Examples include density (mass/volume); fuel efficiency in mpg.

Ratio—things are assigned numbers such that differences and ratios between the numbers reflect differences and ratios of the attribute. Examples include length in centimeters; duration in seconds; temperature in degrees Kelvin.

Absolute—things are assigned numbers such that all properties of the numbers reflect analogous properties of the attribute. The only permissible transformation is the identity transformation. Examples include number of children.

Once a set of measurements have been made on a particular scale, it is possible to transform the measurements to yield a new set of measurements at a different level. It is always possible to transform from a stronger level to a weaker level. For example, a temperature measurement in degrees Kelvin is at the ratio level. If we convert the measurements to degrees Celsius, the level is interval. If we rank the measurements, the level becomes ordinal. In some cases, it is possible to convert from a weaker scale to a stronger scale. For example, correspondence analysis can convert nominal or ordinal measurement to an interval scale under appropriate assumptions.

Number Paradoxes

Think about two students, A and B, from two different universities or from two different majors in the same university. Suppose that student A has a GPA higher than student B, say, 3.2 and 2.8, respectively. Then one may interpret this as A being a better student. But thinking about it carefully, one may wonder what these numbers present and whether they are directly comparable. For example, what if we knew that the average GPA in their universities or majors were respectively 3.4 and 2.7. If so, this implies that A is a below-average student; whereas, B is an above average one. I hope you agree that this could easily lead to confusion.

Simpson's Paradox. Consider treating the T (for "terrible") disease. Say there are two methods, A and B, of treatment. Suppose 390 patients with this disease were given one of the two treatments: 160 were given A, and 230 were given B. The recovery rates were $60/160 = 0.38$ using treatment A and $65/230 = 0.28$ using treatment B. While chances of recovering from the T disease are extremely poor and hopefully much more research will be done on it, it seems from these data that it is more favorable to use treatment A than B.

However, by using one classifier—namely, sex—the total picture changes as follows.

Fraction Recovering

Treatment Sex	A	B
Male	20/100 = 0.2	50/210 = 0.24
Female	40/60 = 0.67	15/20 = 0.75
Total	60/160 = 0.38	65/230 = 0.28

Now, it is clear from this summary that both men and women would prefer—treatment B, which is different from our earlier conclusion. The lesson here is that you must be extremely careful in reporting averages unless the groups under consideration are fairly homogeneous. Note, in this example, the experiences of men and women are quite different.

Who Was a Better Free-Throw Shooter?

Consider two players A and B. Suppose that in the first half of the season, player A made only 20 of his 100 free-throw attempts. In the same period, player B made 50 of his 210 free throws. So in the first half of the season, B performed better (24 percent versus 20 percent). Suppose that in the second half of the season, A made 40 of his 60 and B made 15 of his 20 attempts, respectively. Thus, again, B performed better (75 percent versus 67 percent). Now, if we combined these statistics for the season, we have 60/160 = 0.38 for A and 65/230 = 0.28 for B—that is, A did better for the season. This means that although B performed better in both the first and second half of the season, his performance was worse than A in the whole season. This is, of course, a paradox known as Simpson's paradox. The table below summarizes these statistics.

	First Half	Second Half	Season
Player A	20/100 = 0.20	40/60 = 0.67	60/160 = 0.38
Player B	50/210 = 0.24	15/20 = 0.75	65/230 = 0.28

Now, it is clear from this summary that different conclusions may be drawn from looking at whole or parts of players' statistics. The lesson here is that we must be extremely careful in reporting averages (percentages) unless the groups under consideration are fairly homogeneous.

To stress this point further, let us consider a type of problem that is so often reported in today's newspaper—namely, the difference in the salaries of men and women. Suppose that one hospital in Pennsylvania employs fifty men and fifty women, and their respective average salaries are $16,000 and $14,000. This seemingly supports the argument that there is discrimination in salaries because of sex. However, by using one classifier, length of employment, that in some sense is a measure of qualification, we obtain the following averages.

Sex	*Average Salaries*			
	Male		Female	
Length of Employment	number	average	number	average
less than 5 years	10	10,000	40	12,500
less than 5 years	40	17,500	10	20,000
Total	50	16,000	50	14,000

From this summary, the women's salaries look better than those of the men. Now, it is not the purpose of this example to argue that women have or have not been discriminated against in the matter of salaries (for the record, it is my personal opinion that there has been some discrimination), but only to emphasize that qualifications (amount of experience, education, and so on) of the individual must be taken into account in arguing that discrimination does or does not exist. Yet frequently in the newspaper, we find one salary figure representing each of the groups, men and women. With only those two numbers, as in this example, it is impossible to say anything about discrimination in this situation.

Simpson's paradox can occur because collapsing can lead to inappropriate weighting of the different populations. Treatment 1 was given to 80 males and 120 females, so the marginal table is indicating a success rate for treatment 1 that is a weighted average of the success rates for males and females with slightly more weight given to the females. Treatment 2 was given to 150 males and only 40 females, so the marginal success rate is a weighted average of the male and female success rates with most of the weight given to the male success rate. It is only a slight oversimplification to say that the marginal table is comparing a success rate for treatment 1, which is the mean of the male and female success rates, to a success rate for treatment 2, which is essentially the male success rate. Since the success rate for males is much higher than it is for females, the marginal table gives the illusion that treatment 2 is better.

The moral of all this is that one cannot necessarily trust conclusions drawn from marginal tables. It is generally necessary to consider all the dimensions of a table. Situations in which marginal (collapsed) tables yields valid conclusions are discussed in, for example, section 4.2 of a book by Christenson (1990).

9/11

Conspiracy theories based on numbers have been around for a relatively long time. Here is a relatively recent one related to the 9/11 disaster that involves number eleven. First, recall that the beautiful Twin Towers looked like number eleven. The number 911 is used for emergency, and $9 + 1 + 1 = 11$. That is a good start. The Twin Towers looked like the number eleven. So perhaps all that happened had something to do with the number eleven. Oh, yes, here are more evidence. The first flight to hit the Twin Towers was flight number

11 with ninety-two people on board, $9 + 2 = 11$. September 11 is the 254th day of the year, $2 + 5 + 4 = 11$ and $365 - 254 = 111$. There are eleven letters in "New York City," "Afghanistan," "the Pentagon," and "George W. Bush." It seems that we are getting somewhere. New York was the eleventh state admitted to the union. The area code 119 $(1 + 1 + 9 = 11)$ is for both Iraq and Iran, and flight 77 that crashed in Pennsylvania had sixty-five people on board, $6 + 5 = 11$. Finally, here is another confirmation. Recall the March 11 (2004) attack in Spain. There are exactly 911 days between this and the September 11 (2001) attack.

Data Mining

Data analysis is based on data collected or to be collected. In recent years, because of the availability of high-speed computers, many large databases have become available for analysis. When using such databases, one problem faced by the researchers is to separate useful information or data from the junk. This has led to a relatively new area known as data mining.

In the literature, the name data mining is used in two different contexts. One process refers to the process of analyzing large databases and others to data grubbing, data dredging, or fishing. It is probably no exaggeration to say that most statisticians are concerned with primary data analysis—that is, data are collected with a particular question or set of questions at mind. Data mining, being different in this sense, is entirely concerned with secondary data analysis. In fact, data mining may be defined as the process of secondary analysis of large databases aimed at finding unsuspected relationships that are of interest or value to the database owners.

Also, statistics, as is taught in most textbooks, might be described as being characterized by data sets, which are small and clean. This does not apply to data mining, whose concern is the secondary analysis of large databases. In this area, new problems arise, partly as a consequence of the sheer size of the data sets involved and partly because of issues of pattern matching. In fact, it is a new discipline lying at the interface of statistics, database technology, pattern recognition, machine learning, etc.

Now, the data mining as fishing expeditions refers to is a process where a researcher takes a body of data, tests countless theories, and reports the one that is the most statistically significant, since she believes that only statistically significant results are worth publishing. Most statisticians know very well that if only worthless theories are examined, the researcher will eventually stumble on a worthless theory that seems to be supported by the data. For example, in 1989, the *New York Times* reported that if the Dow Jones Industrial Average increased between the end of the November and the time of the Super Bowl, the football team whose city comes second alphabetically will probably win the Super Bowl. Thus, the Dow increased from 2115 to 2235 between November 30, 1989, and the day of the January 1990 Super Bowl, and San Francisco beat Cincinnati. Clearly this is a coincidence uncovered by data mining. The behavior of the Dow and the names of the cities have no effect on the outcome of a football game. The end of November was selected because this is the starting time that worked the best.

Counting the Possibilities

Suppose that you have two books, A and B, and a small shelf that takes two books. To place the books in the shelf, you have two choices, AB and BA. With three books, your choices are ABC, ACB, CAB, ACB, BCA, and BAC; $3 \times 2 \times 1 = 6 = 3!$ (3 factorial), three choices for placing the first book, two choices for the second book, and one for the third. With four books, it is $4 \times 3 \times 2 \times 1 = 24 = 4!$ Moreover, with only fourteen books, your choices are $14! = 87,178,291,200$, more than twelve times the population of world, amazing. Think about a big shelf, a library. The number of choices is unbelievably large and mind-boggling. Alternatively, think about a room with only fourteen seats. The first person arriving has fourteen choices, the next thirteen, and so on. Therefore, there are total of 14! permutations. Think about a large class, a movie theater, or a football stadium. What is even more surprising is the fact that here we are talking about discrete variables. Imagine what happens with continuous situations like human face and the number of possibilities. Simply countless. I am not sure if I could explain this to my mom and, more importantly, if she would believe it. However, I am still fascinated by the fact that there are infinitely many possibilities even in discrete case.

Addition principle: If an event A1 can occur in a total of n^1 ways and if a different event A2 can occur in a total of n^2 ways, then the event A1 or A2 (A1 $\cup$ A2) can occur in $n^1 + n^2$ ways provided that A1 and A2 are disjointed—that is, they cannot occur simultaneously.

Multiplication principle: If an event (operation) A1 can occur in a total of n^1 ways and if after that a different event (operation) A2 can occur in total of n^2 ways, then the event A1 and A2 (A1 $\cap$ A2), the whole operation, can occur in $n^1 \times n^2$ ways.

Sir Arthur Eddington, an astronomer, once wrote in a satirical essay that if a monkey were left alone long enough with a computer and typed randomly, any great novel could be replicated. What he did not probably know was the fact that Earth's life is not enough to replicate even one book. To see that, consider the following example:

What is the probability that this passage could have been written by a monkey? Ignore capitals and punctuation and consider only letters and space.

"I cannot think about life without sports"

Recall that there are, twenty-six letters plus space, twenty-seven choices for each position. Thus, the number of possibilities is 27^{40}. What we see is just one option. So the probability is $5.564798376768 \times 10^{-58}$.

We shall now use the second of these principles to enumerate the number of arrangements of a set of objects. Let us consider the number of arrangements of the letters a, b, c. We can pick any one of the three to place in the first position; either of the remaining two may be put in the second position, and the third position must be filled by the unused letter. The filling of the first position is an event that can occur in three ways, the filling of the second position is an event that can occur in two ways, and the third event can occur in one way. The three events can occur together in a $3 \times 2 \times 1 = 6$ ways. The six arrangements or permutations, as they are called, are abc, acb, bac, bca, cab, cba. In this simple example, the elaborate method of counting was hardly worthwhile because it is easy enough to write down all six permutations. But if we had asked for the number of permutations of six letters, we should have had $(6)(5)(4)(3)(2)(1) = 720$ permutation to write down. It is obvious now that, in general, the number of permutations of n different objects is

$$n(n-1)(n-2)(n-3)...(2)(1) = n!$$

Example. There are fifteen teams in a tournament. Each team will play with every other team once. How many games are there?

Solution: $(15)(14)/2 = 105$ because, here, AB is the same game as BA. A different solution may be obtained by noting that team 1 will play with other fourteen teams, resulting in fourteen games. Team 2 will play thirteen games (excluding the one with team 1), and so on. So we have $14 + 13 + 9 + \cdots + 2 + 1 = 105$ games.

Example. In table tennis (using the old rules when I played), to win a standard game, a player must either be the first to reach twenty-one points with a margin of two or win two consecutive points following a tie at twenty all before the opponent does. Considering the first case, show that for the game to finish 21–0, 21–1, . . ., 21–19, the number of possibilities are those given in the following table. Note that for the game to end, for example, 21–4, the winner should take the last point plus twenty points from the first twenty-four points played. If you like challenge, try to calculate the number of possibilities for the second case (going through 20–20).

Score	Possibilities
21–0	1
21–1	21
21–2	231
21–3	1771
21–4	10626
21–5	53130
21–6	230230
21–7	888030
21–8	3108105
21–9	10015005
21–10	30045015
21–11	84672315
21–12	225792840
21–13	573166440
21–14	1391975640
21–15	3247943160
21–16	7307872110
21–17	15905368710
21–18	33578000610
21–19	68923264410

Classroom Seats

Consider a classroom with only thirty seats. In the first day of class, students entering the class can choose where they like to sit. The first student has thirty choices, the second student has twenty-nine choices, and the last (thirtieth) student has only one choice. The total number of possible selections for thirty students (30) (29) ... (2) (1) = 30! is astronomical. Thus, the probability of any particular set of selections is practically zero. This is another example of the daily occurrence of highly improbable events.

Bridge Hands

In the card game, bridge, there are 635,013,559,600 different thirteen-card hands possible. This number of possible hands could be realized if all the people in the world played bridge for a day. For an individual, it would take several million years of continuous playing to be dealt each of these hands. Yet any given hand held by a player is equally probable, or rather, equally improbable, as its probability is 1/635,013,559,600. In other words, any hand is just as improbable as thirteen spades. Bridge hands are an example of the daily occurrence of very improbable events. Recall, once again, that some of these hands will surprise us more than the others.

Counting and Physics

Given n particles and $m > n$ boxes, we place at random each particle in one of the boxes. We wish to find the probability p that in n preselected boxes, one and only one particle will be found.

Since we are interested only in the underlying assumptions, we shall only state the results. We also verify the solution for $n = 2$ and $m = 6$. For this special case, the problem can be stated in terms of a pair of dice: the $m = 6$ faces correspond to the m boxes, and the $n = 2$ dice to the n particles. We assume that the preselected faces (boxes) are 3 and 4.

The solution to this problem depends on the choice of possible and favorable outcomes. We shall consider the following three celebrated cases:

Maxwell-Boltzmann statistics. If we accept as outcomes all possible ways of placing n particles in m boxes distinguishing the identity of each particle, then

$$p = \frac{n!}{m^n}$$

For $n = 2$ and $m = 6$, the above yields $p = 1/21$. This is the probability for getting 3, 4 in the game of two dice.

Bose-Einstein statistics. If we assume that the particles are not distinguishable—that is, if all their permutations count as one—then

$$p = \frac{(m-1)!\,n!}{(n+m-1)!}$$

For $n = 2$ and m = 6, this yields $p = 1/21$. Indeed, if we do not distinguish between the two dice, then $n = 21$, because the outcomes 3, 4 and 4, 3 are counted as one.

Fermi-Dirac statistics. If we do not distinguish between the particles and also we assume that in each box we are allowed to place at most one particle, then

$$p = \frac{n!\,(m-n)!}{m!}$$

For $n = 2$ and m = 6, we obtain $p = 1/15$. This is the probability for 3,4 if we do not distinguish between the dice and also if we ignore the outcomes in which the two numbers that show are equal.

One might argue, as indeed it was in the early years of statistical mechanics, that only the first of these solutions is logical. The fact is that in the absence of direct or indirect experimental evidence, this argument cannot be supported. The three models proposed are actually only hypotheses, and the physicist accepts the one whose consequences agree with experience.

Numbers and ASL

Interactive Sign Language (ASL): Fingerspelling and Numbers is the fast, fun way to gain proficiency in finger spelling. From learning the basics to improving comprehension and increasing signing speed, Fingerspelling and Numbers acts as your personal tutor. With the help of its lifelike photographs and animated hands, you'll be able to master fingerspelling at your own speed and pace quickly and easily. You can also create custom vocabularies, such as medical and legal terms, as well as common word endings and letter groups. Whether you are a beginner or an advanced student, Fingerspelling and Numbers is the best way to learn, improve, and master fingerspelling. My major is sign language, and when using numbers with this is only when you are talking about people's age, money, prices, and other things that people can think of in any other language.

Being able to give numerical information in ASL opens many doors. You can give someone your phone number, make an appointment, and warn a potential guest that you have twelve—yes, twelve cats.

When you're indicating quantity and counting things, sign the numbers one through five and eleven through fifteen with your palm facing you, and the numbers six through ten and sixteen through nineteen with your palm facing the person to whom you're signing.

Just as in English, there are exceptions to every rule, especially the one about which way your palm faces. To tell time in sign, let your dominant (active) index finger touch your other wrist—where you would wear a watch. Then use your dominant hand to sign the appropriate hour (number) with your palm facing the person you're signing to; the same goes for addresses and phone numbers. For quantity, though, the numbers one through five have your palm facing you; six through ten have your palm facing the addressee. To sign decade numbers—thirty, forty, fifty, and so on—you sign the first number (3, 4, 5) followed by the sign for the number zero. You sign "hundreds," such as six hundred, seven hundred, eight hundred, and so on, by, first, signing the number (6, 7, 8), and then signing for "hundred."

Cardinal (counting) and ordinal (ordering) numbers will get you through everyday situations, such as counting the millions you won on the lottery, giving your address and phone number to the movie star who wants to get to know you better, telling your mom that you won the first Pulitzer Prize for hip-hop poetry, and telling Cinderella that it's midnight.

When you want to specify that there's more than one item—plural—you sign the item first, followed by the quantity. Unlike English, you don't have to change the item to a plural by adding *s*. A good way to remember this is to keep in mind that you need to show what the item is before you can tell someone how many.

Japanese Sign Language (JSL) is a family of complex visual spatial languages used by deaf communities in Japan. There is no single standard JSL, although the Tokyo form does have some hegemonic force, since many of the TV broadcasts and meetings are sponsored by Tokyo deaf groups. The national sign languages in Taiwan and Korea apparently have incorporated some JSL signs and forms from the colonial occupation of these countries by Japan prior to World War II.

Japanese Sign Language is distinct from spoken/written Japanese in both grammar and lexicon, although many deaf signers will use manually coded Japanese/pidgin signed Japanese when signing to hearing or non-native JSL signers. The grammatical system shares many similarities with other national sign languages in its use of the complex visual space available, classifiers, and other complex forms.

Interestingly, JSL, ASL, and spoken Japanese all use a topic-comment grammatical system. This makes JSL and spoken Japanese more compatible than ASL and spoken English. This is one explanation for widespread use of MCJ/voiced JSL/pidgin JSL forms present in Japan.

JSL appears to be a much "younger" language form than many other national sign languages. The first school for the deaf was established in Kyoto in 1878, and we have very little evidence for sign language communities before that time (although they no doubt existed in small pockets).

Finger spelling was introduced in the early twentieth century and is based on the finger spelling used in Spain, France, and the United States. However, many older deaf do not know the finger spelling forms or numerals, and most deaf born before the end of World War

II (1948) did not attend school, since it was only after the war that compulsory education for the deaf was instituted.

Social Security Numbers

The Social Security program began as a type of insurance policy for America's elderly after the Great Depression. During the 1930s, poverty levels among senior citizens escalated to nearly 50 percent. In 2004, only 12.7 percent of all people living in America were below the poverty level threshold. By simple virtue of this comparison, it becomes clear why Pres. Franklin Roosevelt felt the Social Security program was necessary for the economic survival of the nation.

Social Security was enacted into law in 1935 as part of Franklin Roosevelt's New Deal plan. The law was primarily designed to allow retirees and unemployed persons to begin receiving benefits within two years. The program was, and currently still is, funded by the collection of payroll taxes based on a worker's wages. The exact financial ratios have been amended over the years, but the basics of the program have remained the same. Some of the changes that were made over the years include the addition of survivor's benefits and benefits for retiree's spouses and children. The way that benefits were paid also changed over time. Initially, the program allowed for a onetime lump sum payment but has now been revised to serve as a monthly fund. These benefits are paid out according to the universal adaptation of the social security number.

The social security number is a unique form of identifying United States citizens.

A treasury regulation was issued in 1936 that required an account number for each employee covered by the Social Security program, thus the creation of the social security number. A social security number is a nine-digit number assigned to individuals upon birth (and subsequent application to the Social Security Administration) in the United States. The nine digits of the number are broken into three sections, each separated by a hyphen. The first three numbers are area numbers, which are based upon the ZIP code of the city where application for a social security number was first made. The second two numbers (ranging from 01 to 99) are group numbers that serve to break social security numbers with the same area numbers into more manageable groups. The final four digits are serial numbers that range from 0001 to 9999. These numbers run consecutively within the group number designation.

During the period between 1936 and 1937, approximately thirty million applications for social security numbers were processed. Since that time, the social security number has come to be used as an identifying factor by countless groups. In 1961, the Civil Service Commission adopted the social security number as an official federal employee identifier. The Internal Revenue Service requires use of the social security number as the taxpayer identification number. The Department of Defense adopted the social security number in lieu of the military service number for purposes of identifying armed forces personnel. These government agencies are not the only ones to embrace use of the social security number. Private agencies such as banks, creditors, health insurance companies, and employers now use the social security number as a form of identification.

Approximately 360 million social security numbers have been issued to date, of which 211 million are still active. There have been concerns voiced about duplication of social security numbers and the ill effects that would follow such duplication, but since there are nearly one billion different social security numbers available, this is not an immediate problem. A more timely concern would be worrying about running out of Social Security funds before running out of numbers.

The Social Security program is truly a creation of genius. It is not only a vehicle with which to collect retirement checks but also an avenue that creates choices. It is much, much more than a simple paycheck. The Social Security program allows for vocational rehabilitation, disability pay, Medicare health insurance, and Supplemental Security Income. The individual number that each

American is assigned provides for a financially stable future of which the generations prior to its existence could only dream.

ZIP Code

The ZIP, or Zoning Improvement Plan, code was produced because of the volume of mail, the revolution in transportation, and the rise in manpower costs.

In the 1930s, the post office was still moving the bulk of the mail by train.

However, in 1963, fewer trains were carrying the mail. Furthermore, from the 1930s to 1963, the United States had undergone significant changes. It suffered an economic depression, fought in the Second World War, and went from an agricultural economy to an industrial economy. Also, migration from state to state was becoming more popular as people were searching for work. Because of these changes, the mail had also changed in volume, character, and transportation. Mail had also changed from a primarily social correspondence to more business mail.

In 1963, the Advisory Board of the Post Office Department selected a coding program, and Gen. John A. Gronouski announced that the ZIP code would begin on July 1, 1963.

Some larger cities had existing postal zones set in motion in 1943, but where the old zones failed to fit within the delivery areas, new numbers were assigned. By July 1963, a five-digit code was assigned to every address. The first digit represents a broad geographical area of the United States ranging from zero, for the most far east, to nine for the west.

The second two digits identify a central mail-distribution point known as a sectional center. Although Utah has just four centers for the entire states, there are six of them to service New York City.

The last two digits indicate the town or local post office. This is done alphabetically. For example, towns that begin with A usually have low numbers. However, larger cities are given the digits of 01.

In 1983, the U.S. Postal Service added four digits to the code, called the ZIP+4 code. This allows mail to be eligible for cheaper bulk rates and easily sorted with automatic equipment.

Roman Numerals (Revisit)

During the years I taught numbers, I wrote several times about them, especially Roman numerals. Here, I add them all in case you prefer one to the other.

1. My first exposure to Roman numerals was in elementary school. My teacher used I-II-V to organize her notes and to place events of our stories in chronological order on the board. At the time, I credited my teacher for thinking up such a great idea, but upon studying the history of Roman numerals, I realized that Roman numerals were around long before I encountered them in fifth grade. In this paper, an overview of origins regarding the Roman numeral system will be discussed.

The Roman numeral system was first devised in ancient Rome. The Romans were involved in commerce and trade; in order to express numerical values, they would have to develop a numerical system. Today, Roman numerals are represented as letters from the alphabet; however, in the beginning, Roman numerals were separate symbols that represented the numerical value. The Roman numeral system was actually an adaptation of the Etruscan numeral system. The Etruscan system used symbols to indicate value. The Etruscan system is complex from my point of view, but as I read further, it explained that the shape, size, and other characteristics of the symbols indicated the value it represented.

So with the Etruscan system, the Romans made symbols that suited their need to indicate value. During the Middle Ages, the Roman numeral system underwent a revision. During this time, the symbols used became the letters of the alphabet, and their value was assigned for use. The following table is an example of assigned value to letters.

I = 1 V = 5 X = 10 L=50 C=100 D=500 M=1000

As numbers became more complex, the Roman numeral system increased in complexity as well. For numbers above five thousand, a horizontal line was placed above the letter to indicate its increase in value.

Today, the Roman numeral system can still be found, but the Arabic number system is more versatile and common, because it incorporates the use of the zero and allows for more complicated calculations. The Roman numeral system does not. For these reasons and more, the Roman numeral system was gradually replaced by the Arabic numeral system in the early second millennium.

The Roman numeral system is used in organizational outlines, genealogy, the Olympics, and other disciplines.

2. Roman numerals is defined as a system of writing numbers that is based on the ancient Roman rather than the Arabic used in English-speaking countries today. In today's society, we don't realize the history of Roman numerals. Since Arabic is the number system we use, we take advantage of Roman numerals and only notice them in places like the Super Bowl (e.g., Super Bowl XXXIV). The difference between the two are Romans did not have a symbol for zero and that certain placement within the numbers can determine whether or not you are subtracting or adding. Arabic numbers look like this: 1, 5, 10, 50,100,500, 1,000; compared to Roman numbers, which are constructed out of the basic letter symbols that look like this: I, V, X, L, C, D, and M.

Thus, I means 1, II means 2, and III means 3. Four strokes, however, seemed like too many, which brings us to our next symbol, V for 5. Placing I in front of the V (or placing any smaller number in front of any larger number) indicates subtraction. So IV would represent 4. After V comes VI, meaning 6; VII, meaning 7; and VIII meaning. The next symbol is X, which means 10. Numbers in the teens, twenties, and thirties follow the same form as the first set, only with Xs indicating the number of tens, which means XXXI is 31, and XXIV is 24. L stands for 50. And thus 60, 70, and 80 are

LX, LXX, and LXXX. C (which stands for *centum,* the Latin word for "one hundred") is obviously 100. This is a word we still use today in words like "century" and "cent." D stands for 500. So CDXLVIII is 448. M is 1,000. You see a lot of Ms because Roman numerals are used a lot to indicate dates.

Romans started off with a number notation. For example, 673 is represented by the symbols DCLXXIII, and that is $500 + 100 + 50 + 10 + 10 + 1 + 1 + 1 = 673$. This additive writing of the numbers made the execution of the operations and arithmetics complex, but with the years was been affirming the beginning of position. As for perform the calculations they were traced of the lines for earth and between a line and the other one they put of the small stones that constituted the unities you count". (Nova)

"Each system is differentiated from the others for the base, the more used numeration is that to base 10. In most ancient system of numeration the base was 5, because 5 is the fingers of the hand. The numbers, for written, they were suitable with of the dots or of the lines engraved on tablets of clay or papyruses. The Roman notation was base 5 and 10. These two numbers are pointed out with the V signs and X and the single unity, suitable with the sign I, additions were considered or subtracted one had positioned to their right or left. I Il III IV V VI VII VITI IX X XI XII."

Which brings us to why and how the Romans came up with this symbolic letter representation for the letters L, C, D, M, the superior groups to 50, 100, 500, 1000 unit. "If we define those suitable from the small numbers of the base like unity of I order, the same base is an unity of 2 order. The ancient considered unity of order also superior and they pointed out it with special signs" (Net).

Roman numerals have been replaced by Arabic numerals; however, they are still widely present in our everyday lives. As stated before, we have seen Roman numerals represented for the Super Bowl. We also use them when we are making an outline and to indicate chapters in a book. We have seen Roman numerals in the index of books and in papers. We may not realize it because we are used to Arabic notation, but Roman numerals play a big part in how we read numbers today.

3. The system of Roman numerals was slightly modified in the Middle Ages to produce the system we use today. It is based on certain letters that are given values as numerals: Roman numerals are a system of numerical notations used by the Romans. They are an additive (and subtractive) system in which letters are used to denote certain "base" numbers, and arbitrary numbers are then denoted using combinations of symbols. Unfortunately, little is known about the origin of the Roman numeral system.

The following table gives the Latin letters used in Roman numerals and the corresponding numerical values they represent. For example, I 1, V 5, X 10, L 50, C 100, D 500, M 1000.

Although the Roman numerals are now written with letters of the Roman alphabet, they were originally separate symbols. The Etruscans, for example, used I A? X? 8 ®? for IV XL CM.

They appear to derive from notches on tally sticks, such as those used by Italian and Dalmatian shepherds into the nineteenth century. Thus, the I descends from a notch scored across the stick. Every fifth notch was double cut(?,?,?,?, etc.), and every tenth was cross cut (X), much like European tally marks today. This produced a positional system. Eight on a counting stick was eight tallies, IIIIA?III, but this could be written A?III (or VIII), as the A? implies the four prior notches. Likewise, number four on the stick was the I-notch that could be felt just before the cut of the V, so it could be written as either IIII or IV. Thus, the system was neither additive nor subtractive in its conception, but ordinal. When the tallies were later transferred to writing, the marks were easily identified with the existing Roman letters I, V, X.

(A folk etymology has it that the V represented a hand and that the X was made by placing two Vs on top of each other, one inverted.)

The tenth V or X along the stick received an extra stroke. Thus, 50 was written variously as N, H?, K, 1:P?, ?, etc., but perhaps most often as a chicken-track shape like a superimposed V and I. This had flattened to _1_? (an inverted T) by the time of Augustus, and soon thereafter became identified with the graphically similar letter L. Likewise, 100 was variously)K?, ?, 1><1?, H, or as any of the symbols for 50 above plus an extra stroke. The form)K? (i.e., a superimposed X and I) came to predominate, was written variously as >I< or ?IC, was then shortened to ? or C, with C finally winning out because, as a letter, it stood for centum (Latin for "hundred").

The hundredth V or X was marked with a box or circle. Thus, 500 was like a ? superposed on a? or f--7 (that is, like a I> with a cross bar), becoming a struck-through D or a E> by the time of Augustus, under the graphic influence of the letter D. It was later identified as the letter D. Meanwhile, 1000 was a circled X: ?, @?,®?, and by Augustinian times was partially identified with the Greek letter <I>F. It then evolved along several independent routes. Some variants such as 1:P? and CD (more accurately a reversed D adjacent to a regular D) were historical dead ends (although folk etymology later identified D for 500 as half of <I>F for 1000 because of the CD variant), while two variants of ? survive to this day. One, CI?, led to the convention of using parentheses to indicate multiplication by 1000 (later extended to double parentheses as in ?, ?, etc.); in the other, ? became 00 8 and 1><1?, eventually changing to M under the influence of the word "mille" (thousand).

The number 1732 would be denoted MDCCXXXII in Roman numerals. However, Roman numerals are not a purely additive number system. In particular, instead of using four symbols to represent a 4, 40, 9, 90, etc. (i.e., IIII, XXX:X, VIIII, LXXX:X, etc.), such numbers are instead denoted by preceding the symbol for 5, 50, 10, 100, etc., with a symbol indicating subtraction. For example, 4 is denoted IV, 9 as IX, 40 as XL, etc. However, this rule is generally not followed on the faces of clocks, where IIII is usually encountered instead of IV. Furthermore, the practice of placing smaller digits before large ones to indicate subtraction of value was hardly ever used by Romans and came into popularity in Europe after the invention of the printing press (Wells 1986, p. 60; Cajori 1993, p. 31).

The history of Roman numerals is not well documented, and written accounts are contradictory. It is likely that counting began on the fingers, and that is why we count in tens. A single stroke I represents one finger, five or a handful could possibly be represented by V, and the X may have been used because if you stretch out two handfuls of fingers and place them close, the two little fingers cross in an X. Alternatively, an X is like two Vs, one upside down. Although the Latin for 100 is centum and for 1000 is mille, scholars generally do not think that C is 100 and M is 1000 because they are the initial letters of centum and mille. The use of D could be a representation of a C with a vertical line through it representing half. My own view is that M arose out of the use of () symbols to multiply by 1000. This theory is supported by the use of (I) for 1000 and I) for 500. These could easily become corrupted or abbreviated into M or D, which they resemble.

4. The Romans were active in trade and commerce, and from the time of learning to write, they needed a way to indicate numbers. The system they developed lasted many centuries and still sees some specialized use today. Roman numerals traditionally indicate the order of rulers or ships who share the same name (i.e., Queen Elizabeth II). They are also sometimes still used in the publishing industry for copyright dates and on cornerstones and gravestones when the owner of a building or the family of the deceased wishes to create an impression of classical dignity. The Roman numbering system also lives on in our languages, which still use Latin word roots to express numerical ideas. A few examples: unilateral, duo, quadriceps, septuagenarian, decade, milliliter.

The big differences between Roman and Arabic numerals (the ones we use today) are that Romans didn't have a symbol for zero and that numeral placement within a number can sometimes indicate subtraction rather than addition.

Roman numerals were first discovered in 500 BC from the Greek nation and their alphabet symbols, which are not translated into Latin. There was one example that the Greek used when dealing with numbers. The number five indicates that there would be a shorter way for the Romans to use it. Their way of thinking was when they look at the fingers when putting the number five, you can see in between the fingers there would be the letters Vs. So what they did was made the V a symbol for the number five, and that's just an example.

Another example is the theory of how originally the X used to be 50 but was switched to L and X became 10. Another theory is that ten 1s were written in a row and then crossed with an X to simplify counting. Then X alone became the shorthand for 10. Others believe the V looks like the top half of the X, as to say 5 is half of 10, or put one V doubled with an upside down V, as to say 5 x 2.

Generally, according to the Roman numerals, they are pretty much straightforward. The largest numerals always start from the left to the right but only when adding; whereas in subtraction, the smaller numeral would be in front. The Roman numeral system did not include zero, and Romans had no concept of it in their arithmetic. This is one reason why Roman numerals are so clumsy for calculation, though it is possible. They tended to use an abacus for arithmetic, and that device does have the concept of zero built in—it is represented by an empty row. But it was the Indian and Arab mathematicians after the end of the Roman empire who invented our present system where we have the concept of "place" and have a distinct symbol to represent zero or an empty column. So when we write 10, for example, the zero tells us that the 1 is worth ten times as much as it would be if the number was just 1. The value of this system for arithmetic and calculation and for depicting numbers of any size is so great that the IndoArabic way of writing numbers is now almost universal, and Roman numerals are confined exclusively as "counting" numbers rather than as calculating numbers.

History

The number zero did not have a symbol in the Roman numbering system, but the concept was brought about in the medieval times. According to Wikipedia, the first known person to use zero to compute was Dionysius Exiguus in 525, but I believe that the concept of zero was used earlier. This lack of zero prevented Roman numerals from becoming a positional notation and led to the replacement by Arabic numerals in the beginning of the second millennium.

Throughout history, there has been alternate forms of Roman numerals used in different parts of the world. In the Middle Ages, for example, Latins placed a horizontal line above selected numerals; this means that the numeral was one thousand times larger than the symbol. They also placed vertical lines on both sides of the numerals to signify that the numeral was one hundred times larger than the symbol. Off of the basis, John Wallace is sometimes credited for introducing the symbol for infinity, because one thousand times a number was very large in this period.

The notation of Roman numerals has varied throughout history. In the beginning of the system, it was more common to use IHI than IV, but this changed throughout the history of the system. It seems that at the beginning of the numeral system, they were reluctant to subtract numerals. Then as they got more comfortable with subtraction, they began to simply by using subtraction to keep the numerals smaller in number of characters. The Roman numeral system was kept in everyday use until the fourteenth century, when it was replaced by Arabic numerals.

I will now discuss the modern usage by English-speaking people. According to my source, "The use of Roman numerals today is mostly restricted to ordinal numbers, such as volumes or chapters in a book or the numbers identifying monarchs or popes"(4). Today, we sometimes use lowercase Roman numbers to distinguish certain sections or books or other things. Today, Roman numerals are often used in books and movies that have more than one sequel. This is done to show that the movies are happening in a timeline that is in a certain sequence. Wikipedia also writes that "in music theory a scale degrees or diatonic functions are often identified by Roman numerals (as in chord symbols)" (4).

In conclusion, Roman numerals were a very early and primitive numbering system that was used for a long while. Throughout the history of the system, there were many changes and variations to make the system a little more practical. It is still used today, but it is reserved mainly for books, movies, music, and sporting events.

Applications

Roman numerals were invented over two thousand years ago by the Romans, and they are still used today. Up until the eighteenth century, Roman numerals were used in Europe for bookkeeping even though the Indo-Arabic numerals we use today were known in Europe and widely used in Europe from around AD 1000. Today, they have many uses.

You find them used to show the year when a building was built, a film was made, or a book was published. They are also used on the pages of books before the main part starts or to show subheadings, where they normally use lowercase letters: i, ii, iii, iv, v, xi, . . . etc. Sports events are often designated by a Roman numeral. The Athens Olympics in 2004, the twenty-eighth games in modern times, was called the Games of the XXVII Olympiad. In the USA, the American Football Championship is called Super Bowl. In 2005, the thirty-ninth championship was Super Bowl XXXIX, and the 2006 event will be Super Bowl XL. They are still used to show the hours on some analogue clocks and watches, including Big Ben, one of the most famous clocks in the world, located in the Clock Tower of the Palace of Westminster, England, where the UK Houses of Parliament are located. Intel, the computer chip maker, called the new version of its Pentium processor the Pentium II in 1997, the next version was Pentium III, but in 2000, Intel unveiled its latest chip as the Pentium 4. Perhaps they thought that Pentium IV was too difficult for people to cope with. They are also used to describe the First and Second World Wars, gravestones, and to name monarchs such as King Edward VII of England.

The history of Roman numerals is not well documented, and written accounts are contradictory. Although they are not used very often for arithmetic, they are great when the numbers are simply used for counting. It is likely that counting began on the fingers, and that is why we count in tens. A single stroke I represents one finger, five or a handful could possibly be represented by a V, and the X may have been used because if you stretch out two handfuls of fingers and place them close together, the two little fingers cross in an X. C means 100, and M means 1000. Roman numerals were very popular, and still are today because adding and subtracting are easy with them. Multiplying, dividing, and fractions, however, are not done with Roman numerals because they are more advanced mathematics, and since Roman numerals lack a zero, it is a particular disadvantage.

Even though Roman numerals were invented two thousand years ago, they still have a great impact on our everyday mathematical lives. They are used for hundreds of different reasons and can been seen all over the world, whether on a clock or in a book. They will surely continue to be used and implemented for quite some time.

What Else?

1. Throughout the course of history, mathematics developed out of practical needs in areas of life such as business, industry, agriculture, and technological advances. The early Egyptians and Babylonians of the third millennia BC developed numbering systems for all whole positive numbers; this is the first history of mathematical development and application. The Babylonians developed a method of multiplication using a chart of squares and division using a table of reciprocals. A Babylonian by the name of Ptolemy published a book about the chords of a circle, which proved to be the beginning stages in the development of trigonometry.

The Greeks began to make developments to the world of mathematics in the sixth century BC. Intellectuals such as Pythagoras, Plato, and Aristotle were the early leaders of Greek contribution. The Pythagorean theorem was named and developed during this period; development of this method had begun in early Mesopotamia with the Babylonians, but the Greeks were responsible for rediscovering, refining, and recording it. The Greeks are also credited with realizing that real numbers cannot accurately describe all values, such as the chord of a square, whose sides are of length equal to one. They called these numbers "irrational" but did not develop a way to express them. Much work was also done to express volumes, areas, and relationships between points, lines, and numbers. The most significant contribution from the Greek scholars was that of Euclid in approximately 300 BC. Euclid's book *Elements* gave the fundamental beginning of geometry. Geometry as we know it today is still largely based on his work. Euclid defined many terms such as a point and line, and gave many postulates, without proof, that relate these ideas to other applicable areas. Menelaus was another notable Greek mathematician; he did much work in the development of spherical triangles (early trigonometry) and published six lost books on chords around the time of AD 100.

After AD 300, much development in mathematics came from the East. The areas of China and India were responsible for a new numeral system and advancements in geometry. The areas of algebra and trigonometry as well as other numeral systems were further developed in the Middle East in the fifth and sixth centuries. Numbering systems took another leap forward in the Middle Ages with the development of ways to notate the number zero as well as negative numbers. The properties of the "real number system," including all rational and irrational numbers, were being explored and defined between the 1600s and 1800s in Europe. It was during this time that calculus began to be understood. The 1800s brought research into the areas of polynomials and the process of finding their roots. Mathematicians Gauss and Lagrange did much work in area of solving polynomials; they developed a method of substitution for the variable in a quadratic equation that made solving the equation a much easier task. Many notable mathematicians did extensive work during the seventeenth and nineteenth centuries throughout Europe; a few of these mathematicians and their areas of study and development are as follows: Leonardo da Pisa (arithmetic, algebra, and geometry), Simon Steven (invention of decimal fractions), John

Napier and Henry Briggs (logarithms), Pierre de Fermat (number theory), Pascal (probability), Rene Descartes (analytic geometry), and Isaac Newton (calculus). Calculus has been able to provide a very useful tool for obtaining solutions to physical problems pertaining to various areas of life.

As stated before, mathematics have developed over time as people continued to seek solutions to everyday problems and situations as well as theoretical applications. As technology continues to thrive and progress, so will the use of mathematics in problem solving, designing, and modeling of physical situations. The applications of mathematics will never become useless.

2. The foundation of mathematical and numerical systems traces back to over 2000 BC. The first civilized mathematical system was perfected by the Babylonians, who were believed to have inherited ideas from both the Sumerian and Akkadian cultures. The system they perfected called the sexagesimal system consisted of sixty base symbols. Although there is no solid explanation as to why the number system consisted of sixty symbols, Theon of Alexandria attempted to answer this question in fourth century AD with the following: Sixty, being the only number divisible by 1, 2, 3, 4, and 5, maximized the number of divisors. By developing a base 60 system, we also allow ourselves to measure into thirds.

There are also theories that the base 60 system was founded because of astronomical events, such as sixty being the number of moons per year with the number of planets. That the year was thought to have 360 days was suggested as a reason for the number base of 60 by Moritz Cantor. Another hypothesis is the fact that the sun moves through its diameter 720 times during a day, and with twelve Sumerian hours in a day, it equals to sixty.

Other theories base the foundation of a base 60 number system upon geometry. Since the fundamental geometrical building block for the Sumerians was considered to be an equilateral triangle, and an equilateral triangle is sixty degrees, by dividing sixty into ten, an angle of six degrees would become the basic angular unit. Now, there are sixty of these basic units in a circle, again providing us with a solid possible reason for the base 60 numbering system.

Although all of these reasons may or may not be true, it is hard to consider any of them as a solid reason as to why a base 60 numbering system was chosen. When you think of different civilizations, it is easy to see that each civilization's numbering system was developed almost hand in hand with the way the culture learned to count. For example, civilizations that learned to count on their ten fingers developed into a base 10 number system, or civilizations that learned how to count on their hands and toes developed into a base 20 numbering system. In the end, it is impossible to say for sure just why a base 60 numbering system was developed.

Another numbering system with very early roots was the binary system. It is rumored that the binary system was created upon studying an ancient Chinese system known only to have come from a fish man named Fuxi. This system was called the I Ching.

The I Ching was a coin system where coins are used to randomly generate hexagrams, a group of six lines where each line represents either "yin" or "yang." This would cause each line to represent a six-digit number in the binary system. What is fascinating about this system is that with only a few coins, a very large number of equations and possibilities are created.

As to whether Fuxi was actually a fish man, there are some who say that Fuxi was actually in relation to the first emperor of china, and others who trace Fuxi to be a group of mathematical advisors in ancient northwestern china.

The binary system comes into play later when a man by the name of Leibniz proposed that in order for a numbering system to work, it need only to have two numbers. Leibniz theorized that with a base system of two numbers, binary arithmetic would handle any problem that decimal arithmetic could not.

These are just a few of the many hundred early numbering systems developed across the world. Many of these early systems in turn helped to evolve into new and more refined numbering systems. I have chosen to list these, though, to show the basic principles from where our culture has come. Without the most basic of systems, our most advanced mathematics would never have come to be.

3. Numbers are used every day, everywhere. Numbers are used for security, for problems to solutions, for fun, and more. Students all over the nation study math and learn different theorems for reasons of further study. Some people hold professions that use math, such as math professors, doctors, people in business, and more. There are many times in life you may think numbers are not important or you do not use them. However, if you open your eyes, numbers are all over.

The ancient Egyptians where the first ones to use math. The ancient Egyptians used special symbols known as pictographs to write down numbers over three thousand years ago. The Egyptians had bases of a 10 system hieroglyphs for numerals. They used separate symbols for one unit, one ten, one hundred, and one thousand, one ten thousand, one hundred thousand, and one million. From the Egyptians using this, the use of numbers was able to expand and grow as a process. Eventually, the Romans picked up on this as well.

Years passed and the Greeks and Romans started to develop math. Greeks had their own currency, weights and measures, etc. These in turn led to small differences in the number system between different states, since a major function of a number system in ancient times was to handle business transactions. The ancient Greeks had different systems for cardinal numbers and ordinal numbers.

Later, the Romans developed a system of numerals that used letters from their alphabet rather than special symbols. Today, we use numbers based on the Hindu-Arabic system. We can write down any number using combinations of up to ten different symbols. As mentioned before, the Romans developed the Roman numerals, which is still used highly today. Authors use the numerals for chapters in a book or the table of contents, and we as students use it for outlines.

Even before all these communities started developing a use for numbers, most historians say that finger counting was one of the first times people had a use for numbers and started seeing that the use for counting is used everywhere. The development of ordinal numbers from cardinal numbers (one, two, three, etc.) may have happened as a result of finger counting. Fingers, and sometimes toes, can be used to count on. If this is always done in the same order, and then the same finger will always be used to represent, for example, seven. This provides a link between the seven items being counted and the seventh finger in the sequence.

Numbers basically continued to develop with the development of society. As society grew and times changed, there was more of a need for intelligence. The economy got better and more money was coming in, and because of the use of numbers, math skills developed and we were able to use more. Basically, a good example is a social security number. Everyone has one, and those nine numbers make someone's identity. People use their social security numbers daily—some examples can be at their work or for email; even at school as students we use them on tests so that professors can identify students easier.

4. The ancient Egyptians used special symbols known as pictographs to write down numbers over three thousand years ago. The Romans developed a system of numerals that used letters from their alphabet rather than special symbols. Today, we use numbers based on the Hindu-Arabic system. We can write down any number using combinations of up to ten different symbols. The ancient Egyptians developed number systems to keep accounts of what was bought and sold.

The use of a number system can be dated back to 30,000 BC by an animal bone found with notches carved into it, signifying some type of tally system. Tallying is one of the oldest forms of numbers used to keep track of a flock of sheep, for example. The development of ordinal numbers from cardinal numbers may have happened as a result of finger counting. Fingers, and sometimes toes, can be used to count on.

Pictures were also used to represent numbers. For example, a picture of five men would mean a man five times. Strokes were also used for counting, and five could be represented by hand and ten was represented by two hands. The Egyptians used the demotic system as their everyday form of number use. The numbers went from one to one million and were written from right to left using pictures and symbols.

The Babylonians used clay to form their numbers, which evolved into the writing system known as cuneiform. They used base 10 and base 6 in their counting. They also developed a way to represent zero. They left a large space between numbers to represent zero. The Chinese used monogram form, which consists of various sticks and strokes to represent numbers. The Mayans and Aztecs use base 20 to in their number system, and their symbol for zero resembled a shell.

Some civilizations used the alphabet in representing numbers. The Hebrews used the twenty-two letters of their alphabet to represent numbers up to four hundred. The later Greek number system used their alphabet similar to the way the Hebrews used theirs.

Roman numerals are one of the most substantial systems used. It only requires the memorization of a few symbols and is still used today. This system used subtraction of numbers. For example, four is N (five minus one) instead of !III. Pi is known as one of the oldest mathematical quantities in numbers. The origins of pi are so ancient that they are untraceable.

Arabic numbers comes from Indian numerals and are commonly used today. Early forms of the numerals were developed in India between the second century BC and the sixth century AD. In the eleventh century, a way of calculating these numbers was developed called algorism, as well as the abacus. Arabic numbers soon were used instead of Roman numerals, which were inadequate for writing complex numbers used in astronomy and other sciences.

5. While everyone uses numbers on a daily basis, not many people take the time to understand the origin of numbers. The earliest evidence of counting dates back to 30,00 BC because of a series of notches carved in a wolf bone, which may have represented a tally.

Many different cultures have different origins of their own number system. Ancient Egyptians used pictographs to write numbers over three thousand years ago. Egyptians have three different systems of writing numbers: the hieroglyphic, the hieratic, and the demotic writings. The Romans developed a system of numerals based on their alphabet. While the origin of the Roman number system is still questioned, one theory is that it could be based on the number five. The Sumerians are believed to have invented writing in the fourth to second millennia BC using symbols in clay to represent letters and numbers. Their number system only had two numbers, one and two, as well as a place-value system. In the Mayan number system, the numbers one, two, three, and four were shown by dots. And five, ten, and fifteen were lines or bars. It is possible that the moon was used to represent twenty. The Mayans used a symbol similar to a shell to represent zero. The Greeks had two number systems, one of which was used from the fifth to the first centuries BC. Their number systems were called Herodian numbers or Attic numbers. These systems used letters of the Greek alphabet for numbers. Indian Arabic numbers are difficult to trace because only a few of their early examples still remain. Their number system began between the second century BC and the sixth century AD. They discovered that a place-value system with a specific symbol for zero would make allow them to make calculations by simply writing down numbers, as opposed to using abacus.

Pi, which is 3.14, may be one of the oldest mathematical quantities known. The origin of pi is not traceable because of its ancient history, but it is thought to have been first known in Egypt because of a reference in the Egyptian Rhind Papyrus to pi dated approximate 1650 BC.

6. What is a number? A number is any figure or group of figures identifying somebody or something. In the ancient years, about three thousand years ago, Egyptians used special symbols called pictographs to keep accounts of what was bought and sold and were able to write numbers. Eventually, the Romans developed a different system of numerals that used letters from the alphabet. The Hindu-Arabic system is what we use today, using ten different symbols: 0,1,2,3,4,5,6,7,8,9.

The earliest method of keeping a record of quantities has been called tallying. This only compares two objects and does not give any idea of the actual number of items. Another idea was using different number words depending on the context. There are cultures that have number systems consisting of only one, two, and many, such as the Australian aboriginal tribes.

Roman numerals are still used today such as for numbering appendices in books, denoting dates in credits of films and television, or on watches or clocks. The widespread influence of the Roman Empire, the force of tradition, and e, because the system had many advantages over other European systems, was the purpose behind the long use of the Roman numerals. One of the advantages that the Roman numerals had was the fact that they were easier to memorize, only having to remember a few symbols and their worth. There is a theory behind the Roman number system that states that the system may have been based on the number 5, which equals V in the Roman numeral system and represents a hand held flat with the fingers together. The symbol X, for 10, could be a double of 5, and three of the other symbols could have come from the Greek letters such as C, for 100, from theta, M, for 1000, from phi, and L, or 59, from chi. Theta and phi were thought to have changed gradually under the influence of the initials of the number words centum, for 100, and mille, for 1000, to C and M.

Roman numerals use a subtractive principle such that when four is written as IV, it is five minus one, rather than IIII; and 900 is written as CM, rather than DCCCC. The only symbols used today in the Roman numeral system are I, II, III, IV, V, VI, VII, VIII, IX, X, L, C, D, and M (1, 2, 3, 4, 5, 6, 7, 8, 9, 10, 50, 100, 500, and 2000).

Without the use of numbers, the world we live in would not be as simple as everyone makes it out to be. We are a fast-paced society and rely on computers, cell phones, money, calendars, social security numbers, and so on; the list could go on forever with items that we use in daily life that involve the use of numbers. It is interesting that the Roman numeral system is still around and has not died out completely, although it is similar to the ten-symbol system that we use now. As we continue to grow a society, so will the use of numbers, and as devastating as it is, many will remain oblivious to the fact that numbers are all around us.

7. Every day, we go about our lives without ever really taking the time to appreciate the mathematics that has been the foundation of our easy lives. We use transportation, technology, and other fundamental things needed for the comfortable society we live in today. Where did it all come from, and why should we know this?

Thousands of years ago, people began to use numbers for elementary purposes. Using "tally stick," people would use numbers as a representation for something. In Africa, bones were found that had notches in them. The thing that was remarkable about the notches is that it was a representation of days until the moon cycle, similar to the modern-day calendar. Another big thing they found about the tally sticks is that it represented the hunting schedule and how long they have been in that location. The tally stick used base numbers and was imperative for the survival of both hunting and gatherers.

Empires such as the Romans were very big in commerce and trade, and with such a growing dynasty, they needed numbers. I am sure everyone has used Roman numerals before; that's just how important numbers are and how powerful they can be. What is great about numbers is that using them enables other things to come about, similar to the Romans using numbers for expressing numerically in Latin. They also used numbers for keeping track of ships and other items important to them.

In relation to the contribution of Roman numerals, Italian Leonardo Fibonacci realized how strenuous and time consuming they were to use and began to make his own numbering system. Daring to go against the "prestigious" Roman numerals, Leonardo wrote a book called *Liber Abaci* in AD 1202. Although looked down on compared the Roman numerals, people began to realize just how imperative Leonardo's number system was in the fifteenth century because it was much less time consuming and problematic.

Another big culture that had an amazing impact in mathematics and the numbering system are the Babylonians. They took the numbers and began to show forms of what we know today as geometry and algebra. In the 1600s, Isaac Newton took the knowledge passed down from the Babylonians and began to work on numbers with all different possible variable ranges. In doing so, his work brought misunderstanding to original problems and numbers. The Babylonians sometimes are looked down upon for the mere fact that they never incorporated a zero into their numbering system; however, the system was used for over a thousand years and never created a problem for them.

However, Georg Cantor was a big help with the numerical system as he cleared up some of the confusion people had with Newton. Although his work cleared up some confusion, many people found his work "mathematical curiosities" instead of practical mathematics because of other findings such as the chaos theory and fractal geometry.

If you think about it, numbers has been around before writing has. In the Fertile Crescent of southwest Asia, where farming may have developed first, many farmers had a surplus of crop. To store these bins of crop, they created records of what bin and what type of crop they had. Then they would make clay tokens or coins, similar to how people make pottery today, and they would keep a record of what they had.

I think that numbering is a sophisticated a way of classifying certain things. Take a child, for instance; he can look at the fingers on his hand, but he doesn't know what they mean. The baby knows that he has several fingers but must learn to count to know how to classify the information he got from his observation.

After researching the beginning of mathematics and numbers, I really have gained a new appreciation for it. It appears that many cultures developed their own form of numbering, and they all can be useful. I guess I never really took the time to respect where it all came from; it's ignorant to think that it's just always been here. I mean, numbers are all over the place; I can't imagine my life without them. I praise the person who started numbering.

Egyptian Mathematics

The way that the Egyptians made a calendar to mark where they were in the year was by the astronomy. The main reason they needed to know this was they had to know when the Nile River was going to flood. This also helped the farmers decide when they had to harvest their crops and when to plant them. The beginning of the year was made when Sirius, the brightest star in the sky (O'Connor), rose. This occurred in July but shortly after the Nile would flood, so this was marked as the beginning of the year in history. This calendar was 365 days, so it ran in stride with the civil calendar, which came later to help taxes, army support, etc. Eventually, the calendars got divided into twelve-month periods, which are like the calendars in America today.

The first number system used by the Egyptians was a system called hieroglyphs. This system used pictures to different numbers. They used a system of base 10. Since they did not have paper to write on at first, there are still stone carvings that represent the kind of system they used (O'Connor). Since it was hard to show fractions of a number, it was written underneath the number hieroglyph; in the form of l/n. n was the integer that was going to be used (O'Connor). The way this was written was with the highest number first followed in descending order.

The Egyptians knew they needed to find a quicker/easier way to write things, so they developed the dried papyrus (paper) and a pen, which was the tip of a reed (O'Connor). With the development of the paper came along a new type of numbering for them to use. This type was hieratic numbers, which was written in a more compact way but still used symbols. This was also a lot more symbols, because all numbers had separate symbols. Even with this system, arithmetic operations were not very easily performed. Historians also believed that they could not think of abstract numbers, so if they were using the number seven, they actually thought of seven different objects to be used for the problem (O'Connor).

There are many things that are being examined about the

Egyptians. One thing is how remarkably close the secant of one of the faces on the Great Pyramid (51° 50' 35"), which is 1.61806, compared to the golden ratio used, 1.618034 (O'Connor). It is not believed that Egyptians knew about this rule, but the closeness of these numbers is outstanding. Another close aspect with the Great Pyramid is how the slope of 51° 50' 35" is close to n/4. This is a cotangent, which is also believed not to be known to the Egyptians (O'Connor). From all of what is known about Egyptians, it is believed that they are possibly one of the first places to used scientific art. This is a good theory, since the word "chemistry" is derived from the word *alchemy*, which means "Egypt" (O'Connor).

Especially now more than ever, numbers have such an influence. Numbers are so important for the world to run daily. You cannot think of an aspect of life today without looking deeper and seeing the use of numbers. Math only progresses with new theorems and something new to learn.

A Comparison

In our society today, there are many uses for numbers. Throughout the millennia, we've discovered different ways to apply numbers to everyday life. One way we use numbers is through Roman numeral system. This system was invented by the Greeks.

At first, the Greeks used notation mirroring Roman numerals. They used a vertical line to represent the number one. For the number five, the Romans used the letter A, C for ten, C for one hundred, and x for one thousand. After a while, the Romans decided to give up on this method and changed the idea around. They decided to use the first nine letters of the alphabet so they could stand for the numbers one through nine. They used the next nine numbers for ten through ninety, and so on. After nine hundred, the Romans used suffixes to symbolize the numbers up to 100,000,000. The Romans found out that they needed twenty-seven letters to make this system work. Their alphabet only contains twenty-four letters, so they had to provide the extra letters. There were numbers from the original Greek alphabet that were taken out of the one they decided to use. The letters that were discarded were used to replace the missing letters they needed to make their system work. These two letters were digamma and koppa. For the last one, the Romans borrowed a letter off of the Phoenician alphabet.

The biggest difference between the Arabic system and the Roman system is the use of the zero. The Romans did not use the number zero in system. They did not believe it existed. The Romans also participated in trade and commerce. They used the system for many centuries. It helped them out a lot with all kinds of problems. It helped their trading businesses. Most of the numerals are used today. For example; the Roman numerals help us indicate the order of which kings and queens ruled (King George I, King Arthur II, and King Phillip III, etc.). Other examples for using the Roman system would be for books. The numerals are used for the copyrights and dates. We also use them for writing outlines for essays and carving them into gravestones.

In conclusion, the Arabic system was created to improve our society. The Roman system was good, but the Arabic system is a huge improvement. The Romans started a great thing by inventing the numerical numbers. It helped not only their society but ours as well.

Mathematics Learning Disorder

As pointed out earlier, some students having issues with numbers showed concern about having what is called mathematics disorder.

Mathematics disorder is a condition in which a student's mathematics ability is considered far below normal for their age, intelligence, and education. The disorder often interferes with academic achievement of the student and sometimes with activities of daily living.

One author of this article has taught mathematics for more than forty years and in several different countries. During this period, he has dealt with many students who, despite their efforts, were disappointed with their performance and test scores. To learn more about such students and their problems, sometimes he presented the same problem in a context more familiar or more interesting to the student. He found that in many instances, and to his surprise, a high percentage of students were able to deal with the questions and manage to find answers. Contexts often used were sports, games, and daily life situations.

The other author worked as a psychological counselor for years. During this period, she dealt with many situations involving students of this type of problems seeking help and advice.

Based on our experience, students who have mathematics disorder often have trouble with writing, understanding mathematical symbols, and word problems. Some were unable to understand graphs and what they represent. Few even had problems in class, including behavior problems and loss of self-esteem. Some seemed anxious or afraid when given mathematics problems, making the problem even worse.

We sometimes wondered if there are genetic factors associated with a mathematics disorder. According to the some research, it is possible that some people have problems with mathematics because of their genetic makeup. In contrast to some families whose members have great difficulty solving mathematics problems, some mention that there are some in their families who tend to have members that consistently have a very high level of mathematics functioning.

Diagnosis

Mathematics disorder may be discovered when a student shows poor functioning in several mathematics skills—for instance, if a student has difficulty understanding and working with various mathematical terms and concepts or identifying mathematical symbols and signs.

Also, mathematics disorder may be indicated when the student is unable to attend to details such as carrying numbers or counting and memorizing the basic things such as multiplication tables.

Although standardized, group testing is important; it alone should not be relied on in making this diagnosis. It is very important that special psychoeducational tests be individually administered to the student to determine if this learning disorder is present. In administering the test, the examiner should give special attention to the student's ethnic and cultural background. Additionally, as we discovered, some students have difficulty with specific subjects or concepts such as fractions or irrational numbers.

In our view, treatment for mathematics disorder should include individual tutoring, placement in special mathematics classrooms with expert mathematics teachers, and other educational aids that focus on mathematics skills.

It is interesting to note that the course of mathematics disorder is varied. Some students go on to do well in other mathematics classes. Others, even with remedial attention, continue to struggle with mathematics during the years in college.

Dyscalculia

A. Difficulties in production or comprehension of quantities, numerical symbols, or basic arithmetic operations that are not consistent with the person's chronological age, educational opportunities, or intellectual abilities.

Multiple sources of information are to be used to assess numerical, arithmetic, and arithmetic-related abilities, one of which must be an individually administered, culturally appropriate, and psychometrically sound standardized measure of these skills.

B. The disturbance in criterion A, without accommodations, significantly interferes with academic achievement or activities of daily living that require these numerical skills.

CHAPTER 4

Mathematics and Sports

S PORTS PROVIDE AN inexhaustible source of fascinating and challenging problems. According to authoritative opinion, modern sports have been getting more and more intellectualized, and as such, their analysis requires more advanced scientific methods. Also being profitable, sports has long been an object of attention for media and business owners.

Most sports can be studied from a mathematical perspective to yield a more valid quantitative results. For example, mathematical methods are applied to estimate an athlete's chances of success, to identify the best training conditions for him/her, and to measure their effectiveness. Information theory makes it possible to estimate the amount of eyestrain in mountain skiing, table tennis, etc. Mathematical physics is used to identify the best shape of rowboats and oars. Applied probability and statistics has been instrumental in analysis of vast amount of sport data available. Probabilistic Monte Carlo method is used for simulation model.

In sum, athletic competitions provide the researcher with a wealth of material that is registered by coaches, carefully preserved and continuously piled up. There is plenty of opportunity out there to experiment and test mathematical models and optimal strategies for situations occurring in sports. Only a tiny part—quite possibly not the most intriguing one—of the problems arising in sports has been described in the pages of books and journals. Think how many yet unsolved problems arise in different sports. Because of this, over the past few decades, the distinctive characteristics of traditional and high-level sports competitions have attracted the interest of the scientific community. Some of these studies have been directed at the modeling and analysis of the characteristics of different sports, and some to analysis of records.

Other than research problems, the universal popularity of sports has inspired a gold mine of interesting examples for the teaching of mathematics. It is generally recognized that the use of sports marks an exciting new direction in teaching and learning mathematics and statistics. With the present state of education, publications that connect mathematics to popular activities like sports are much needed. The importance and practicality of this approach have also been recognized and emphasized by the three major professional organizations: American Mathematical Society, Mathematical Association of America, and American Statistical Association.

Classification of Sports

As can be anticipated, a large number of sports are played around the world. The eighty-three well-known sports may be classified into three groups: combat sports, object sports, and independent sports. A combat sport is one in which each competitor tries to control the opponent by direct confrontation (e.g., boxing, wrestling, and fencing). An object sport is one in which each competitor tries to control an object while the other competitor is in direct confrontation (e.g., soccer, baseball, and chess). An independent sport is one in which one competitor may not interfere with the other competitor (e.g., swimming, shooting, and golf). In an independent sport, the competitors may perform at different times and even in different places, and it might be said that each competitor tries to control himself or herself.

There are three ways in which performance is evaluated in the sports: judged (e.g., all combat sports, diving, and gymnastics), measured objectively (e.g., weightlifting and swimming), and scored objectively (e.g., baseball, archery, and golf).

Sport in Schools

It has been argued that schools in the United States spend too much time and money on sports, much more than academics. Some argue that if what schools do here was good for students, other countries specially European countries would have adopted it long ago. Although these arguments have some relevance, there is also a positive side to sports in school. For example, my son played soccer in school and, while in school, took courses in college and earned fifty-six credits before graduation. He also finished high school as valedictorian. He just went to graduate school and received his PhD from Harvard. When I asked him about high school and its best part, he said without any doubt playing soccer. He then talked about the feeling of being a part of a team, winning, losing, sharing the happiness and sadness, feeling of representing a school, making friends for life, and many other things. What I like to add to this relates to students who are not accidentally strong. My son always felt good about himself simply because he was the best in the classroom. Participating in sport provides opportunity for students to experience being good or even the best. In addition, it is an avenue for students to release the extra energy and become calm.

Teaching Values of Sports

The widespread interest in sports in our culture provides a great opportunity to catch students' attention in mathematics and statistics classes. Many students, whose eyes would normally glaze over after few minutes of algebra, will happily spend hours analyzing their favorite sport.

For the past fifteen years, I have taught several courses using sports as a theme. I have come to conclusion that initiatives of this type increases students' interest in mathematics and statistics and public's interest in college education. Many students and members of the public think that learning mathematics is one thing and going in for sports is quite another. I have also developed a large number of lesson plans that include some novel, stimulating, and interesting applications of mathematics and statistics in sports. They expose students to the process of mathematical and statistical reasoning that is vital for analyzing data correctly. They also provide a source of material that has an intrinsic interest for most students. For those who plan to become mathematics teachers, both at secondary and in higher education, it provides a source of material that could enrich their teaching of many mathematical topics. Here are more reasons to use sports as a teaching theme:

1. Sports data offer a unique opportunity to teach mathematics and test methodologies offered by statistics. In fact, it is hard to find a discipline other than sports where one could collect reliable data with the highest precision possible. In sports, in addition to the quality measurements, one has access to the names, faces, and life history of the participants and their coaches, trainers, and everybody involved. Almost all other data-producing disciplines are susceptible to "data mining" and error, since, unlike sports, they are not watched by millions of the followers and also media representation. Clearly, a theoretical result can only be tested when data is reliable and satisfies the required conditions. If validity of

data produced or collected by an individual or an organization cannot be confirmed, one may end up disbelieving the methodology used. In fact, historically, this has been the case with statistics. Whenever something goes wrong or results do not make sense, statistics gets the blame, not the researcher or the presenter. This is unlike medical sciences where doctors get the blame, not the medical science.

Consider, for example, a sport such as track and field. The nature and general availability of track-and-field data have resulted in their extensive use by researchers, teachers. and sports enthusiasts. The data are unique in that they (1) possess a meaning that is apparent to most people; (2) are collected under very constant and controlled conditions, and thus are very accurate and reliable; (3) are recorded with great precision (e.g., to the hundredth of a second in races), and thus permit very fine differentiation of change and/or differences; (4) are both longitudinal (ninety years for men's records) and cross-sectional (over different distances and across gender); and (5) are publicly available at no cost. Thus, they provide wonderful data sets to test mathematical and statistical models of change.

2. The difficulties faced by educators teaching mathematics and statistics are recognized by the community of scholars involved. To help this matter, many textbooks try to motivate students by introducing varied applications. This addresses both students' apparent desire to see the relevance of their studies to the outside world and also their skepticism about whether mathematics and statistics have any value. This is, of course, a great idea. However, it is noticed that it works mostly with students who are deeply committed to a particular academic or career field. For typical students, applied examples may fail to motivate because they are not usually of immediate concern to them.

So, is the prospect of motivating students entirely bleak? Despite many voids in student interests and despite many ideas that fail to motivate them, students have some common interests that we can build on as we try to teach them mathematics and statistics. Based on my experience, connecting their studies to a familiar theme with which they have interest or immediate concern almost always works better.

We can utilize a motivational strategy different from those that may be found in popular textbooks. I think motivation of students to study mathematics and statistics may be accomplished by linking their study to subjects such as familiar games and sports. With this focus, we can help students to build their studies on a foundation—an understanding of a sport—already possessed by most of them. This approach is adaptable to all levels at which mathematics and statistics is taught, from junior high to graduate schools.

Other reasons for using sports to teach include the following:

1. Sports have a general appeal, and it is an area to which modern scientific methods are increasingly applicable.
2. Sports have become a part of everyday life, especially for young people.
3. Students usually enjoy sports and show a great deal of interest in mathematics and statistics applied to them.
4. A major part of the calculus and statistics sequences offered at high school and college levels can be taught using a chosen sport.
5. Most students relate to sports and understand the rules and meanings of the different statistics presented to them.

Among the popular sports, tennis lends itself to mathematical analysis better than other popular sports. To see this, let us do such analysis. To see a summary of what is covered here, please read the tennis-related short articles in chapter 1 first.

Do Rules of Tennis Favor the Better Player?

Retired British major Walter Wingfield invented lawn tennis in 1873. He borrowed heavily from court tennis, a French game that had been played by the British aristocracy for centuries, and badminton, a game that had more recently come to Britain from India. Court tennis was played in special indoor courts. Major Wingfield's primary innovation was to move tennis outside, a la badminton. Those who said, "I could have invented this game," criticized him. Still, his "tennis sets" (balls, four rackets, and a net) sold well, and lawn tennis caught on quickly in Britain. The first "world tennis championship" was held at Wimbledon in 1877, and soon the game spread to the United States and throughout the British Empire. Today, tennis is popular worldwide, and Wimbledon is truly a world championship. Perhaps one reason tennis is so popular is players of all ages and abilities can enjoy the game. Even weekend players can experience the satisfaction of serving an ace, "putting away" a volley at the net, or watching a well-placed lob sail over the head of a helpless opponent.

Tennis has also become popular with mathematicians. Calculating the odds of winning a tennis match is a fun exercise in mathematical modeling. The British mathematician Ian Stewart wrote an amusing dialog along these lines called *The Drunken Tennis-Player* for *Pour la Science*, the French version of *Scientific American*. Stewart's article was later reprinted in his book *Game, Set and Math*. Another nice discussion of mathematical modeling of tennis appears in *Mathematics of Sports* by Russian mathematicians L. E. Sadoski and A. L. Sadoski.

Stewart's "drunken tennis-player" makes some calculations on a napkin in the pub after a game and exclaims, "The rules of tennis favor the better player." Is this true? As his opponent responds, "But they should, shouldn't they? I mean, the better player ought to have a better chance of winning?" A better way to phrase the question we will examine is "Do the quirks of the rules of tennis inflate the stronger player's chance of winning?" We will attempt to answer the question by building several mathematical models of a hypothetical tennis match between two players: Reza and Bill. By changing the rules of tennis in some of our models, we can investigate how rule changes would affect the odds of winning. Millionaire tennis player, tennis organizer, and tennis innovator James Van Alen promoted many of the alternative scoring systems we discuss.

In all of our models, we will assume that Reza has a 60 percent probability of winning any given point. We assume that the 60 percent probability remains constant throughout the match. This assumption is clearly an oversimplification, particularly since it is likely that Reza's probability of winning a point is higher when he is serving than when he is receiving. However, as we will discuss later, these models behave in essentially the same way as more complicated models that allow the probability of winning a point to vary from point to point. The advantage of using a simple model is that it makes it easy to calculate probabilities. Most of the calculations needed for the results in this paper can be done on a hand calculator. We used the computer software Mathematica, since it makes it easy to do even the more complicated calculations.

The Quirks of Scoring

Major Wingfield proposed that lawn tennis use badminton scoring: matches of fifteen points with players scoring only when serving. This innovation was largely ignored. During the early years, tennis players used a variety of scoring systems. By the time of the first championship at Wimbledon, the All England Croquet Club had settled on a scoring system based on the traditions of court tennis. This system remained unchanged until the introduction of tiebreakers in 1970.

One quirk of tennis scoring is that strange names are used for points in scoring a game: love, fifteen, thirty, and forty, game. Although no one knows the origin of this odd system, it has been proposed that fifteen, thirty, forty-five, sixty were originally used to represent the four quarters of an hour. Over the years, the score forty-five became abbreviated as forty. (In informal play, fifteen is sometimes abbreviated as five.) It would be simpler to score the game zero, one, two, three, and four. Still, the weird point names give no advantage to either player.

A more important quirk is that two points must win a game. If Reza and Bill have each scored three points, the score is called deuce rather than 40:40. If Reza wins the next point, the score becomes advantage Reza. If Reza wins again, he wins the game; otherwise, the score returns to deuce. This feature of tennis scoring does increase the chance that the stronger player will win, as we shall see.

Game versus No-Ad Game

To see how the "win by two rule" affects the probability of the stronger player winning a game, we will construct two simple mathematical models. First, we will model an alternate version of a tennis game, proposed by Van Alen, called a no-ad game. The winner of a no-ad game is the first player to win four points. It is not necessary to win by two. For comparison, we will also model a standard tennis game.

What is Reza's probability of winning a no-ad game? Figure 1 shows all scores that are possible in a no-ad game.

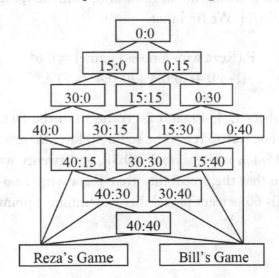

Figure 1

We always list Reza's score first. For example, 15:30 means Reza has scored one point and Bill has scored two points. As each point is played, we descend one level of the chart with a 60 percent probability of moving to the left and a 40 percent probability of moving to the right. After the first point, the probability is 60 percent that the score is 15: love and 40 percent that the score is love: 15. After the second point, there are three possible scores: 30: love, 15:15, and love: 30. The probability that Reza will win both of the first two points is the product of the probabilities of winning each point. We will write P(30: love) for the probability that the score will be 30: love after two points. In this notation, P(30: love) $=0.6 \cdot 0.6=36\%$. Similarly, P(love: 30) $=0.4 \cdot 0.4=16\%$. There are two ways of reaching a score of 15:15—Reza scores first and Bill scores second, or vice versa. So P(15:15) $= 0.6 \cdot 0.4+0.4 \cdot 0.6=48\%$. The following table summarizes our results for the three possible scores after two points.

Score	30:love	15:15	love:30
Probability	36%	48%	16%

By continuing to calculate in this way, we may find the probability of each possible score in a no-ad game and, ultimately, the probability that Reza will win. We find that

$$P \text{ (Reza wins a no-ad game)} = 0.6^4$$
$$(1+4 \cdot 0.4+10 \cdot 0.4^2+20 \cdot 0.4^3)=71\%.$$

The numbers 1, 4, 10, and 20 come from the fact that, while there is only one way to reach the score 40:love, there are four ways to reach 40:15, ten ways to reach 40:30, and twenty ways to reach 40:40. Notice that the probability that Reza wins a no-ad game is *higher* than his 60 percent probability of winning a point.

Deuce Problem

How can we model the "win by two" rule used in the standard game? We start by replacing 40:40 in figure 1 by deuce, as shown in figure 2.

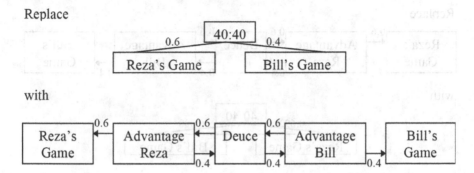

Figure 2

Now, there is no limit to how long the game could go on, alternating between advantages Reza, deuce, and advantage Bill. In principle, the game could go on *forever*, but the probability of that is zero. How can we calculate infinitely many probabilities? Fortunately, we do not have to: algebra comes to the rescue. The infinitely many probabilities we need to calculate form a geometric series: an infinite sum with a constant ratio between the terms. A little algebra allows us to find the sum of our geometric series. Let $x = $ P(Reza eventually wins, starting from deuce). If Reza wins the next two points, he will win the game. If Bill and Reza split the next two points, the score will be deuce again, and Reza's probability of winning the game will be x again. It follows that x satisfies the equation

$$x = \text{P (Reza wins the next two points)} + \text{P (Reza and Bill split the next two points) } x$$

This may seem to be a circular definition of x, but it allows solving for x algebraically. We have

$$x = 0.6 \cdot 0.6 + (0.6 \cdot 0.4 + 0.4 \cdot 0.6)x.$$

Algebraically simplifying the equation, we find that $0.52x = 0.36$. Hence, $x = 32/56 = 69.2\%$.

We can now simplify the picture of deuce as shown in figure 3.

Replace

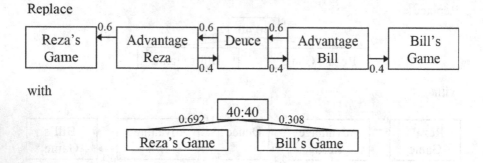

with

Figure 3

Notice that the "win by two rule" favors Reza, since it increases P(Reza wins, starting from 40:40) from 60 percent to 69.2 percent. However, the effect of the "win by two rule" does not appear as large when one looks at the probabilities of winning a game. Reza's probability of winning a standard game turns out to be 73.6 percent, while his probability of winning a no-ad game is 71 percent, a difference of only 2.6 percent. The "win by two" rule *does* help Reza, but only if the score reaches 40:40. Dropping the rule would only change the outcome of 2.6 percent of games between Bill and Reza.

Tradition versus Predictability

The longest match in professional tennis history took place at Wimbledon in 2010. It lasted eleven hours and five minutes (played over the course of three days), easily eclipsing the previous record of six hours and three minutes set during the 2004 French Open. The Wimbledon match was decided after 980 points, and, most amazingly, the final set (which does not end until one of the players obtains a two-game lead) reached a score of 70–68. The improbability of this score motivates our investigation of win-by-two games.

Traditionally, the first player to win six games and lead by two games wins a *set*. After 5:5, the set goes on until one player gets ahead by two, sometimes for a very long time. To avoid such lengthy sets, James Van Alen promoted two variations on traditional scoring.

Of Van Alen's many proposals, the one that found greatest acceptance is the tiebreaker. A tiebreaker is an extended "game" used to decide the set if the score reaches 6:6. Van Alen originally proposed a best-of-nine-point sudden-death tiebreaker. The United States Tennis Association used this version in the early '70s but later switched to what Van Alen disparagingly called a "lingering-death" tiebreaker—the first player to win seven points *and lead by two points* wins. The acceptance of the tiebreaker was the first change in the official rules of tennis scoring in almost a century. Tradition bowed to the need to make the length of televised tennis matches more predictable.

Suppose that Bill and Reza are tied at 6:6. Would Bill have a better chance of winning if they play a tiebreaker or if they follow the traditional rules? In the traditional game, the score of 6:6 is analogous to deuce, as we see in figure 4.

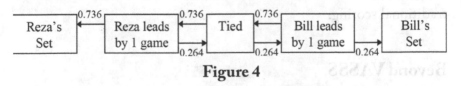

Figure 4

The only difference is that we must replace the probability of winning a point by the probability of winning a game. We can find $x = $ P(Reza wins, starting from 6:6) in the same way we found P(Reza wins starting from deuce). We solve the equation $x = 0.736^2 + 2 \cdot 0.736 \cdot 0.264x$ and find that $x = 88.7\%$.

What happens if Bill and Reza use a tiebreaker? We may calculate P(Reza wins the tiebreaker) in the same way we found P(Reza wins a game). The result is that P(Reza wins the tiebreaker) $= 78.7\%$.

Thus, the tiebreaker gives Bill a somewhat better chance of winning. However, if we now use these results to determine the probability of winning a set, we find that Reza's probability of winning a traditional set is 96.6 percent, while his probability of winning a set with a tiebreaker is 96.3 percent. Using a tiebreaker will change the outcome of only about 0.3 percent of the sets between Reza and Bill.

The Van Alen Streamlined Scoring System

A more radical variation on the rules for scoring sets is the Van Alen Streamlined Scoring System (VASSS), devised by Van Alen in 1958. To simplify the scoring and speed up matches, Van Alen scrapped games entirely. The first player to score thirty-one points wins the set, with a best-of-nine sudden-death tiebreaker at 30:30. The serve alternates every five points as in table tennis. To calculate the probabilities of winning a VASSS set, we use the same kind of calculations we used for the no-ad game. The result is that Reza has a 94.7 percent probability of winning a VASSS set. Although VASSS is not likely to please tennis traditionalists, it does not usually change the outcome of a set. Only about 2 percent of sets between Bill and Reza will have different outcomes under VASSS from under traditional scoring.

Beyond VASSS

Traditionally, there are two kinds of tennis matches: a three-set match and a five-set match. The winner is the first to score two sets in a three-set match or three sets in a five-set match. If scrapping games doesn't make much difference, why not *really* streamline tennis scoring by scrapping sets as well? A match could simply be scored in points. Play would continue until one player reaches a certain number of points, say 150. We will call such a match a 150-match. The graph below produced using Mathematica shows the probability of winning a five-set match (using traditional scoring) as a function of the probability of winning a point.

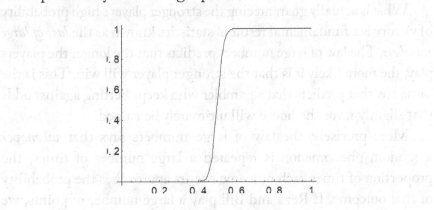

Since Reza has a 60 percent probability of winning a point, his probability of winning the match is 99.96 percent! The steepness of the graph seems to support the thesis that the rules of tennis inflate the stronger player's chance of winning. Is there really something special about the rules of tennis that is working in the stronger player's favor? In the examples we have examined so far, Reza's strong advantage has been robust to a number of variations in the scoring. In the next graph, we plot the probability of winning a 150-match along with the probability of winning a traditional five-set match.

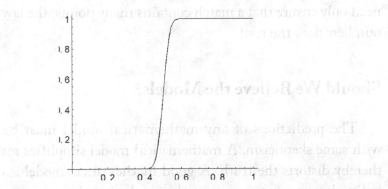

The results are almost identical, indicating that it is not really the rules of tennis that gives the stronger player such a high probability of winning. Reza's probability of winning a 150-match is 99.98%.

The Law of Large Numbers

What is actually guaranteeing the stronger player's high probability of victory is a fundamental result of statistics known as the *law of large numbers*. The law of large numbers predicts that the longer the players play, the more likely it is that the stronger player will win. This is the same law that predicts that a gambler who keeps betting against odds that slightly favor the house will ultimately be ruined.

More precisely, the law of large numbers says that *whenever* a random phenomenon is repeated a large number of times, the proportion of times each outcome occurs approaches the probability of that outcome. If Reza and Bill play a large number of points, we can expect Reza to win close to 60 percent of the points. As the number of point's increases, the probability approaches one that Reza will win nearly 60 percent of the points.

Thus, in any kind of match in which Bill and Reza play a large number of points, Reza will have a high probability of ending with a significantly higher number of points than Bill. Let us define a *reasonable* scoring system to be one that rarely allows a player to win if he is behind in points. Then *any* reasonable scoring system will award Reza the victory almost all the time. Therefore, it is not the quirks of tennis scoring that "favor the stronger player." The rules need only ensure that a match contains many points; the law of large numbers does the rest!

Should We Believe the Models?

The predictions of any mathematical model must be viewed with some skepticism. A mathematical model simplifies reality and thereby distorts the truth. A good mathematical model is one that is simple enough to use to make predictions but close enough to reality that the predictions of the model are useful. How well do our models reflect what actually happens in a tennis match? One glaring omission is that we have failed to take the effect of the serve into account. Although having to serve is sometimes a negative factor for a beginner, experienced players have a higher probability of winning the games they serve. Usually, the player who is serving will win the game. Sets are close because the players alternate serving. The turning point in a set is often a service break, where the receiving player wins a game. Thus, the alternation of the serve would appear to somewhat reverse the advantage of the stronger player and give the weaker player a better chance. As the "drunken-tennis player" says,

> The way the scoring amplifies any advantage means that each player has a chance rather close to one of winning his service game—provided his chance of winning a point is above 1/2. That tends to act the *opposite* way, which evens the game out again!

It is not terribly hard to redo our analysis of a set, taking into account changing probabilities of winning a point as the player's alternate serving. For instance, we can look at a model where Reza has an 80 percent probability of winning a point when he is serving, but only a 40 percent probability when he is receiving. In this model, Bill has a 60 percent probability of winning a point on his serve and a 73.6 percent probability of winning his service games. This model predicts that Reza's probability of winning the match is 96.2 percent, a little less than in the model in which we ignored the serve. However, the difference is small enough that it confirms the conclusions we made using the simpler model. Since the serve alternates, the 80 percent and 40 percent balance out in the end, giving us results consistent with a fixed probability of 60 percent. Below are two graphs plotted on the same axes. For one of the graphs, the horizontal axis represents Reza's fixed probability, p, of scoring a point. The vertical axis represents his probability of winning a set. The second graph shows the result of considering the serve. In this model, Reza's probability of scoring a point is $p + 0.2$ when he is serving and $p - 0.2$ when he is receiving. The fact that the graphs are almost identical confirms that we can safely ignore the serve.

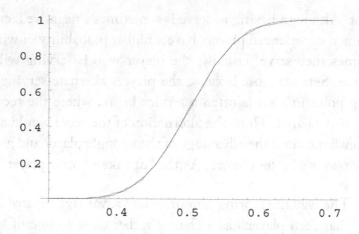

Perhaps a more serious problem with the model is that the results do not seem to be consistent with experience. According to the model, Bill will essentially never be able to beat Reza. Even if Bill could improve his probability of winning a point from 40 percent to 45 percent, the model predicts that he would only win one out of every twenty-five matches. According to the model, the usual situation is for one player to dominate the other. However, it is not uncommon for a player who loses a match to win a rematch. According to our model, this will only happen if the players are extremely well matched. It seems unlikely that the players are always so well matched.

A better explanation of the fact that different players win on different occasions is that the probability of winning a point does not stay fixed each time the players play but fluctuates as the players' level of practice or concentration varies from day to day. For instance, suppose Bill and Reza play every day of the week. Bill's probability of winning a point is 55 percent on Monday but 37.5 percent the other days of the week. Then, on average, Bill will win (55%+6*37.5%)/7 = 40% as before. (For simplicity, we are assuming that each match contains the same number of points.) However, Bill's winning points are now not evenly distributed. He has a high probability of winning the Monday match and losing the others. Thus, he would win about 1/7th of the matches instead of less than 1/1000th of them.

Still, even if the probability of winning a point fluctuates from match to match, it is not unreasonable to suppose that the probability of winning a point is constant throughout a given match. Even when we incorporated the fluctuation of the probabilities from game to game, as the serve alternates, we saw very little change in the probabilities of winning. Therefore, we have a good deal of confidence that our comparisons of different scoring systems are valid.

Conclusions

The answer to the question of whether the "drunken-tennis player" is correct in saying that the "rules of tennis favor the better player" is yes and no. Yes, in that the rules call for the players to play a large number of points in a match. No, in that any reasonable scoring system with a large number of points will produce the same results.

What can we say about changing the rules? Our models predict that the use of a tiebreaker changes the outcome of a set very rarely. The tiebreaker helps keep matches from running too long. It is no wonder that it has been widely adopted. We could go further and replace games with no-ad games or sets with VASSS sets or even matches with 150 matches without altering the outcome of the match in most cases. Doing so would simplify the scoring and probably speed up play, but there is something to be said for tradition. Deuce and tiebreakers do make tennis more exciting to play and to watch, even if they do not often change the outcome of the match. Traditional scores are also easier to remember. We imagine players of a 150-match would often have to stop to ask questions such as "Is it 127:109 or 126:109?"

Finally, some comments about winners and losers. Our models show that it is unlikely for the weaker player to win a tennis match purely by chance. Thus, if Reza wins the match, he is very likely the better player, at least for that day. However, Bill can take consolation in the fact that the loser of a tennis match may be only slightly weaker than his opponent. Even though Reza has beaten Bill consistently in the past, Bill only needs to win one more point in ten to be even with Reza. With a little practice, Bill may be able to win the next match!

Table Tennis

As mentioned, most sports can be related to mathematics in general and several specific subjects in particular. Many can be utilized to enhance teaching of mathematical concepts. In fact, sports and the data related to them offer a unique opportunity to teach mathematics and statistics and to test the methods they offer. This section presents a probabilistic analysis of a table tennis game with a view toward its possible use as an aid for demonstrating steps of mathematical modeling. The materials presented are suitable for further analysis and study. We begin by describing the history and the rules of the game.

History and Rules

Table tennis, also known as ping-pong, has its origins dating back to medieval tennis. Beginning as a mere social diversion, table tennis became popular in England during the latter part of the nineteenth century. It did not take long for the popularity of this game to spread. As early as 1901, table tennis tournaments were being organized, books about table tennis were written, and even local table tennis associations were formed. In 1902, an unofficial "World Championship" was held. Table tennis was widely popular in Central Europe from 1905 to 1910. A slightly altered version was introduced prior to this to Japan, which then spread to China and Korea.

Today, table tennis has become a major worldwide sport, with approximately thirty million competitive players and countless others who enjoy playing the sport at a leisurely level. Presently, the International Table Tennis Foundation represents 140 countries. In 1988, it became an Olympic sport.

Table tennis is an object sport in which each competitor tries to control an object, while the other competitor is in direct confrontation. The object of the game is to hit the ball into the opponent's table. If the ball is not returned after one bounce, a point is won. The only other way to win a point is when the opposing player commits an error by striking the net or by hitting the ball off the table. The table must be 2.74 meters (9 feet) long and 1.525 meters (5 feet) wide. The top of the net must be 15.25 centimeters (6 inches) above the playing surface, while the posts that secure the net are 15.25 centimeters beyond the sideline. The spherical ball must weigh 2.7 grams (0.09524 ounces) with a diameter of 40 millimeters (1.574 inches). The ball may be white or orange, in either case not glossy, and it can be made of celluloid or similar plastic material. While most rackets seem to be the same shape, it can actually be of any size, shape, or weight as long as the blade is flat and rigid. Natural wood must comprise 85 percent of the blade by thickness. Neither can be thicker than 7.5 percent of the total thickness or 0.35 millimeters, whichever one is smaller. A game is won by the player who either reaches eleven points by a margin of two or gains a lead of two points after both players have scored ten points. This scoring scheme was implemented in 2001. A match is simply the best of any odd number of games. A new serving system was also put into place in 2001. Each player serves for two points and then alternates with the other player until the end of the game. If the game reaches ten all, then the number of consecutive serves for each player is reduced to one. Concerning the serve, another new rule was put into effect in 2002. This rule states that the ball must be visible at all times, meaning the server cannot block the receiver's view of the ball in any way (e.g., body, clothing, and table).

Figure 1 displays possible outcomes (states) of a game of table tennis under the new rules. Prior to 2001, each game consisted of twenty-one points instead of eleven, and five serves instead of two.

Figure 1. Table tennis scoring illustration (new rules)

```
                        0-0
                     1-0 0-1
                  2-0 1-1 0-2
               3-0 2-1 1-2 0-3
            4-0 3-1 2-2 1-3 0-4
         5-0 4-1 3-2 2-3 1-4 0-5
      6-0 5-1 4-2 3-3 2-4 1-5 0-6
   7-0 6-1 5-2 4-3 3-4 2-5 1-6 0-7
8-0 7-1 6-2 5-3 4-4 3-5 2-6 1-7 0-8
9-0 8-1 7-2 6-3 5-4 4-5 3-6 2-7 1-8 0-9
10-0 9-1 8-2 7-3 6-4 5-5 4-6 3-7 2-8 1-9 0-10
11-0 10-1 9-2 8-3 7-4 6-5 5-6 4-7 3-8 2-9 1-10 0-11
11-1 10-2 9-3 8-4 7-5 6-6 5-7 4-8 3-9 2-10 1-11
11-2 10-3 9-4 8-5 7-6 6-7 5-8 4-9 3-10 2-11
11-3 10-4 9-5 8-6 7-7 6-8 5-9 4-10 3-11
11-4 10-5 9-6 8-7 7-8 6-9 5-10 4-11
11-5 10-6 9-7 8-8 7-9 6-10 5-11
11-6 10-7 9-8 8-9 7-10 6-11
11-7 10-8 9-9 8-10 7-11
11-8 10-9 9-10 8-11
11-9 10-10   9-11
```

A's game Advantage A Tie Advantage B B's game

Probability of Winning a Game

Table tennis can be related to many mathematical topics. We start exploring these relationships by first focusing on some probability questions.

Suppose that a game is played by two players, A and B. Let the probability of A winning the point be denoted by x, and the probability of B winning the point by y, where $x + y = 1$ and both are independent of the present score. In practice, x and y may be estimated from players' past games against each other.

Before proceeding further, we need to comment on the question of possible advantage associated with serving and its effect on probability of winning or losing a game. The assumption of fixed probability of winning a point is a standard one and is usually made for simplicity. However, it is not a serious one, and its effect on probability of winning or losing a game is not as pronounced as it is for tennis, since the server has to hit her own side of the table first. In fact, as is pointed out in an article titled "How Long Is an 11 Point Game?", djmarcusetasc.com (Wed. Feb. 10 10-39-01 1993), in most cases, the assumption of fixed probability has no effect on conclusion. Also, in a more detailed study titled "Does It Matter Who Serves First?", djmarcusetasc.com (Wed. Feb. 10 10-39-02 1993), which appeared in page 31 of Jan/Feb 91 *TT Topics*, the answer to the question posed in the title is stated as no based on the following argument. If the game goes tie (deuce), then it does not matter who served first, since no matter who wins, each player will have served the same number of times. For other cases, consider the following modifications of the rules. Rather than stopping when one player reaches eleven, keep playing until twenty points have been played. If player A wins the game under modified rules, then she must win at least eleven of the twenty points and hence would have won the game under standard rules. Similarly, if A loses under the modified rules, she also would have lost under the standard rules. But under the modified rules, both players serve ten times, and so it doesn't matter which one served first.

In the rest of this article, we will assume that x is fixed. However, we will return to this question later and make further comments before closing. For simplicity, we will only calculate the probability of A winning a game. Following the new rules of the game, A can either win by reaching a score of eleven with a margin of two (case 1) or if players tie at 10:10 by winning extra points to have a margin of two points (case 2). Thus, the probability of A winning a game is the sum of these two probabilities—that is,

P(A winning a game) = P(Case 1) + P(Case 2).

Let us discuss each case in more detail.

Case 1: In this case, A can win the game by any of the following scores: 11:0, 11:1, 11:2, 11:3, 11:4, 11:5, 11:6, 11:7, 11:8, or 11:9. For these, A must win ten out of the respective ten, eleven, . . ., nineteen points played, plus the last point. Using binomial distribution, it is easy to see that the probability of A winning with the score of 11:j is $\binom{10+j}{10} x^{11} y^j$, $j = 0, 1, \ldots, 9$. Thus, the probability of A winning the game by reaching a score of eleven with a margin of two is

$$P(\text{case 1}) = \sum_{j=0}^{9} \binom{10+j}{10} x^{11} y^j = \sum_{j=0}^{9} \binom{10+j}{10} x^{11}(1-x)^j.$$

Case 2: In this case, A and B must first reach the score of 10:10. Then A must gain a lead of two points to win the game. The probability of this event is, therefore,

$$P(\text{Case 2}) = P(\text{A and B both reach 10})$$
$$\times\ P(\text{A wins by a margin of 2}).$$

First, we calculate the first term in the right-hand side. Since to reach 10:10, a total of twenty points must be played, we have

$$P(\text{A and B both reach 10}) = \binom{20}{10} x^{10} y^{10}.$$

Next, to win the game, A must gain a lead of two points. This can be done in infinitely many ways until A finally gains a lead of two. Let p denote the probability that A wins the game after reaching the score of 10:10. Then we have

$$p = x^2 + 2xyp.$$

This is because starting from 10:10, A can either take the next two points with probability x^2, or each player takes a point with probability of $2xy$ and restart from the score of 11:11. The probability of A winning a game restarting from 11:11 is the same as starting from the score of 10:10. Solving for p, we get

$$p = \frac{x^2}{1=2xy}$$

Putting these together, the probability that A wins a game after reaching a score of 10:10 is

$$P(\text{Case 2}) = \binom{20}{10} x^{10}y^{10} \frac{x^2}{1-2xy}.$$

Replacing $1-x$ for y, and $x^2+(1-x)^2$ for $1 - 2xy$, the probability of case 2 is

$$P(\text{Case 2}) = \binom{20}{10} \frac{x^{12}(1-x)^{10}}{(x^2+(1-x)^2)} = \frac{184756(1-x)^{10}x^{12}}{1-2x + 2x^2}.$$

Finally, $g(x)$ the probability of A winning a game, is obtained by adding the probabilities for the two cases—that is,

$$g(x) = P(\text{A wins}) = P(\text{Case 1}) + P(\text{Case 2})$$

$$= \sum_{j=0}^{9} \binom{10+j}{10} x^{11}(1-x)^j + \binom{20}{10} x^{12}(1-x)^{10}/(x^2+(1-x)^2)$$

Under the old rules, the probability of A winning a game denoted by $g_0(x)$ is

$$G_0(x) = \sum_{j=0}^{19} \binom{20+j}{20} x^{21}(1-x)^j + \binom{40}{20} x^{22}(1-x)^{20}/(x^2+(1-x)^2).$$

Table 1 lists values of P(Case 1) and P(Case 2) and the total ($g(x)$) for different values of x for the new rules. Table 2 lists the same values for a game played according to the old rules, where rather than eleven, players had to score twenty-one points to win. Table 3 and table 4 present different probabilities under the new and the old rules, respectively.

Table 1. Values of P(Case 1), P(Case 2) and the total probabilities ($g(x)$) for x = 0.5 – 0.9 (new rules).

Score	0.5	0.505	0.51	0.525	0.55	0.6	0.7	0.8	0.9
11 to 0	0.0005	0.0005	0.0006	0.0008	0.0014	0.0036	0.0198	0.0859	0.3138
11 to 1	0.0027	0.003	0.0033	0.0044	0.0069	0.016	0.0653	0.189	0.3452
11 to 2	0.0081	0.0088	0.0096	0.0124	0.0186	0.0383	0.1175	0.2268	0.2071
11 to 3	0.0175	0.0189	0.0204	0.0256	0.0363	0.0664	0.1527	0.1965	0.0897
11 to 4	0.0305	0.0327	0.0350	0.0426	0.0572	0.0930	0.1603	0.1376	0.0314
11 to 5	0.0458	0.0486	0.0515	0.0606	0.0772	0.1116	0.1443	0.0825	0.0094
11 to 6	0.0611	0.0642	0.0673	0.0768	0.0926	0.1190	0.1154	0.0440	0.0025
11 to 7	0.0742	0.0771	0.0801	0.0886	0.1012	0.1156	0.0841	0.0214	0.0006
11 to 8	0.0835	0.0859	0.0883	0.0947	0.1025	0.1040	0.0568	0.0096	0.0001
11 to 9	0.0881	0.0898	0.0913	0.0950	0.0974	0.0879	0.0360	0.0041	3E-05
Case 1	0.4119	0.4296	0.4475	0.5015	0.5914	0.7553	0.952	0.9974	1
Case 2	0.0881	0.0898	0.0913	0.0945	0.0955	0.0811	0.026	0.0019	6E-06
Total	0.5	0.5194	0.5387	0.596	0.6868	0.8364	0.9781	0.9993	1

Table 2. Values of P(Case 1), P(Case 2), and the total probabilities ($g(x)$) for x = 0.5 – 0.9 (old rules).

Score	0.500	0.505	0.510	0.525	0.550	0.600	0.700	0.800	0.900
21-0	0.000	0.000	0.000	0.000	0.000	0.000	0.001	0.009	0.109
21-1	0.000	0.000	0.000	0.000	0.000	0.000	0.004	0.039	0.230
21-2	0.000	0.000	0.000	0.000	0.000	0.001	0.012	0.085	0.253
21-3	0.000	0.000	0.000	0.000	0.001	0.002	0.027	0.131	0.194
21-4	0.000	0.000	0.000	0.001	0.002	0.006	0.048	0.157	0.116
21-5	0.001	0.001	0.001	0.002	0.003	0.012	0.072	0.157	0.058
21-6	0.002	0.002	0.002	0.004	0.007	0.021	0.094	0.136	0.025
21-7	0.003	0.004	0.004	0.006	0.012	0.032	0.108	0.105	0.010
21-8	0.006	0.007	0.007	0.011	0.018	0.045	0.114	0.073	0.003
21-9	0.009	0.011	0.012	0.016	0.027	0.058	0.110	0.047	0.001
21-10	0.014	0.016	0.017	0.023	0.036	0.069	0.099	0.028	0.000
21-11	0.020	0.022	0.024	0.031	0.046	0.078	0.084	0.016	0.000
21-12	0.026	0.029	0.031	0.040	0.055	0.083	0.067	0.009	0.000
21-13	0.033	0.036	0.039	0.048	0.063	0.084	0.051	0.004	0.000
21-14	0.041	0.043	0.046	0.055	0.069	0.082	0.037	0.002	0.000
21-15	0.047	0.050	0.053	0.061	0.072	0.077	0.026	0.001	0.000
21-16	0.053	0.056	0.058	0.065	0.073	0.069	0.018	0.000	0.000
21-17	0.058	0.060	0.062	0.067	0.071	0.060	0.011	0.000	0.000
21-18	0.061	0.063	0.064	0.068	0.068	0.051	0.007	0.000	0.000
21-19	0.063	0.064	0.065	0.066	0.063	0.042	0.004	0.000	0.000
Case 1	0.437	0.462	0.488	0.564	0.684	0.870	0.0994	1.000	1.000
Case 2	0.063	0.064	0.065	0.066	0.061	0.038	0.003	0.000	0.000
Total	0.500	0.526	0.553	0.629	0.746	0.909	0.997	1.000	1.000

Table 3. Different probabilities under the new rules.

x	Tie	A wins after tie	A wins with tie	A wins without tie
0.000	0.000	0.000	0.000	0.000
0.050	0.000	0.003	0.000	0.000
0.100	0.000	0.012	0.000	0.000
0.150	0.000	0.030	0.000	0.000
0.200	0.002	0.059	0.000	0.001
0.250	0.010	0.100	0.001	0.004
0.300	0.031	0.155	0.005	0.017
0.350	0.069	0.225	0.015	0.053
0.400	0.117	0.308	0.036	0.128
0.450	0.159	0.401	0.064	0.249
0.500	0.176	0.500	0.088	0.412
0.550	0.159	0.599	0.095	0.591
0.600	0.117	0.692	0.081	0.755
0.650	0.069	0.775	0.053	0.878
0.700	0.031	0.845	0.026	0.952
0.750	0.010	0.900	0.009	0.986
0.800	0.002	0.941	0.002	0.997
0.850	0.000	0.970	0.000	1.000
0.900	0.000	0.988	0.000	1.000
0.950	0.000	0.997	0.000	1.000
1.000	0.000	1.000	0.000	1.000

From these tables, it is clear that under the old rules, the game will virtually never reach a tie unless x (or y) is between 0.35 and 0.65. If players are equally good, the probability of reaching a tie takes the maximum value of 0.125.

Similarly, under the new rules, the game will virtually never reach a tie unless the probability that a player wins a point is between 0.25 and 0.75. This means that a tie is more likely to occur under the new rules than under the old rules. If players are equally good, the probability of reaching a tie takes the maximum value of 0.176. Note that, although reaching a tie is more likely under the new rules, it is clear that the probabilities of winning the game after reaching a tie are the same under both rules.

Figure 2 presents the graphs of $g(x)$ and $g_0(x)$. As can be seen, $g(x) \geq g_0(x)$ for $0 < x \leq 0.5$ and $g(x) < g_0(x)$ for $0.5 < x \leq 1$. This means that the probability of winning the game is higher for the weaker player under the new rules, making the game less predictable and more exciting.

Table 4. Different probabilities under the old rules.

x	Tie	A wins after tie	A wins with tie	A wins without tie
0.000	0.000	0.000	0.000	0.000
0.050	0.000	0.003	0.000	0.000
0.100	0.000	0.012	0.000	0.000
0.150	0.000	0.030	0.000	0.000
0.200	0.000	0.059	0.000	0.000
0.250	0.000	0.100	0.000	0.000
0.300	0.004	0.155	0.001	0.002
0.350	0.019	0.225	0.004	0.017
0.400	0.055	0.308	0.017	0.074
0.450	0.103	0.401	0.041	0.213
0.500	0.125	0.500	0.063	0.437
0.550	0.103	0.599	0.061	0.684
0.600	0.055	0.692	0.038	0.870
0.650	0.019	0.775	0.015	0.964
0.700	0.004	0.845	0.003	0.994
0.750	0.000	0.900	0.000	0.999
0.800	0.000	0.941	0.000	1.000
0.850	0.000	0.970	0.000	1.000
0.900	0.000	0.988	0.000	1.000
0.950	0.000	0.997	0.000	1.000
1.000	0.000	1.000	0.000	1.000

Figure 2. Graphs of $g(x)$ (dashed curve) and (solid curve).

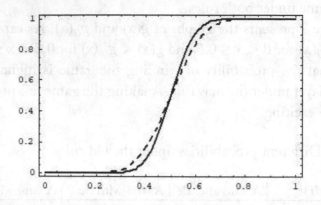

Analysis Using Difference Equations

Let us once more consider the event that A and B reach score of 10:10. We can use difference equations to find the probability that A wins the game starting from this score or, in fact, any other possible score. Here, we use the tie situation for demonstration.

Suppose that $g(i,j)$ represents the probability of A winning the game starting from a score of $i{:}j$ —that is, A has won i points and B has won j points. Since either player can win the next point, this implies that

$$g(i,j) = xg(i+1,j) + yg(i,j+1).$$

Now, consider the case of a tie—that is, the score of 10:10. Using this relationship, we get

$$g(10,10) = xg(11,10) + yg(10,11),$$
$$g(11,10) = xg(12,10) + yg(11,11),$$

and

$$g(10,11) = xg(11,11) + yg(10,12).$$

Substitution for $g(11,10)$ and $g(10,11)$ in the first equation yields

$$g(10,10) = x[xg(12,10) + yg(11,11)] + y[xg(11,11) + yg(10,12)]$$
$$= x^2 g(12,10) + 2xyg(11,11) + y^2 g(10,12).$$

If the game reaches a score of 12:10, then A has won the game. Therefore, $g(12,10) = 1$. Also, if the game reaches a score of 10:12, then B has won the game. This means that $g(10,12) = 0$. Finally, if the game reaches a score of 11:11, the situation is no different from the score of 10:10. Therefore, we have $g(11,11) = g(10,10)$. Replacing for these, we obtain

$$g(10,10) = x^2 + 2xyg(10,10)$$

or

$$g(10,10) = x^2/(1 - 2xy).$$

The value of $g(10,10)$ represents the probability that A wins the game starting from a score of 10:10. Note that this is the same as what we obtained in the previous section. Following the same lines, we can obtain the probabilities for other starting scores. For example,

$$g(11,10) = xg(12,10) + yg(11,11)$$
$$= x + yx^2/(1-2xy) = x(1-xy)/(1-2xy)$$
$$g(10,11) = xg(11,11) + yg(10,12) = xg(10,10) = x^3/(1-2xy)$$
$$g(9,9) = g(10,10) = x^2/(1-2xy)$$
$$g(10,9) = g(11,10), \qquad g(9,10) = g(10,11)$$

Markov Chains

A Markov chain is a mathematical system that experiences transitions from one state to another according to certain probabilistic rules. The defining characteristic of a Markov chain is that no matter how the process arrived at its present state, the possible future states are fixed. In other words, the probability of transitioning to any particular state is dependent solely on the current state and time elapsed. For example, think of a tennis game. Suppose that in a game currently played the score is 30–30. The probability of moving to the state of 40–30 does not depend on how the game has arrived in the state of 30–30. The state space, or set of all possible states, can be anything: letters, numbers, weather conditions, tennis scores, or stock performances.

Markov chains may be modeled by finite state machines, and random walks provide a prolific example of their usefulness in mathematics. They arise broadly in statistical and information-theoretical contexts and are widely employed in economics, game theory, queuing (communication) theory, genetics, and finance. While it is possible to discuss Markov chains with any size of state space, the initial theory and most applications are focused on cases with a finite number of states.

Transition Matrix

As pointed out, Markov chain is used to describe the "change" of state of a system as time passes. For example, suppose that John, who loves outdoor sports, keeps track of the weather in the area he lives. For simplicity, he only considers two states—namely, sunny and rainy. He is planning to play tennis tomorrow, but today the weather is in state "rainy." By keeping the records, John is trying to figure out the chance that tomorrow's weather would be in state "sunny." Here, John is dealing with a system that has only two states, and changes of states may happen from one day to the next. As another example, the result of the latest game played by two teams must be one of three possible states: state 1, team A won the game; state 2, tie; or state 3, team B won the game. In a playoff series or when considering games from one season to the next, it is possible that team A wins the first game and then team B wins the second one. Such a situation happens in sports like tennis or volleyball when several games or sets are played one after another or in basketball, ice hockey, etc., during the playoffs. In such situations, we may observe a "change of state"—that is, as "time" passes and we move from the first game to the second game, the state may change from state 1 into state 3. Note that in some sports, there is no tie, and therefore we need to consider only two states. In general, from the past records, it is possible to estimate the conditional probability of moving to one state given that in the previous period "time," we were in that or in another state. For example,

P(Tomorrow is Sunny | Today is Rainy)
P(Lisa will be in a good mood tomorrow
| She is in a good mood today)
P(Having "hot hands" in tomorrow's game |
Had a "hot hand" in today's game),

or

P(Team B wins the second game | Team A wins the first one)

The latter is exactly the probability for the state to change from 1 into 3 and is conventionally denoted by p_{13}. Similarly, all the probabilities can be estimated from the past records. Here, p_{ij} denotes the probability that the state changes from i to j, in unit of time or period, etc.

In general, if a process can be in one of the k possible states, labeled $1,2,...,k$, and if the probability p_{ij} that the state changes into state j at any time after it was in state i at the preceding time can be determined, then the process is called a Markov chain. The probability p_{ij} is called the transition probability from state i to state j. The matrix $P = [p_{ij}]$ is called the transition matrix of the Markov chain. Note that we can also think of Markov chain as a sequence of stochastic moves such that the probability for the next state is completely determined by the present state.

For example, suppose that John has a record of the last ten days as

SSRSRRRSSR.

Then since S(sunny) followed S twice, R(rainy) followed S three times, S followed R twice, and R followed R twice, we have the following (estimate) for the transition matrix:

$$
\begin{array}{cc}
 & \begin{array}{cc} S & R \end{array} \\
\begin{array}{c} S \\ R \end{array} & \begin{bmatrix} 0.4 & 0.6 \\ 0.5 & 0.5 \end{bmatrix}
\end{array}
$$

A Practice Problem. Many people believe in "hot hands" in sport. This thinking gives rise to the belief that streaks or "hot hands" are unusual. In fact, streaks are to be expected in random sequences, and long streaks need not be amazing coincidences. As an example, here is the 1999 win-loss record for the Baltimore Orioles:

WLLLWLLLWLLLLLLWLLLWLLWLWWWWW
LWLLLLLLWWLLWLWLWWL WLLWLLWLLLWWW
WWWWLWWWWLLLLLLLLLLWWLLLLWWWWW
LLWWWWWWLWLLLL1LWWLWLLWWLLLWW
WLLWWWLLLLWLW WLWLLWWWWWW
WWWWWWWLWWLW

The team's overall average was almost exactly 0.500. Use this to find a transition matrix. Can you draw any conclusion?

Steady State Solution

We start by noting that if all elements of the transition matrix or its powers are positive, the corresponding Markov chain is called a regular Markov chain.

If a Markov chain is regular, then the process approaches to a fixed state vector q. The vector q is called the steady-state vector of the regular Markov chain. Note that this vector does not require the knowledge of an initial state vector. In other words, it is the same for any arbitrary initial state vector.

Absorbing States

In this section, we use two examples from sports to present the concept of the absorbing Markov chain. The first example involves tennis.

In a tennis game played by players A and B, after the state "deuce" is reached, the game becomes a Markov chain involving five states: A's game, B's game, deuce, advantage A, and advantage B. Let us number these state as 1, 2, 3, 4, and 5, respectively. If we further assume that the probability of winning a point by either players is fixed and equals to x and $y = 1 - x$—that is,

$$x = P(\text{player A wins the point})$$

and

$$y = 1 - x = P(\text{player B wins the point}),$$

then we have the following transition matrix.

$$
\begin{array}{c}
 \\
1 \\
2 \\
3 \\
4 \\
5
\end{array}
\begin{array}{ccccc}
1 & 2 & 3 & 4 & 5 \\
\left[\begin{array}{ccccc}
1 & 0 & 0 & 0 & 0 \\
0 & 1 & 0 & 0 & 0 \\
0 & 0 & 0 & x & y \\
x & 0 & y & 0 & 0 \\
0 & y & x & 0 & 0
\end{array}\right] = P
\end{array}
$$

Note that in this example, both p_{11} and p_{12} equal 1, reflecting the fact that once we reach these states (A's game or B's game), everything is over and the process will never change its state again. Thus, once in such a state, the process stays in that state forever.

Analysis Using Markov Chain

In this section, we use matrices and a Markov chain to calculate the probability that A wins a game after reaching the score of 10:10. The analysis presented here addresses the first passage characteristics of the Markov chain. First, we set up a transition matrix with elements representing the probabilities of moving from a state to other states. Let state 1 indicate that A has won the game, and state 2 indicate that B has won the game. Let state 3 refer to tie. Let state 4 indicate that A has one-point advantage over B, and state 5 indicate that B has a one-point advantage over A.

State 1	State 4	State 3	State 5	State 2
A's Game	Advantage A	Tie	Advantage B	B's Game

The transition matrix denoted by T is then

$$T = \begin{array}{c} \\ 1 \\ 2 \\ 3 \\ 4 \\ 5 \end{array}
\begin{array}{ccccc} 1 & 2 & 3 & 4 & 5 \end{array}
\left[\begin{array}{ccccc}
1 & 0 & 0 & 0 & 0 \\
0 & 1 & 0 & 0 & 0 \\
0 & 0 & 0 & x & y \\
x & 0 & y & 0 & 0 \\
0 & y & x & 0 & 0
\end{array} \right]$$

Note that, here, states 1 and 2 are absorbing states; whereas, the other three states are transient.

Suppose now that the game is in state 3. We can represent this by a state vector taking the form

$$(0 \quad 0 \quad 1 \quad 0 \quad 0).$$

Multiplication by transition matrix T yields

$$(0 \quad 0 \quad 0 \quad x \quad y).$$

This is the state vector after playing one point. It is easy to show that the repeated multiplication leads to a final state vector of the form

$$(x^2/(1-2xy) \quad y^2/(1-2xy) \quad 0 \quad 0 \quad 0)=$$
$$(x^2/(x^2+y^2) \quad y^2/(x^2+y^2) \quad 0 \quad 0 \quad 0).$$

This means that, eventually, one of the players will win the game. However, depending on the value of x, the time it takes for this to happen will vary. In fact, it is possible to determine the average length of the time it takes for each player to win starting from one of the three nonabsorbing states (3, 4, and 5). For this, we divide the transition matrix into four matrices, I, 0, R, and Q—that is,

$$T=\begin{bmatrix} I & 0 \\ R & Q \end{bmatrix}$$

where I and 0 are identity and zero matrices, respectively. Here, I, 0, R, and Q are matrices of dimensions (2x2), (2x3), (3x2), and (3x3), respectively. Next, we find the matrix $F=(I-Q)^{-1}$ known as the fundamental matrix. This matrix determines how long it would take, on average, to reach an absorbing state starting from a non-absorbing state. Here, we have

$$(I-Q)=\begin{bmatrix} 1 & -x & -y \\ -y & 1 & 0 \\ -x & 0 & 1 \end{bmatrix}$$

$$F=(I-Q)^{-1}=\begin{array}{c} 3 \\ 4 \\ 5 \end{array}\begin{pmatrix} \dfrac{1}{1-2xy} & \dfrac{x}{1-2xy} & \dfrac{y}{1-2xy} \\ \dfrac{y}{1-2xy} & \dfrac{1-xy}{1-2xy} & \dfrac{y^2}{1-2xy} \\ \dfrac{x}{1-2xy} & \dfrac{x^2}{1-2xy} & \dfrac{1-xy}{1-2xy} \end{pmatrix}$$

This indicates, for example, that starting from state 3 (a tie score), it takes an average of $= (1 + x + y)/(xy)$ points for the game to end.

The sum of the other two rows provides similar information for games starting from states 4 and 5.

Finally, we calculate the matrix FR. This matrix provides probability of A or B winning the game starting from a non-absorbing state. It is

$$
FR = \begin{array}{c} 3 \\ 4 \\ 5 \end{array} \begin{pmatrix} \dfrac{x^2}{1-2xy} & \dfrac{y^2}{1-2xy} \\[2ex] \dfrac{x(1-xy)}{1-2xy} & \dfrac{y^3}{1-2xy} \\[2ex] \dfrac{x^3}{1-2xy} & \dfrac{x(1-xy)}{1-2xy} \end{pmatrix} \begin{array}{c} 1 \qquad\qquad 2 \end{array}
$$

For example, when the score is tied at 10:10, the probability that A wins the game equals $x^2/(1-2xy)$. This is exactly what we obtained using the other approaches. Also, the probability that B wins starting from state 3 is $y^2/(1-2xy)$. Other entries are the corresponding probabilities for the games starting from states 4 or 5. Table 5 presents numerical values of some of these quantities for different values of x.

Table 5. Conditional probability of A winning and expected number of points given a tie or advantage situation.

x	Probability that A wins starting at state 3	Probability that A wins starting at state 4	Expected number of points before game ends starting at state 3	Expected number of points before game ends starting at state 4
0.00	0.000	0.000	2.000	3.000
0.05	0.053	0.000	2.210	3.099
0.10	0.111	0.001	2.439	3.195
0.15	0.176	0.005	2.685	3.282
0.20	0.247	0.012	2.941	3.353
0.25	0.325	0.025	3.200	3.400

0.30	0.409	0.047	3.448	3.414
0.35	0.496	0.079	3.670	3.385
0.40	0.585	0.123	3.846	3.308
0.45	0.671	0.180	3.960	3.178
0.50	0.750	0.250	4.000	3.000
0.55	0.820	0.329	3.960	2.782
0.60	0.877	0.415	3.846	2.538
0.65	0.921	0.504	3.670	2.284
0.70	0.953	0.591	3.448	2.034
0.75	0.975	0.675	3.200	1.800
0.80	0.988	0.753	2.941	1.588
0.85	0.995	0.824	2.685	1.403
0.90	0.999	0.889	2.439	1.244
0.95	1.000	0.947	2.210	1.110
1.00	1.000	1.000	2.000	1.000

Further Analysis

In section 3, we calculated the probability for A to win a game as a function of x. In this section, we study certain aspects of this function. Recall that

$$G(x)=\sum_{j=0}^{9}\binom{10+j}{10}x^{11}(1-x)^{j}+\binom{20}{10}x^{12}(1-x)^{10}/(x^2+(1-x)^2).$$

Let us examine the graph of g(x) over the interval [0,1] shown in figure 2.

It is clear that g(x) is an increasing function over the interval [0,1]. This means that as x increases, the probability of A winning the game also increases. The minimum occurs at x = 0, where the probability of A winning a point is 0. The maximum occurs at g(x) = 1, where the probability of A winning a point is 1. The first derivative of g(x) is

$$g'(x) = 369512(1-x)^{10} x^{10}(5-9x-9x^2)/(1-2x-2x^2)^2.$$

Figure 3 depicts a representation of this function and $g'_0(x)$. Values of $g'(x)$ for few x's are shown in table 6.

Figure 3. Graph of $g'(x)$ (dashed curve) and $g'_0(x)$ (solid curve).

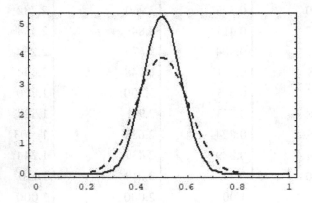

Table 6. Selected values of $g'(x)$

x	$g'(x)$	x	$g'(x)$
0.5000	3.8763	0.8000	0.0313
0.6000	2.4607	0.9000	0.0001
0.7000	0.5698	1.0000	0.0000

Since $g'(x) > 0$ for all, $0 \leq x \leq 1$, $g(x)$ is increasing on this interval. This indicates that as x increases, the probability of A winning a game also increases. The rate of increase for the probability of A winning a game is maximum at $x = 0.5$, where $g'(x)$ equals 3.876. This indicates that around $x = 0$, for every 1 percent change in x, $g(x)$ changes by 3.876 percent. Thus, whenever one player is slightly better than her opponent, her chance of winning a game changes a great deal. So around $x = 0.50$, a small advantage for either player will result in a significant advantage in a game. On the other hand, for x values close to 0 or 1, the rate of change for her chance of winning will not change a great deal because $g'(x)$ has low values at these extremes. This means that for these values, the chance of A winning a game is already very high or very low, and thus any small change in x will not affect A's chance of winning significantly. Finally, since g(x) decreases over the interval [0.5,1], it follows that the rate of change for the probability of A winning a game is decreasing.

We also have

$$g''(x) = \frac{-369512\,x^9(1-x)^9(-50+279x-699x^2+1006x^3-810x^4+324x^5)}{(1-2x+2x^2)^3}$$

Figure 4. Graph of g''(x) (dashed curve) and g''$_0(x)$(solid curve).

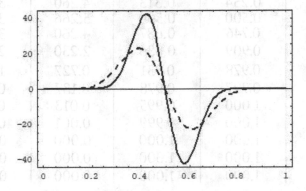

from which it can be determined that there is an inflection point at x = 0.50. The graph of this function and g''$_0(x)$are given in figure 4. When x < 0.5, g''(x) > 0, indicating that g(x) is concave up on (0, 0.5). Since g'(x) > 0, this means that the rate of increase for the probability of A winning the game increases on this interval. However, when x > 0.5, g''(x) < 0, meaning that g(x) is concave down on (0.5, 1). This indicates that while A's chances of winning the game is increasing on this interval, the rate of increase decreases as it approaches its maximum of 1.

The information provided so far can be used to compare the probabilities and rate of changes under the new and old rules. Table 7 provides a summary. As expected, the weaker player has a better chance under the new rules.

Table 7. Probabilities and rates of changes under the new and old rules.

x	$g_O(x)$	$g_N(x)$	$g'_O(x)$	$g'_N(x)$
0.000	0.000	0.000	0.000	0.000
0.050	0.000	0.000	0.000	0.000
0.100	0.000	0.000	0.000	0.000
0.150	0.000	0.000	0.000	0.003
0.200	0.000	0.001	0.001	0.031
0.250	0.000	0.005	0.013	0.168
0.300	0.003	0.022	0.137	0.570
0.350	0.022	0.069	0.727	1.364
0.400	0.091	0.164	2.230	2.461
0.450	0.254	0.313	4.260	3.465
0.500	0.500	0.500	5.266	3.875
0.550	0.746	0.687	4.260	3.465
0.600	0.909	0.836	2.230	2.461
0.650	0.978	0.931	0.727	1.364
0.700	0.997	0.978	0.137	0.570
0.750	1.000	0.995	0.013	0.168
0.800	1.000	0.999	0.001	0.031
0.850	1.000	1.000	0.000	0.000
0.900	1.000	1.000	0.000	0.000
0.950	1.000	1.000	0.000	0.000

We end this section by noting that the analysis presented may help teach some mathematical concepts. For example, calculation of $g'(x)$ can be related to the idea of one player having a slight advantage (edge) over the other player in one point and her edge in one game. The instructor can ask the students to calculate $g'(x)$ for values close to $x = 0.5$ by calculating the difference between the $g(x)$ values and also by using the definition directly.

Additional Studies

The analysis of the game can be expanded in many different directions. This section includes few suggestions for this.

A.   It is possible to determine the size of a handicap to make a game a fair game. For this, we first ask the following question; how many points should be accorded the weaker player for the game to be fair? Here, it is possible to construct a table to determine the number of handicap points that will nearly equalize player's chances of winning using a variable such as $x / y = r$.

We can also consider the inverse problem. Again, we ask the following questions: if B has been given a handicap of j points to make the game fair, what does that mean for the relative strength of the players? Here, we need to solve the equation $g(0,j)=1/2$ with respect to $r = x / y$ for a given value of j. This leads to an algebraic equation in r.

As pointed out, in practice, it is also possible to make the game fair by moving the net away from the middle of the table. To find out how much net should be moved, one can try different amounts and let players play games until several successive games end with close scores.

B. In the above analysis, we assumed that the probability of winning a point is fixed for each player throughout the match. Suppose that the probability of winning the point is not constant during the match. For example, a player may have a higher probability of winning the point when serving. If a game is played under a set of conditions that remains unchanged during the game, then it is possible to carry the analysis using an equivalent game with a fixed probability of winning a point. Here, equivalent means that the two games have the same chance of being won or lost. To clarify, consider the possible advantage associated with serving. Suppose that the probability for player A to win a point is p when serving and is q when receiving. We can describe this situation using the following transition matrix.

$$\begin{bmatrix} p & 1-p \\ q & 1-p \end{bmatrix}$$

Here, the fixed probability of winning a point, x, in the equivalent game may be taken as the stationary solution of a Markov chain having above transition matrix—that is, $q/(1-p+q)$. If players are equally good servers—that is, $p = 1 - q$, then $x = 0.50$. Also, if $p>1 - q$, then $x > 0.50$. For example, for $p=0.60$ and $q = 0.45$, we get $x = 0.529$. This means that if A and B play according to the above matrix and C and D play with fixed probability of winning a point (0.529 and 0.471), then A and C have the same chance of winning the game. Note that this solution is also applicable to other situations. For instance, suppose that A after winning a point wins the next point with probability p, and after losing a point wins the next point with probability q, then the equivalent game with a fixed probability of winning a point is the one with $x=q/(1-p+q)$.

Conclusion

Elementary concepts of probability, calculus, and linear algebra are used for analysis of a table tennis game. The modeling is carried out using several different approaches. Comparison of the results under the new and old rules of the game reveals that the outcome is less predictable under the new rules, making the game more exciting. The analysis can be expanded in many different directions and illustrates the steps of modeling.

Teaching Mathematics/Statistics Using Tennis

The difficulties faced by educators teaching mathematics and statistics are recognized by the community of scholars involved.

To help this matter, many textbooks try to motivate students by introducing varied applications. This addresses both students' apparent desire to see the relevance of their studies to the outside world and also their skepticism about whether mathematics and statistics have any value. This is a great idea that works mostly with students who are deeply committed to a particular academic or career field. For typical students, varied applied examples may fail to motivate, as they are not of immediate concern to them or they do not occur in their daily lives.

Fortunately, students have some common interests that we can build upon when teaching mathematics and statistics. It seems that connecting their studies to a familiar theme with which they have interest or immediate concern almost always works better. Unfortunately, it is not always easy to find a motivating theme of interest to the majority of students. I have tried a few and have come to the conclusion that games and sports are the best way to accomplished this. With this focus, we can help them to build their studies on a foundation—an understanding of a sport—already possessed by most of them. I think this approach is adaptable to all levels at which mathematics and statistics are taught from junior high to graduate schools. In what follows, we discuss other advantages of using sports.

General

- Sports have a general appeal, and it is an area to which modern scientific methods are increasingly applicable.
- Sports have become a part of everyday life, especially for young people.
- Students usually enjoy sports and show a great deal of interest in mathematics and statistics applied to them.
- A major part of the calculus and statistics sequences offered at college level can be taught using a chosen sport.

- Most students can relate to sports and can understand the rules and meanings of the different statistics presented to them.

Specific

Sports data offer a unique opportunity to test methodologies offered by mathematics and statistics. I believe it is hard to find an area other than sports where one could collect reliable data with the highest precision possible. Here, in addition to the quality measurements, one has access to the names, faces, and life history of the participants and their coaches, trainers, and everyone involved. Almost all other data-producing disciplines are susceptible to "data mining" and error, since, unlike sports, they are not watched by millions of fans and media. Clearly, a theoretical result can only be tested when data is reliable and satisfies the required conditions. If the validity of data produced or collected by an individual or an organization cannot be confirmed, one may end up being suspicious of the methodology used.

Consider, for example, a sport such as track and field. The nature and general availability of track and field data have resulted in their extensive use by researchers, teachers, and sports enthusiasts. The data are unique in that they

1. possess a meaning that is apparent to most people;
2. are collected under very constant and controlled conditions, and thus are very accurate and reliable;
3. are recorded with great precision (e.g., to the hundredth of a second in races), and thus permit very fine differentiation of change and/or differences;
4. are both longitudinal (one hundred years for men's records) and cross-sectional (over different distances and across gender); and

5. are publicly available at no cost, thus they provide wonderful data sets to test mathematical and statistical models of change.

Illustration

The focus of a lesson could be a single concept based on examples from several different sports or several different concepts based on a single sport. In this article, we illustrate how a single sport, tennis, may be used to teach certain mathematical and statistical concepts.

The Quirks of Scoring. During the early years, tennis players used a variety of scoring systems. By the time of the first championship at Wimbledon in 1877, the All England Croquet Club had settled on a scoring system based on the traditions of court tennis. This system remained unchanged until the introduction of tiebreakers in 1970.

One quirk of tennis scoring is that strange names are used for points in scoring a game: love, fifteen, thirty, forty, and game. Although no one knows the origin of this odd system, it has been proposed that fifteen, thirty, forty-five, and sixty were originally used to represent the four quarters of an hour. Over the years, the score forty-five became abbreviated as forty. (In informal play, fifteen is sometimes abbreviated as five.) It would be simpler to score the game zero, one, two, three, and four. Still, the weird point names give no advantage to either player.

A more important quirk is that a game must be won by two points. If players each score three points, the score is called deuce rather than 40:40. If the server wins the next point, the score becomes advantage in. If the server wins again, she wins the game; otherwise, the score returns to deuce. If server loses the next point, the score becomes advantage out. If the server loses again, she loses the game; otherwise, the score returns to deuce. This feature of tennis scoring has the virtue of increasing the chance that the stronger player will win, as we shall see.

Consider a game of tennis between two players, A and B. The

progression of the game can be used to teach many statistical concepts and critical thinking. Throughout for any event E, we use P(E) to denote the probability that E occurs.

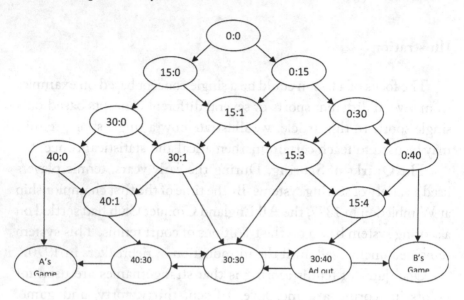

1. Let

$$x = P \text{ (A wins a point)}$$
$$y = 1-x = P \text{ (B wins a point)}$$

To simplify the modeling and analysis of the game, we may first assume that x is fixed—that is, it does not change during the game. Do you think this is a reasonable assumption?

Objective: To teach critical and logical thinking versus practical significance.

2. Starting from 0:0, what are the possible outcomes after one exchange (one point), two exchanges, and so on?

Objective: To teach concepts such as sample space (all possibilities), events (a collection of outcomes), and their algebra.

3. How do you assign probabilities to the outcomes of the sample space after one exchange? Note that possible outcomes are 0:15 and 15:0.

Objective: To teach concepts such as quantification of uncertainty, probability (classical, objective, and subjective), and odds.

4. Do you think winning a point will affect the probability of winning the next point?

Objective: To teach concepts such as conditional probability and independence.

5. Suppose that the answer to question 4 is no. Let $x = 0.60$. How do you find the probabilities such as P (15:15) or P (30:30) or P (40:15), etc.? How do you find these probabilities if the answer to question 4 is yes?

Objective: To teach combinations, multiplication rule, Bernoulli, binomial, and Poisson distributions.

6. Find (a) P (A wins the game without deuce),
   (b) P (A wins the game after reaching deuce),
   (c) P (A wins the game)

Objective: To teach addition rule, infinite series, and geometric progression.

7. Find (a) P (A wins the set without tiebreak),
   (b) P (A wins the set after a tiebreak),
   (c) P (A wins the set)

Objective: To teach modeling and problem solving.

8. Find P (A wins the match)

Objective: To teach pattern identification and model building.

9. Find general formulas for probabilities in questions 6, 7, and 8.

Objective: To teach functions, graphs, and function of functions (composite functions).

10. Let $e = x - y$ represent the edge in one point. For example, $x = 0.51$ means player A has a 0.02 edge over player B in one point. Find edges in one game, one set, and the match.

Hint: call the edge in one point Δx and edge in a game Δy

Objective: To teach the derivative, chain rule, and differential equations.

11. Let $r = x/y$. Since $0 \leq x, y \leq 1$, it follows that $0 \leq r < \infty$ and $r = 1$ for $x = y = 0.5$. Express the probabilities in question 9 in terms of r.

    Objective: To teach transformations and homogeneous polynomials.

12. Let 0, 1, 2, 3, and 4 represent the scores 0, 15, 30, 40, and the game, respectively. Let $g(i,j) = P$ (A wins the game starting from the score (i,j)). Show that

$$g(i,j) = xg(i+1,j) + yg(i,j+1).$$

    Also show that $x^2/(x^2 + y^2)$.

    Objective: To teach recursions and difference equations.

13. Think about a game that has reached the state deuce (or 30: 30). Recall that there is no limit to how long the game could go on. From this point, the game could reach one of the five possible states. Let 1, 2, 3, 4, and 5 denote the states A's game, B's game, deuce, advantage A, and advantage B, respectively. Recall that the game moves from state to state until one player wins. The probabilities of moving from one state to another can be summarized as

	1	2	3	4	5
1	1	0	0	0	0
2	0	1	0	0	0
3	0	0	0	x	y
4	x	0	y	0	0
5	0	y	x	0	0

Objective: To teach matrices, Markov chains, and states of a Markov chain.

14. Suppose that the game is now in state deuce (state 3). This can be expressed as the state matrix:

$$1 \quad 2 \quad 3 \quad 4 \quad 5$$

$$[0 \quad 0 \quad 1 \quad 0 \quad 0]$$

Show that after one and two exchanges, the state matrices are respectively

$$1 \quad 2 \quad 3 \quad 4 \quad 5 \qquad 1 \quad 2 \quad 3 \quad 4 \quad 5$$
$$[0 \quad 0 \quad 1 \quad x \quad y] \qquad [x^2 \quad y^2 \quad 2xy \quad 0 \quad 0]$$

Objective: To teach matrix algebra.

15. Starting from deuce
    a. How many exchanges (points) are expected to be played before the game ends?
    b. How many times is each state expected to be visited/revisited before the game ends?

Objective: To teach stationary solution, inverse of a matrix, and the fundamental matrix.

16. Suppose now that x_1, x_2 represent respectively the probabilities in part *a* and x_3, x_4 represent respectively the probabilities in part *b*.
    a. P(A wins a point when serving) and P(A wins a point when receiving)
    b. P(A wins a point after winning a point) and P(A wins a point after losing a point)

Find the probabilities of winning a game, a set, and the match for player A.

Objective: To teach basic concepts of modeling.

17. Consider a tournament like the Davis Cup. Suppose that countries A and B each have three players represented as

A_1, A_2, A_3 and B_1, B_2, B_3, respectively. Suppose that the following matrix represents their chances of winning or losing against each other.

$$
\begin{array}{c c c c}
 & B_1 & B_2 & B_3 \\
A_1 & \begin{bmatrix} 40\% & 52\% & 50\% \\ A_2 & 40\% & 41\% & 30\% \\ A_3 & 55\% & 45\% & 60\% \end{bmatrix}
\end{array}
$$

For example, using this matrix, we have $P(A_1$ beats $B_1) = 40\%$. Recall that in the Davis Cup, each team decides which player plays first, second, etc., game without knowing about the selection of the other team. How do you think teams should make their selection?

Objective: To teach game theory.

18. Recall that in tennis, the server gets a second chance to serve after missing the first one. Ordinarily, players go for a speedy (strong) but risky first serve and a slow but a more conservative second serve. Analyze all the possible serving strategies and their consequences.

Objective: To teach basic concepts of decision analysis and its role in the game theory.

19. How do you summarize statistics related to a tennis player, a team, and a tournament?

Objective: To teach descriptive statistics.

20. Suppose that you have data for the speed of player A's first serve. How do you calculate the probability that in the next match, the average speed of A's first serves would exceed a certain value?

Objective: To teach sampling distribution and central limit theorem.

21. How do you compare two tennis players? How do you rank tennis players?

Objective: To teach performance measures, measures of relative standing, z-score, etc.

22. A claim is made about the performance of a tennis player. Using the player's statistics, how do you validate the claim?

Objective: To teach hypothesis testing, type I and type II errors, and P-value.

23. How can you use the past statistics of a player to predict his/ her future performance?

Objective: To teach estimation (prediction), confidence intervals, regression, time series, and forecasting.

24. Suppose that you have statistics on the speed of player A's first serves. How do you predict the next record speed and perhaps the maximum possible speed of A's serves?

Objective: To teach theory of records, asymptotic theory of order statistics, extreme value theory, and threshold theory.

25. How do you organize a tennis tournament?

Objective: To teach planning and scheduling.

26. Recall that the winner of men's tennis match must win three out of five sets. Each set has six games. Do you think the present scoring system is fair? For example, player A could win two sets, six games to none (6–0) and lose three tiebreak sets (6–7). In this case, A could win thirty games and lose only twenty-one games and yet lose the match. Do you have any suggestion to make the match more balanced?

Objective: To teach methods for adaptive modeling.

Binomial Distribution, Matrices, Markov Chain, and Derivatives

Consider a match between two players, A and B. Suppose that player A has a 10 percent edge over player B in one point—that is, the probability that player A will win a point is 55 percent and the probability that player B will win a point is 45 percent.

1. Show that the probability calculations before reaching deuce can be carried out using a binomial distribution.
2. Find the probability that player A wins a game, a set, and the match given the edge A has in one point. Also, calculate the edge in a game, a set, and a match given the edge in one point.

Suppose that the information regarding the players A and B is summarized in a rectangular array given below.

$$
\begin{array}{c@{\quad}ccccc}
 & 1 & 2 & 3 & 4 & 5 \\
1 & 1 & 0 & 0 & 0 & 0 \\
2 & 0.55 & 0 & 0.5545 & 0 & 0 \\
3 & 0 & 0.55 & 0 & 0.45 & 0 \\
4 & 0 & 0 & 0.55 & 0 & 0.45 \\
5 & 0 & 0 & 0 & 0 & 1
\end{array}
$$

This is called a transition matrix. It includes the probabilities of moving from one state to another after one point. Here, state 1 represents A won the game, state 2 represents the advantage A, state 3 represents the deuce, state 4 represents the advantage B, and state 5 is B won the game.

3. Apply matrix algebra and interpret the results in the context of a tennis game.
4. If we look at the tennis game as a Markov chain, what are the states? Which states are non-recurrent? Which states are recurrent? Which states are absorbing?
5. Let x denote the probability that player A wins a point, and $y = 1 - x$ denote the probability that player B wins that point. It can be shown that

$$P(\text{A wins the game}) = x^4 \left[1 + 4y + 10y^2 + \frac{20xy^3}{x^2 + y^2}\right].$$

Replace y by $1 - x$ in this formula and find its derivative with respect to x. Calculate the value of the derivative at the point $x = 0.50$. For players close in ability (small edge in one point, e.g., 1 percent), the resulting value provides the edge in one game. Compare the value obtained using derivative with the actual value of the edge.

6. Consider the formula in problem 5. Replace y by $1 - x$. The resulting function has several properties. For example, it is symmetric with respect to $x = 0.50$. Study the other properties of this function.

7. Consider the formula in problem 5. Replace y by $1 - x$. Suppose that P (A wins the game) $= 0.60$. Use numerical methods to find x.

8. Find the probability of winning a set as a function of x and show that this is an example of a function of a function. Use this to find the edge in a set both directly and by using the derivative (chain rule) as in problem 5.

Calculations Based on Normal Distribution

1. The average speed of a well-known tennis player's first serve is 117 mph with a standard deviation of 5 mph. What is the probability that this player's next first serve will
   a. be slower than 115 mph?
   b. be faster than 120 mph?
   c. have a speed between 116 and 122 mph?

2. Suppose that tennis balls are produced to have COR $= 55.5\%$ (target value). To see if the process is on target, once a day, fifty balls are tested. If the average COR falls outside the interval 53.5 percent and 57.5 percent, the process is judged out of control. What is the probability that the process will be judged out of control incorrectly? Assume that the standard deviation is 1.5 percent.

Pythagorean Theorem and Baseball

The original Pythagorean theorem of baseball was devised by Bill James in the 1980s. The "theorem" predicts the winning percentage of a baseball team based on how many runs the team scores as well as how many runs it allows. When a team scores fewer runs that it allows, the model suggests that the team should have a losing record; and when a team scores more runs than allowed, they should have a winning record. In the 2001 season, the New York Mets allowed more runs than they scored but still had a winning record, so they were considered an overachieving team. The Colorado Rockies, on the other hand, scored more runs that they were allowed but still had a losing record, so they were considered an underachieving team.

Now, about twenty years later, Michael Jones and Linda Tappin of Montclair State University in New Jersey have devised mathematically simpler alternatives to the Pythagorean theorem of baseball.

(a) To predict the winning percentage of a team, one new model simply uses addition, subtraction, and multiplication. It starts with the total runs scored by the team in all of its games (R_S) and subtracts the runs that it allows (R_A) and then multiplies it by a number Beta (β), which is chosen to produce the best results. For the 1969–2003 seasons, the optimal values of B range from 0.00053 to 0.00078 with an average of 0.00065.

Adding 0.5 to the result gives the predicted winning percentage of the team. The resulting linear formula looks like this:

$$P = 0.5 + \beta(R_S - R_A).$$

This equation can be used to teach simple linear regression. If an equation takes the form $y = b + mx$, then here we are using x to predict y, and the equation is said to be linear. In the above equation using the baseball variables, the R_S and R_A are the values that are used in trying to predict what P, the winning percentage, will be. One advantage of using this formula is we can find a range for P.

(b) In contrast, the original Pythagorean theorem of baseball is a little more complex. It uses the square of the previously used variables. The resulting formula is

$$P = R_S^2 / (R_A^2 + R_S^2).$$

This equation gets its name because of its similarity to the Pythagorean theorem in geometry, which relates the lengths of the sides of a right triangle as $a^2 + b^2 = c^2$, where a and b are the shorter sides of the triangle and c is the hypotenuse.

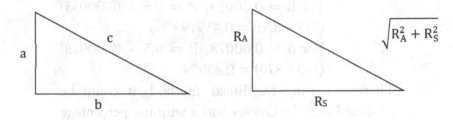

Examples using both methods (a) and (b):

1. In the 2003 season, the Philadelphia Phillies scored 791 runs and allowed 697 runs to be scored against them. According to the 2003 MLB relative power index, the winning percentage (PCT) for the Phillies was 0.531. Test to see how close these methods predict the actual PCT.

    a.   The range of values for the winning percentage can be predicted for the different values for Beta (β).
         For $\beta = 0.00053$, P = 0.5 + (0.00053)
         (791 – 697) = 0.54982
         For $\beta = 0.00078$, P = 0.5 + (0.00078)
         (791 – 697) = 0.57332

    Therefore, using the linear method, it could be predicted that the Phillies had a winning percentage between 0.54982 and 0.57332.

    b.   Using the Pythagorean theorem,
         P = $(791)^2$ / $[(697)^2 + (791)^2]$ = 0.56292

In this case, both strategies predict a very similar result.

2. The Baltimore Orioles scored 743 runs in the 2003 season, while allowing 820 to be scored upon them. According to the 2003 MLB relative power index, the winning percentage (PCT) for the Orioles was 0.490. Test to see how close these methods predict the actual PCT.

    a. The range of values for the winning percentage can be predicted for the different values for Beta (β).

        For $\beta = 0.00053$, $P = 0.5 + (0.00053)$ $(743 - 820) = 0.45919$

        For $\beta = 0.00078$, $P = 0.5 + (0.00078)$ $(743 - 820) = 0.43994$

Therefore, using the linear method, it could be predicted that the Orioles had a winning percentage between 0.43994 and 0.45919.

    b. Using the Pythagorean theorem,

$$P = (743)^2 / [(820)^2 + (743)^2] = 0.45085$$

In this case, both strategies again predict a very similar result.

The question is whether the Pythagorean theorem was needlessly more complicated. It was found that for the baseball seasons from 1969 to 2003, the new linear formula predicts the percentages almost as well as the old theorem. The one real exception was the 1981 season in which there was a baseball strike. Therefore, the new linear method may be a better and simpler solution to predicting the winning percentages of baseball teams.

Sports and Exercise

Regular Exercise: Several sports are used by many as both exercise and social activity. These days, most people get their physical activities by going to gyms. Considering this, let me present a fun reading related to exercise and its benefits.

Recently, there has been some research on the possible association between physical activities and life expectancy for adults. To determine the number of years of life gained from physical activity, researchers have examined data on more than 650,000 adults. One study that appeared in *PLOS Medicine*, November 6, 2012, reports that people who engaged in physical activity had life expectancy gains of up to 4.5 years. Based on such studies, physical activities are recommended for anyone who can afford the time, energy, and cost.

While reading this and few similar studies, I was wondering about possible drawbacks of regular exercise, and I think I have found some. Of course, needless to say that this is not to encourage you to give up such activity. However, pointing out them may provide a few "comforting" arguments for those who, for any reason, do not or cannot exercise regularly.

Since so many different kinds of physical activities could be classified as exercise, to simplify the discussion here, I only consider more serious ones that require traveling to a gym. Consider an individual who devotes ten hours of his/her weekly life to exercise (exercise plus travel to and from a gym). Suppose also that this individual does this through his/her adult life of fifty years, say, and with the goal of increasing his/her life expectancy. Simple calculation shows that for this person, time devoted to exercise would take almost three years of his/ her life. Moreover, since we cannot exercise while sleeping, it takes, in fact, 4.5 years of our day lives. So the question that arises at this point is the following: would exercise increase person's life by more than three years already invested? The answer is it may or may not, and we cannot be sure about that. So here we are spending a sure period of our young life for an unsure gain of the end of life when we have less energy, fewer desires, and maybe less health to enjoy what life has to offer. Of course, exercise could improve the quality of life, but for most people, regular exercise has its own physical, emotional, and mental stress. Just imagine spending three years of your life traveling to a gym or running on a treadmill. Other issues to consider are

- time taken from other activities;
- cost, especially for those who live in big cities;
- time to shower and to wash the gym clothes;
- possible injuries during the exercise;
- cost of buying gear such as shoes, etc.

Now, clearly, we can extend the discussion to many other related factors. For example, most of us enjoy casual recreations such as walking, biking, or swimming. These are probably the best forms of physical activity that are helpful to both body and mind.

Here is another good news and interesting research finding. According to a recent research, drinking red wine is better than going to the gym. Jason Dyck and his colleagues in University of Alberta in Canada found that red wine, nuts, and grapes have a complex called resveratrol, which improves heart, muscle, and bone functions, the same way they're improved when one goes to the gym. Details of this research can be found in the article by Natalie Roterman | Sep 15 2014, 04:51PM EDT.

So to my friends who have a busy life: maybe regular exercise is not always what we need to do, but as they say, we all need to get up, dress up, and show up, even if we do it as slow as a turtle, remembering that turtles with a minimum physical activity live significantly longer than many apparently active creatures.

Hot Hand in Sport "Hot hand" in basketball is a phenomenon that most basketball fans accept and consider well understood. It is a topic that has generated a great deal of interest among researchers in areas such as psychology, statistics, and physical education. For a player who is hot, basket appears as if it is so wide that he/she will make shot after shot with ease. The opposite situation may be called "cold hand."

Although hot hand is meaningful to basketball players and fans, several publications have demonstrated that statistically such phenomenon may not even exist. Amos Tversky, a psychologist who studied every basket made by the Philadelphia 76ers for more than a season, has concluded the following:

1. The probability of making a second basket (hit) did not rise following a successful shot (success did not breed success).
2. The number of "runs" or baskets in succession was no greater than what a coin-tossing model would predict.

To clarify item 2, consider a good player who makes half of his shots. For such a player, four hits in a row (HHHH) is expected to occur once in sixteen sequences of four shots, as this is one of the sixteen equally likely outcomes. The chance that this player hits at least three shots in a row (HHHH, HHHM, MHHH) is 3/16, assuming that trials are independent. Does this mean that the player has a hot hand? To add to this, suppose that I flip a coin four times and get four heads in a row. Could I claim that I have a hot hand, say, for heads?

Clearly, a great player will have more sequences of five hits than an average player—but not because he has greater will or gets in that magic rhythm more often is his/her talent. He/she has longer runs because his/her average success rate is so much higher and has a much better chance of having more frequent and longer sequences. Suppose that we simulate a great player's game whose chance of making a field goals is 0.6. Such player will get five in a row about once in every thirteen sequences, since the chance of this event is 1/13 or $1/(0.6)^5$. If another player's chance of making a field goals is 0.3, he will get his five hits in a row only about once in 412 times. In other words, we need no special explanation for the apparent pattern of long runs, except perhaps when the performance is far beyond what is expected from a player based on his past statistics.

Similar studies in baseball indicate that nothing ever happened in baseball above and beyond the frequency predicted by coin-tossing models. The longest run of wins or losses are as long as they should be and occur about as often as they ought to. Again, this may also be demonstrated using computer simulation. However, this rule has one exception that should never have occurred: DiMaggio's fifty-six-game hitting streak in 1941. Purcell calculated that to make it likely (probability greater than 50 percent) that a run of even fifty games will occur once in the history of baseball up to now (and fifty-six is a lot more than fifty in this kind of league), baseball's rosters would have to include either four lifetime 0.400 batters or fifty-two lifetime 0.350 batters over careers of one thousand games. In actuality, only three men have lifetime batting averages in excess of 0.350, and no one is anywhere near 0.400. He then concluded that DiMaggio's streak is the most extraordinary thing that ever happened in American sports.

Having stated all the above, the question that still remains is how people define "hot hand" and whether this definition is based on a pattern or reality or is just a perception. Study of Larkey et al. (1989) reveals that different measures for hot hand leads to consider different players hot. It is interesting to note that, although the precise meaning of a term like "hot hand" is unclear, its common use implies a shooting record that departs from coin tossing with probability of success greater than that expected. An equally interesting point relates to the fact that fans who talk about hot hand usually refer to pattern of streak shooting or something that is noticeable, memorable, or unlikely. Examples are observations such as HHHHHM or MHHHHH. But then, why an outcome like HMHMHM, which presents a sequence of hit followed by miss and vice versa is not considered a notable or extraordinary pattern, and why is it not given a name? As pointed out in *The Skeptic's Dictionary*, the clustering illusion is the intuition that random events that occur in clusters are not really random events. The illusion is due to selective thinking based on a false assumption. A good example occurs in lottery. People think that a number like 2145362 is more random than 2222222; whereas, when choosing an even digit number randomly, they have the same chance to be selected.

Here is another example supporting the fact that people only recognize certain patterns and ignore many others. Let us replace hit by 1 and miss by 0. Then we have a sequence of ones and zeros (a number in binary system). Consider, for example,

$$0\,0\,0\,0\,0\,0\,1\,1\,1\,1\,1 \text{ and } 1\,1\,1\,1\,0\,0\,1\,1\,0\,1\,0.$$

The first one may be recognized as having a pattern. But what about the second one? In fact, this is a famous pattern. It is the binary presentation of the number 1946, my birth year. So to me this is a recognizable pattern, but to the others it may not be (in fact, it is not). A similar example happens in poker. Getting a royal flush is surprising even though the chance of any hand in poker, with no particular name, is extremely low.

It is also possible to argue in favor of hot hand. In an interesting paper, Wardrop (1998) presents many discussions concerning an inherent weakness in the methods used by Gilovich et al. (1985) and Tversky and Gilovich (1989a). Hooke (1989) discusses the inherent difficulty of using statistical methods to study complete phenomenon such as a game of basketball. Hale (1999) has discussed this issue and has raised several questions. He has argued that hot hand is an internal phenomenon and that the sense of being "hot" does not predict hits or misses. When a player realizes he is hot, he tends to push the envelope and attempt more difficult shots, which then leads to predictably a failure. Or he raises the question, precisely how unlikely does a streak of success need to be before we are prepared to count it as a legitimate instance of hot hand?

According to Hale (1999), there are three prominent arguments that conclude there are no hot hands in sports. The first argument of the hot hands critics creates a tradition in the very act of destroying it. By making "success breeds success" a necessary condition of having hot hands, the critics have established a previously undefended and barely articulated account of hot hands, only to demolish it. Instead, he has argued that there are good reasons to reject "success breeds success" as a requirement for having hot hands. While it is true that many players believe that future success is more likely when they are already hot, either this is only a belief that their current state has causal efficacy into the future, or is it inductive reasoning that their current high rate of success is evidence of future success? Yet neither disjunction makes "success breeds success" part of the concept of having hot hands.

The next two arguments offered by the critics of the hot hand theory are of a well-known skeptical pattern: set the standards for knowledge of something extremely high, and then show that no one meets those standards. The canonical reply to this strategy of which he availed himself is to reject those standards in favor of more modest ones that charitably preserve our claims of knowledge. The skeptical insistence upon exceedingly rare streaks or statistically remote numbers of streaks as being the only legitimate instances of hot hands is arbitrary and severe. He has then argued that "being hot" denotes a continuum, one that is nothing other than deviation from the mean itself. This obviously comes in degrees.

To summarize the arguments against the hot hand, Tversky and Gilovich (1989), using several data sets, concluded that existing data does not support "hot hand." They have devised a clever experiment to obtain convincing evidence that knowledgeable basketball fans are much too ready to detect occurrences of streak shooting and the "hot hand" in sequences that are, in fact, the outcome of Bernoulli trials. To clarify this argument further, they have considered the data concerning the free throws for nine regular players of Boston Celtics from 1980 to 1982. Then they asked the following question: when shooting free throws, does a player have a better chance of making the second shot after making the first shot than after missing the first shot? To answer this. they have chosen a sample of one hundred Cornell and Stanford students randomly. The responses were 68 percent yes, implying hot hand; and 32 percent no, implying independence or negative association.

After analyzing the data, Tversky and Gilovich (1989) concluded that the data provided no evidence that the outcome of the second shot depends on the outcome of the first one. Adams (1992), using data on eighty-three players, showed that the mean interval from making a field goal (n $= 372$) to making a field goal in nineteen NBA games did not differ from the mean interval from making to missing (n = 394), which further challenges assumptions regarding hot hand.

In a more recent study, Koehler and Conley, in "The Hot Hand Myth in Professional Basketball," which appeared in the *Journal of Sport and Exercise Psychology,* have offered further evidence against hot hand in a unique setting, the NBA long-distance shootout contest. They have concluded that declarations of hotness in basketball are best viewed as historical commentary rather than as a prophecy about future performance.

Having discussed opposing views, the final and perhaps more important question is, why do arguments for and against hot hand seem convincing? In an attempt to answer this question, Wardrop (1995) has performed a very interesting analysis of data for Boston Celtics players. His analysis is based on, as he stated, the fact that the data available to laypersons may be very different from the data available to professional researchers. In addition, laypersons unfamiliar with a counterintuitive result, such as Simpson's paradox, may give the wrong interpretation to the pattern in their data and their analysis. There are many problems of this type in probability theory where the right answer is counterintuitive.

Here, we will borrow the data and its analysis presented in Wardrop for demonstration. To understand Wardrop's analysis, consider the following data presented in Wardrop (1995).

Observed Frequencies for Pairs of Free Throws by Larry Bird and Rick Robey, and the Collapsed Table (Wardrop 1995, Table 1)

	Larry Bird		
	Second		
First	Hit	Miss	Total
Hit	251	34	285
Miss	48	5	53
Total	299	39	338
Rick Robey			

	Second		
First	Hit	Miss	Total
Hit	54	37	91
Miss	49	31	80
Total	103	68	171

Collapsed Table			
	Second		
First	Hit	Miss	Total
Hit	305	71	376
Miss	97	36	133
Total	402	107	509

Let us use the notations

P_h = the proportion of first shot hits that are followed by a P_m = the proportion of first shot misses that are followed by a hit. Then we have,

$P_h = 251/285 = 0.881$ and $P_m = 48/53 = 0.996$ for Larry Bird,

$P_h = 54/91 = 0.593$ and $P_m = 49/80 = 0.612$ for Rick Robey; but for collapsed data, we have

$P_h = 305/376 = 0.811$ (= $(285/376)(251/285) + (91/376)(54/91)$), and $P_m = 97/133 = 0.729$.

Note that, contrary to the hot hand theory in the sense of success breeds success, each player's shot is slightly better after a miss than after a hit, although as is shown by Wardrop, the differences are not statistically significant.

It is possible, of course, to ignore the identity of the player attempting the shots and examine the collapsed data in the table. For example, on 509 occasions, either Bird or Robey attempted two free throws; on 305 of those occasions, both shots were hit; and so on. For the collapsed data, $ph = 0.811$ and $pm = 0.729$. These values support the hot hand theory—that is, a hit was much more likely than a miss to be followed by a hit.

The data from Bird and Robey illustrate Simpson's paradox—namely, $P_h < P_m$ in each component table, but $P_h > P_m$ in the collapsed table. It is easy to verify algebraically that the proportion of successes for a collapsed table proportions equals the weighted average of individual player's proportions, with weights equal to the proportion of data in the collapsed table that comes from the player. For the after-a-hit condition, for example, the weight for Bird is $285/376 = 0.758$, the weight for Robey is $91/376 = 0.242$, and the proportion of successes for the collapsed table, $305/376 = 0.811$, is

$$(285/376)(251/285) + (91/376)(54/91).$$

As a result, even though both Bird and Robey shot better after a miss than after a hit, the collapsed values show the reverse pattern because of the huge variation in weights associated with each player. In short, Wardrop has concluded that Simpson's paradox has occurred because the after-a-miss condition, when compared to the after-a-hit condition, has a disproportionately large share of its data originating from the far inferior shooter Robey.

A Fun Problem

In 1990, they ordered the eleven non-playoff teams by using sixty-six ping-pong balls placed in a hopper. The team with the worst record in the league had eleven balls with its logo painted on them, the second worst team had ten balls, and so on, with the non-playoff team having the best record getting only one ball. A drawing was held to determine the first three draft picks. The first draft pick was awarded to the team whose logo was on the first ball drawn. More balls were drawn until a new logo appeared, and the second pick was awarded to that team. Still, more balls were drawn to determine which of the remaining nine teams would receive the third pick. At that point, the lottery ended, and the remaining nonplayoff teams were assigned draft positions in the reverse order of their records. Analyze this system using the ideas of probability theory.

A Mathematics Course Based on Sports

In recent years, numerous books and articles have been published in the interdisciplinary area of statistics and sports. These activities indicate importance and popularity of the subject. In fact, recent investigations have demonstrated that students show a great deal of interest in mathematics and statistics when applied to sports-related problems.

Considering this, I have developed a course titled "Mathematics and Sports." The object of the course is to provide novel, stimulating, and interesting applications of mathematics and statistics in sports. It is based on problem-solving ideas and can be taken by students who have had a first course in statistics.

It is my belief that courses of this type increase students' interest in mathematics and statistics and public's interest in university activities. In fact, many students and members of the public mistakenly believe that learning mathematics is one thing and going in for sports is quite another.

Description of the Course

The course is about links between mathematics, statistics, and sports. It provides valuable source of material that has an intrinsic interest level for most people. It shows just how mathematical and statistical analysis can be used to explain the structure of many sports and sporting events as well as to indicate how performance can be modeled. Those who plan to become teachers, both in secondary and higher education, will learn about a valuable source of material that will enrich their teaching of many mathematical topics. To include a little history, a few lectures are devoted to the life and work of mathematicians who were keenly interested in sports.

For the sporting enthusiast, this course demonstrates the effective mathematical and statistical model that can help to improve analysis of frequently used techniques. While modeling usually only confirms what intuition tells the expert sportsman, it is nevertheless gratifying to validate one's own interest.

Simpson's Paradox

Consider two players, A and B. Suppose that in the first half of the season, player A made only 20 of his 100 free-throw attempts. In the same period, player B made 50 of his 210 free throws. So in the first half of the season, B performed better (24 percent versus 20 percent). Suppose that in the second half of the season, A made 40 of his 60 and B made 15 of his 20 attempts, respectively. Thus, again, B performed better (75 percent versus 67 percent). Now, if we combined these statistics for the season, we have 60/160=.38 for A and 65/230=.28 for B—that is, A did better for the season. This means that although B performed better in both the first and second half of the season, his performance was worse than A in the whole season. This is, of course, a paradox known as Simpson's paradox. The table below summarizes these statistics:

	First Half	Second Half	Season
Player A	20/100 = 0.20	40/60 = 0.67	60/160 = 0.38
Player B	50/210 = 0.24	15/20 = 0.75	65/230 = 0.28

Now, it is clear from this summary that different conclusions may be drawn from looking at whole or parts of players' statistics. The lesson here is that we must be extremely careful in reporting averages (percentages) unless the groups under consideration are fairly homogeneous.

Penalty Kicks

A penalty kick (penalty) is a free kick taken from the penalty mark thirty-six feet (twelve yards or approximately eleven meters) from the center of the goal and with no player other than the goalkeeper of the defending team between the penalty taker and the goal.

Penalty kicks occur during a normal play. They also occur in some tournaments to determine who progresses after a tie game; though similar in procedure, these are not penalty kicks and are governed by slightly different rules.

In practice, penalties are converted to goals more often than not, even against world-class goalkeepers.

A penalty kick may be awarded when a defending player commits a foul punishable by a direct free kick against an opponent or a handball within the penalty area (commonly known as "the box" or "eighteen-yard box"). It is the location of the offense and not the position of the ball that defines whether a foul is punishable by a penalty kick or direct free kick, provided the ball is in play. The penalty taker (who does not have to be the player who was fouled) must be clearly identified to the referee.

The penalty kick is a form of direct free kick, meaning that a goal may be scored directly from it. If a goal is not scored, play continues as usual. As with all free kicks, the kicker may not play the ball a second time until it has been touched by another player even if the ball rebounds from the posts. However, a penalty kick is unusual in that, unlike general play, external interference directly after the kick has been taken may result in the kick being retaken rather than the usual dropped ball.

Lessons

Objective: To teach basic geometry and trigonometry.

Background information: Penalty kicks are normally taken from the penalty mark, which is a midline spot thirty-six feet from center of the goal. The penalty mark has the same distance from both goalposts.

Opening question: Suppose the penalty kick can be taken from any point in the field as long as its distance from the center of goal is thirty-six feet. Suppose also that the goalkeeper stands in the center of the goal and can protect the area inside a circle of radius eight feet. From which point on the penalty taker has the greatest chance of scoring?

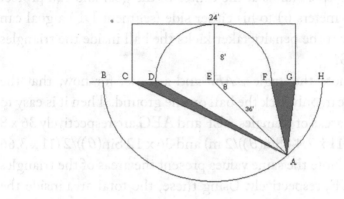

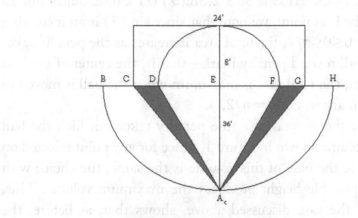

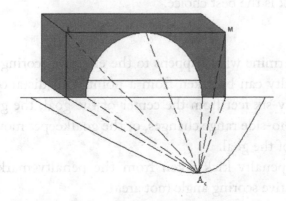

Solution

Consider a kick taken from an arbitrary point A shown in figure 1. The goal width (CG) is 24 feet or 7.32 meters. Since we assume that the goalkeeper stands at the center of the goal and can protect 8 feet or 2.44 meters (r) to his either side (segment DF) a goal can be scored only if the penalty taker kicks the ball inside the triangles ACD or AFG.

Let θ denote the angle $\angle AEF$ and assume, for now, that the penalty taker can only kick the ball on the ground. Then it is easy to show that the areas of triangles AEF and AEG are respectively 36 x 8 x Sin(θ)/2 ft (11 x 2.44 Sin(θ))/2 m) and 36 x 12 Sin(θ))/2 (11 x 3.66 Sin(θ))/2 m). Note the same values present the areas of the triangles ADE and ACE, respectively. Using these, the total area inside the triangles ACD and AFG is 36 x 2 Sin(θ) (11 x 0.61 Sin(θ) m). To determine the best point, we note that since sin (θ) is an increasing function for $0 \leq \theta \leq \pi/2$, the total area increases as the penalty taker moves the ball toward penalty mark—that is, the center of the half circle. In fact, the total area is maximum when the ball is moved to the penalty mark where $\theta = \pi/2$.

Now, for the general case, the penalty taker can kick the ball inside the volume shown in figure 1. Since for any point selected on the half circle the base of this volume is the same, the choice with the largest possible height produces the maximum volume. This, referring to the case discussed above, shows that, as before, the penalty mark is the best choice.

Activity

- Determine what happens to the effective scoring area if the penalty can be taken from a point with distance less than thirty-six feet from the center of the goal, the goalkeeper's side-to-side range changes, or the goalkeeper moves forward out of the goal.
- For penalty kick taken from the penalty mark, find the effective scoring angle (not area).

- Suppose that the penalty taker can kick the ball with the speed of v. Could you think of a way of incorporating this dimension of the penalty kick into the problem? Would this change your answer to the problem of choosing the point with the highest chance of scoring?

Objective: To teach game theory.

Background information: Consider the game played between the opposing goalie and a penalty taker. The penalty taker can elect to kick toward one of the two goalposts or else aim for the center of the goal. The goalie can decide to commit in advance (before the kick) to either one of the sides or else remain in the center until he sees the direction of the kick.

Opening question: Can you think of optimal strategies for the penalty taker and goalie?

Solution: This two-person zero-sum game can be represented as follows, where the payoffs are the probability of scoring a goal.

| | | *Goalie* | | |
		Breaks left	Remains Center	Breaks right
	Kicks Left	0.5	0.9	0.9
Kicker	Kicks Center	0.1	0	0.1
	Kicks Right	0.9	0.9	0.5

If we assume that decisions between the left or right side are made symmetrically (with equal probabilities), then this game can be represented by a two-by-two matrix, where $0.7 = (1/2)(0.5) + (1/2)(0.9)$:

| | *Goalie* | |
	Remains Center	Breaks side

		0	1
Kicker	Kicks center	0	1
	Kicks side	0.9	0.7

The optimal strategies of the kicker and goalie are to

    a.   more often kick center and remain center.
    b.   more often kick side and break side.
    c.   choose center and side equally often.

Activity

    a.   Suppose that a penalty taker kicks rights more often than left. Discuss how that may change the optimal strategies discussed above.
    b.   Make any changes you consider appropriate or interesting and discuss their consequences.

Closing Remarks

How do you summarize the penalty statistics for players, teams, and tournaments? This can be used to teach descriptive statistics.

How do you compare players and teams? This can be used to teach performance measures, measures of relative standing, z-score, etc.

A claim is made about a player. Using his/her statistics, how do you validate the claim? This can be used to teach hypothesis testing.

Using the past statistics of a player, how do you predict his/her future performance? This can be used to teach estimation (prediction), confidence intervals, regression, time series, and forecasting.

Suppose that the outcomes of successive penalty kicks have dependency. How do you model and analyze such data? This can be used to teach Markov chain.

Extra Activity: Hypothesis Testing

Collect stats of a soccer tournament of your choice—for example, number of penalties, goals, shots, corners, etc.—and provide numerical, graphical, and tabular summary of the data.

Test the claim that nowadays most goals are scored (a) from set pieces, (b) during the first or last five minutes of the halves, or (c) by head.

Prediction of Future Records (Details)

Apart from intrinsic interest, there are several medical and physiological reasons why we would like to know how fast a human being could, for example, run a short distance such as one-hundred-meter dash or a medium distance such as four hundred meters. While there is a general agreement among the physiologists and physical educators about existence of, for example, an upper limit for such a speed (or a lower limit for the time needed to run this distance), the limit is not known at the present time. Because of a great interest in this question, apart from physiological research, there have been some attempts to estimate (predict) the limits via mathematical modeling. Recall that the performance of a human being in athletic events is an issue of great popular interest.

Methods

Prediction of the future records can be divided into short-term and long-term different approaches, which can be taken depending on goal of study and the data one may use.

1. Using all the available data, one may use *general trend analysis*.
2. Considering the fact that both records and improvements are decreasing, one may use ideas from population biology

and apply models such as *logistic* that has an asymptote, which makes its tail to level off.

3. Data analysis and modeling may be performed using only the best record of, for example, each year. The modeling may then be carried out using the *extreme value theory*.

There are three type of extreme value distribution of which one has a lower bound to represent the ultimate record.

4. Data analysis and modeling may be performed using only the times below a chosen time such as ten seconds. The modeling may then be carried out using the *threshold theory*. This theory models data on the lower tail of the distribution for times.

Here, also, there are three types of tail models of which one has a lower bound representing a short tail model and the ultimate record.

5. Data analysis and modeling may be performed using only the records. The modeling may then be carried out using the *theory of records*.

This theory models number of records, time of the records, time interval between records, and, finally, size of the records.

Since sport records are more frequent than what this theory predicts, some modification to account for this is needed. Here, one can use population growth or growth in population of participants for this adjustment.

Trend Analysis

Trend analysis seeks to decompose the data into a systematic (deterministic, trend, signal, talent) part and a random (stochastic, noise, chance) part. When the form of the systematic part is known, it is easy to separate that from the random part. In the absence of information pertaining to the either systematic or random parts, smoothing is used for separation. This is usually carried out assuming that the systematic part is smooth and the random part is rough. Smoothing is an exploratory operation, a means of gaining insight into the nature of data without precisely formulated models or hypotheses. Smoothing is often achieved by some sort of averaging (low-pass filtering). Once the smooth part is determined, the difference between the original message and the smooth part is used to present the rough part. The rough part is usually utilized to make reliability statement regarding the systematic part. For data with a time index (time series), one popular smoothing technique is the so-called moving averages. The idea is to average the neighboring values and to move it along the time axes.

The idea is to smooth the data using techniques such as moving average in stock market to recover the trend or replacing the data points by a smooth curve using methods such as regression analysis. When smoothing, we can weight more recent performances (times) or records more than others. Finally, we can assume that improvements are proportional to the last record and use the *logistic model*.

General Method

To analyze sports records, a large number of investigators have proposed models that are made up of a deterministic component $z(t,\theta)$ to account for the trend and a stochastic component $x(t)$ to account for the variation—that is,

$$y(t) = z(t,\theta) + x(t) \tag{1}$$

(see for example Chatterjee and Chatterjee 1982, Morton 1983, Tryfos and Blackmore 1985, and Smith 1988). For deterministic component $z(t,\theta)$ many different forms are suggested. See Blest (1996) for the list.

For example, Smith (1988) considered model (1) assuming that $x(t)$'s are i.i.d. random variables. Particular distributions considered for $x(t)$ were normal, Gumbel, and the generalized extreme value.

For $Z(t,\theta)$ the following linear, quadratic, and exponential-decay models were examined:

$$Z(t, \theta) = \theta_0 - \theta_1 t, \, \theta_1 > 0 \tag{2}$$
$$= \theta_0 - \theta_1 t + \theta_2 t^2 / 2, \, \theta_1 > 0 \tag{3}$$
$$= \theta_0 - \theta_1[1 = (1 - \theta_2)^t] / \theta_2, \, \theta_1 > 0, \quad 0 < \theta_2 < 1 \tag{4}$$

Using the maximum likelihood method and numerical approximation procedures, these models were applied to some data. The normal distribution was found to be the most appropriate among the three distributions, although it was noted that the choice of distribution was not crucial for forecasting purposes. It was also noted that the quadratic or exponential model did not provide a significant improvement over the linear model.

When estimating the limiting time (ultimate records) from exponential-decay model, the standard errors were so large that it made the predictions meaningless. Smith has implied that the choice of model may be inappropriate and acknowledged the wide variability of estimates corresponding to different error distributions and different portions of the series. In fact, he is doubtful that the use of such methods, in general, and model (4), in particular, can produce meaningful performance estimates for the distant future.

Although useful in some cases, additive models of the form (1) may not be appropriate for sports data, as they imply no relation between variation in $x(t)$ and change in $z(t,\theta)$. In fact, it is reasonable to expect decrease in variability in the latter portion of the data because performances are getting closer to the ultimate record and significant improvements are less likely now than, say, fifty years ago. Also, since most world-class runners remain competitive for a number of years (usually between three and six), we may see some dependency between neighboring performance measures.

Noubary (1994) has suggested the following models:

$$\log y(t) = \theta_0 - \theta_1 t x(t), \qquad \theta_1 > 0 \tag{5}$$

$$\log y(t) = \theta_0 - \theta_1 t + \theta_2 \log(t) + x(t), \qquad \theta_1 > 0 \qquad (6)$$

where $\{x(t), t = 1, 2, ...\}$ is a zero-mean stationary process. Note that models (5) and (6) can alternatively be written in multiplicative form as

$$y(t) = \theta_0^* \, e^{-t\theta_1} x^*(t) \qquad \theta_1 > 0 \qquad (7)$$

$$y(t) = \theta_0^* t^{\theta_2} e^{-t\theta_1} x^*(t) \qquad \theta_1 > 0 \qquad (8)$$

where $\theta_0 \log \theta_0^*$ and $x(t) = \log x^*(t)$ These models imply that both means and variances vary with time—that is, unlike the additive model where variation in $y(t)$ is independent of t, here, the variation decreases as t increases. As a result, the standard errors of the future records are smaller than those of an additive model, and, therefore, the likelihood of obtaining a meaningful prediction is higher.

Extremes and Records

There are many instances where extreme values and values with low frequencies are of main concern. Examples include large natural disasters compared to moderate ones, weak components or links compared to their average counterparts, large insurance claims compared to average claims, and extraordinary sport performances.

Extreme values are usually analyzed using one of the three major theories:

1. the extreme value theory, which deals with maxima or minima of the subsamples;
2. the threshold theory, which deals with values above or below a specified threshold; and
3. the theory of records, which deals with values larger (smaller) than all the previous values.

The frequencies of extremes are analyzed using the theory of exceedances.

This theory deals exclusively with number of times a chosen threshold is exceeded.

In most applications, the number of exceedances and values of excesses over a threshold are combined to yield a more detailed analysis. For example, insurance companies analyze both the number of times a "large" claim is made and the amount by which these claims exceed a specific large threshold. Reliability engineers study both the number of extreme loads, such as large earthquakes and their magnitudes.

Exceedances

In sport, extraordinary individuals or teams are those who perform significantly better than expectation and pass a threshold for average and ordinary. For example, in one-hundred-meter race, only male athletes who run the distance in less than ten seconds may be considered exceptional. In basketball, player with a point per minute or assist per minute greater than 2/3 may be referred to as exceptional. In tennis, players having more than forty-five aces in one match may be considered exceptional. If we examine the distribution of times for one-hundred-meter race for the runners or points per minute (PPM) for basketball players, then we find that the exceptional performances constitute the tail of the corresponding distribution (for one-hundred-meter times, the lower tail; and for PPM, the upper tail). Thus, to study the records or to predict them, rather than entire distribution covering the entire range of possible values, we may just study the upper (lower) tail of the corresponding distribution. We can do this either by seeking a model for values above (below) a specified threshold or by considering a past performance (e.g., the third best performance) in the history of the sport of interest and study the number of times it was surpassed or exceeded.

In this chapter, the second approach is discussed. Here, exceedances are used to answer questions such as, how many (basketball players) would it take to produce a player that will perform like or better than, for example, Michael Jordan?

In many situations in sports, instead of records, we are interested in the events associated with the exceedances of certain performance measures of the variable under study. For example, in sports such as long jump, one needs to jump a certain length to qualify for a competition. This leads to deal with the frequencies (number of athletes) instead of the values on the random variable itself. For instance, we may like to know how many baseball players will have more than fifty home runs next season. Or how many of the players will surpass the performance of the third best player in the history of a given sport? Many authors have discussed exceedances and their applications.

The Theory

A theory known as the theory of exceedances deals with number of times a threshold is exceeded. As a result, it is a counting process. Some examples in sports include the following:

1. In track and field, athletes need to exceed specified qualifying levels (e.g., time, distance, or height) to be selected for certain competitions. We are concerned with the number that succeed.
   For instance, in the long jump, a distance of eighteen feet, nine inches will qualify for districts.
2. n sports such as basketball, we may be interested in the number of players who average more than m points per game.

Let Y be a random variable representing, for example, a performance measure in a given sport and x a real number representing the qualifying level. The event $Y = y$ is an exceedance of the level x if $y > x$. Now, assuming independent and identically distributed trials, what is the probability of r exceedances in the next n trials?

This is clearly a Bernoulli experiment with two possible outcomes: "exceedance" or "not exceedance." Since it is repeated n times, the number of exceedances has a binomial distribution with parameters n, $p(x)$, where $p(x)$ is the probability of exceedance of the level x of the variable under study. Note that, here, $p(x) = P(X > x) = 1 - F(x)$, where $F(x)$ is the distribution function of X. Using this, the probability of r exceedances of level x in the next n trials can be calculated from

$$\binom{n}{r} [1 - F(x)]^r \, F^{n-r}(x) \quad , 0 \leq r \leq n.$$

Now, suppose that, rather than a fixed level, we make the level x dependent on n, x_n, say. This means that we change the level with increase in frequency of events such as earthquakes. If we choose x_n in such a way that the following condition is satisfied:

$$\lim_{n \to \infty} n[1 - F(x_n)] = \tau; \, 0 \leq \tau \, \infty,$$

then the probabilities of r exceedances of level n_x can be approximated by those of a Poisson distribution (process) with parameter (rate) τ. This is because exceedances become rare events and the binomial distribution tends to Poisson.

Another interesting problem related to exceedances is the following: Assuming independent and identically distributed trials, determine the probability distribution of number of exceedances in the next N trials of the mth largest observation in the past n trials. This is useful when choosing a design load smaller than magnitude of some that have already occurred.

Suppose that p_m is the probability of exceedance of the mth largest observation in the past n trials, then it can be shown that pm has Beta distribution with density function

$$F(P_m) = \frac{P_m^{m-1} (1-p_m)^{n-m}}{B(m, n-m+1)}; \ 0 \le P_m \le 1.$$

Using this and the fact that the probability of r exceedances in the next N trial is binomial with parameters N and p_m, it can be shown that the mean and the variance of the number of exceedances of the mth largest observation in future N trials are respectively

$$\frac{Nm}{n+1}, \frac{Nm(n-m+1)(N+n+1)}{(n+1)^2(n+2)}.$$

Example. The yearly maximum number of points scored in high school basketball games during the last forty years was seventy. Then the mean and variance of the number of exceedances of seventy of the yearly maximum number of the points during the next thirty years are respectively

$$30/41 = 0.732 \quad \text{and} \quad (30)(40)(71)/(41)^2 (42) = 1.207.$$

If the second largest score was 67, then the mean and variance of the number of exceedances of this value in the next twenty years are respectively

$$(20)(2)/41 = 40/41 = 0.976 \quad \text{and}$$
$$(20)(2)(39)(61)/(41)2(42) = 1.348.$$

Example. Consider the yearly maximum score during the last sixty seasons. As a goal in our school, we want to choose a level to have a mean value of four exceedances in the next twenty years.

Using the formula for the mean, we get $20m/61 \approx 4 \Rightarrow m \approx 12$. This means that the value to be chosen is the twelfth largest score (order statistic) in the data.

Now, suppose that $m=1$. It follows from (1) that

$$f(p_1) = n(1-p_1)^{n-1}, \ 0 \le p_1 \le 1.$$

For this distribution, the mean and variance are respectively

$$\frac{1}{n+1}, \quad \frac{n}{(n+1)^2(n+1)}.$$

Noting that for a relatively large n, the variance is small (e.g., for fifty years of data, it is equal to 0.0003696), we may replace p1 with its expected value so that the probability of exceedances in the next trial is $\frac{1}{n+1}$. Using this, the probability of no exceedances in the next N trials is

$$(1-\frac{1}{n+1})^N.$$

Example. Recall the basketball scores mentioned above. We have $n = 40$ and $N = 30$. Thus, the required probability is $(1-(1/41))^{30} = 0.4767$, and the probability of at least one exceedances is 0.5233.

Note that a good approximation for $(1-\frac{1}{n+1})^N$ can be found if we let $N = h(n+1)$. This gives

$$[(1-\frac{1}{n+1})^{\frac{1}{n+1}}]^h \rightarrow \exp(-h) = \exp(-\frac{1}{n+1}).$$

In the above example, h = 30/41 and exp(-30/41) = 0.481.

The above approximation was based on replacing the expected value for the probability of no exceedance. Since the distribution of p_1 is not symmetric, one may prefer to replace the mean (expected value) with the median—that is, consider $1-2^{-\frac{1}{n}}$ in place of $\frac{1}{n+1}$. This gives $2^{-\frac{N}{n}}$ in place of $(1-\frac{1}{n+1})^N = \exp(-\frac{1}{n+1})$. Continuing with the basketball example, $2^{-\frac{30}{40}} = 0.595$.

Finally, suppose that K is the number of trials up to the first exceedance. The possible values for K are 0, 1, 2, . . ., and we have

$$P(K = k) = \frac{1}{n+1}(1-\frac{1}{n+1})k, \quad k=0, 1, 2, \dots .$$

This is a geometric distribution with expected value and variance equal to respectively

$$n+1, n(n+1).$$

This means that to have an exceedance, we need an average of n+1 trials.

Return Periods

Assume now an event such that its probability of occurrence in a unit of time (normally one year) is p. Assume also that occurrences of such an event in different periods are independent. Then, as time passes, we have a sequence of equally likely Bernoulli experiments (only two outcomes: [a] occurrence or [b] not occurrence, are possible). Thus, the time (measured in unit periods) to the first occurrence is a geometric random variable with parameter p with mean value 1/p. This motivates the following definition.

Definition. Let A be an event and T the random time between consecutive occurrences of A events. The mean value, τ, of the random variable T is called the return period of the event A.

Note that if F(x) is the distribution function of the yearly maximum of a random variable, the return period of that random variable to exceed the value x is 1/[1-F(x)] years. Similarly, if F(x) is the distribution function of the yearly maximum of a random variable, the return period of the variable to go below the value x is 1/F(x) years.

Example. Suppose that the distribution function of the yearly maximum in girls' high school basketball games has the following Gumbel (type I extreme value) distribution:

$$F(x) = \exp\left[-\exp\left(-(x-38.5)/7.8\right)\right].$$

The return periods of yearly maximum score of sixty and seventy are respectively

$$\tau_{60} = 1/(1- F(60)) = 16.25 \text{ years}$$
$$\tau_{70} = 1/(1- F(70)) = 57.24 \text{ years}$$

Exceedances and English Premier League

Astrophysicists at the University of Warwick studying the extreme variability in X-rays emitted from matter falling into black holes have discovered that their research methods also show that the world's top-division football (soccer) matches have an unusually large proportion of high scoring games—so much so that international football (soccer) actually shows a pattern of "extreme events" similar to that seen in the large bursts of X-rays from the accretion discs of black holes. However, analysis of just English Premier League and Cup football (soccer) games showed that English top-division football (soccer) is in fact thirty times less likely to have high-scoring games than the rest of the world taken as a whole and could thus be seen by some people as thirty times more boring.

"External events"—that is, large events that are more likely than would be expected from a random process—can be a signature of complexity in nature. In the case of matter moving in accretion disks around black holes, it tells us about the turbulent flow in the accreting matter.

Whist seeking to compare this distribution of events with other patterns in the world around us, two University of Warwick postgraduate physics students (John Greenhough and Paul Birch), with their supervisor Sandra Chapman and colleague George Rowlands, looked at the number of goals scored by the home and away teams in over 135,000 games in 169 countries since 1999 and found that the results followed the pattern of "external statistics."

However, they also compared their data with an analysis they made of the scores of thirteen thousand English top-division games and five thousand FA Cup matches between the 1970/71 and 2000/01 seasons. They found that these scores contained far less high-scoring games than the world as a whole, and rather than fitting an external statistics pattern, the English games more closely fitted either Poisson or negative binomial distributions.

In summary, their analysis revealed that a total score over one hundred goals in any one game occurs approximately only once in every ten thousand English top-division matches (once every thirty years); but in top-division matches worldwide, such a score is seen once in three hundred games (about once every day).

How Fast

I have worked for years to combine my interests in mathematics, statistics, and sports. After Beijing Olympics and Berlin competitions, I was approached by press around the world because of work I have done on prediction of sport records and their relevance to records set in sprint events recently.

In the last two decades, the athletics performances and related issues have received considerable attention by physiologists, physical educators, and the public. One aspect of interest has been the improvement of the performances over time to address the question of predicting the future performances.

Research in this area includes short-term prediction (e.g., what would be the next Olympic record?) and long-term prediction (e.g., what would be the ultimate record?). I have done some research in both areas and have developed methods for prediction of both. My works have some advantages over most existing methods in two directions:

1. They include margin of error. For example, I have estimated that with a 95 percent confidence, the ultimate time for one-hundred-meter dash is 9.44 seconds.
2. They go far beyond trend analysis and utilize recent developments such as theory of records, threshold theory, and tail modeling, and their modified versions.

The remarkable performances of Usain Bolt in the two-hundred-meter and one-hundred-meter sprints during the Beijing Olympics in 2008 and Berlin competitions in 2009 have prompted many, especially United States Olympic committee, to focus on predicting records. They are very interested to know just how surprising Bolt's performances in one hundred and two hundred meters were. They have found my work interesting and useful. They are working on several projects with the aim of finding (among others) an improved ability to know whether a given USA athlete (or program) is on track to be competitive against top international competitors, and if so, how is it possible to establish more meaningful and realistic goals and performance interventions, and perhaps to be more objective and informed about their funding decisions. Some of these studies aim at finding answer to the following questions:

1. What is the short-term prediction (e.g., next Olympics, for the fastest time in a one-hundred-meter sprint)?
2. What is the long-term prediction or ultimate record for one-hundred-meter sprint? My prediction was much lower than previously held beliefs (he believes that the ultimate time is 9.44 seconds). And, last year, Usain Bolt had a record-breaking one-hundred-meter sprint with a time of 9.69 seconds during the Olympic Games in Beijing, and this year an amazing time of 9.58 seconds during the Berlin competitions.

My methods for calculating this minimum time have been of interest to the United States Olympic committee. My methods are different from most other predictions because they are accompanied with a margin of error. The United States Olympic committee is interested in a program that would compare U.S. runners with runners from other countries and how they can interfere with an athlete's performance to make him more competitive. My methods provide a lower bound for how far they can expect athletes to go. I have been interested in the tail of the distribution for data. I have used the theory of records, threshold data, and tail modeling to make my predictions.

Once the gun sounds, Usain Bolt seems to test the very limits of the race—the human race. The Jamaican sprinting sensation put on another amazing performance at the world championships on Sunday, shattering his own record in the one hundred meters by 0.11 seconds to take it down to an almost inhuman 9.58 seconds. Maybe "inhuman" is a bit too strong, but the man is certainly on another level.

Note that this result was the biggest improvement in the one-hundred-meter record since electronic timing began in 1968. Bolt's not done either. A few days later, he cruised into the semifinals of the two hundred meters, and he also figures to lead his nation's four-hundred-meter relay. At that, who knows how low he can go? He's certainly willing to try.

"Personally, I think I have more work to do," Bolt said after winning the one-hundred title at the same Olympic Stadium where Jesse Owens won four gold medals at the 1936 Berlin Games.

Several researchers have done studies recently to predict how fast a man can run one hundred meters. The latest from Tilburg University in the Netherlands predicts that someone will eventually break the tape at 9.51 seconds. Bolt, who has set three records at the one-hundred-meter distance with times of 9.72, 9.69 and 9.58, is already looking to rip that theory apart.

"I said 9.4," Bolt said. "I think the world records will stop at 9.4."

"You'd have to think he can't keep getting faster, but we wouldn't put it past him," Ladbrokes spokesman Robin Hutchinson said in a statement. Bolt became the premier runner in the world when he ran 9.72 in May 2008—only two and a half months before the Olympics, where he lowered the mark to 9.69. At the Bird's Nest in Beijing, he outdid himself, easing up at the end of the one hundred, mugging for the cameras even before he showboated across the finish line—and also setting records in the two-hundred and four-hundred relay.

NOTES: one indicator of getting close to limit is that amount of improvements should decrease. This is not true both for one-hundred- and two-hundred-meter data.

Conclusion

A procedure is developed for calculation of probabilities of future athletic records based on modeling the tail of the distribution for performance measures. The results obtained for men's long jump and four-hundred-meter run are reasonable. For long jump, the last two records, 8.95 and 8.90, are significantly greater than the third-best record, 8.35, indicating a medium or long tail model. Data regarding the performance of Usain Bolt have a similar characteristic, as his Olympic times are significantly lower than his performances prior to that. Additionally, the use of medium or long-tailed models avoids difficulties posed by intractable likelihood equation and the complications regarding ultimate records and their estimation.

CHAPTER 5

Mathematics of Risk

A traveler who refuses to pass over a bridge until he has
personally tested the soundness of every part of it is not likely
to go far; something has to be risked, even in mathematics.
—Horace Lamb (1849–1934), English mathematician
And one day I realized that not taking a risk was the greatest risk of all.
Insurance, a profession based on mathematics of risk.

WE ALL TAKE risk, not because we want to but because we have to. As such, we all need to know something about how to evaluate the risks involved in our decisions if we wish to make appropriate choices. After all, risk pervades virtually all areas of human endeavor, whether these endeavors be for personal, social, commercial, or national purposes.

Risk is complicated. What makes it even more complicated is the fact that there is no universally accepted definition for it. Here are a couple of commonly accepted definitions:

- Risk is a situation or event where something of human value is at stake and where the outcome is uncertain.
- Risk is an uncertain consequence of an event or an activity with respect to something that people value.

Basically, these definitions express risk in terms of uncertainties and their consequences rather than quantities representing them. There is a general agreement that risk should be assessed using mathematical quantities so that we can use them for evaluation and comparison.

Mathematical Expressions of Risk

Let us start with the most general mathematical expression of risk—namely,

Risk = [Prob.{event occurs}] x [probable cost if event occurs].

With this in mind, three types of risk models can be distinguished:

a. Risk associated with the uncertainty of the occurrence of an undesired event and its fixed or deterministic consequences.

b. Risk associated with the uncertainty of the magnitude of the consequences of the fixed or deterministic occurrence of an undesired event.

$$\text{Risk II} = [\text{Prob.}\{\text{event}\} = 1] \times [\text{P}\{\text{cost/event}\}]$$

c. Risk associated with the uncertainty of the occurrence of an undesired event and the uncertainty of the magnitude of its consequences.

Summarizing these, we may formulate them as

$$Risk = [Prob.\{event\}] \times [P\{cost/event\}]$$
$$Risk = Probability \times Severity$$

From these definitions, it is evident that there are two basic components of risk:

a. a future outcome that can take a number of forms, some of them unfavorable.

b. a nonzero probability indicating that such unfavorable outcomes may occur.

Here are some other definitions involving these two components:

A. Risk is the chance of loss.
B. Risk is the possibility of loss.
C. Risk is uncertainty.
D. Risk is the dispersion of actual from expected results.
E. Risk is the probability of any outcome different from the one expected.
F. Risk is the probability of an event times the cost (or loss) if the event occurs.

Each of us may prefer one over the others depending on our understanding, intuition, and experience, and the type of the media we follow.

Risk and Intuition

Making individual decisions in the face of uncertainty reflects our response to loss and reveals a great deal about our personality as well as our intuition about the risk. Most things we do in life involve some degree of uncertainty. As such, the return of our investments, whether it is money, time, love, etc., may or may not meet our expectation. So, in this sense, it is not unreasonable to think of life decisions as a form of gambling. Some decisions such as investing in education usually have a long-run positive return. A few decisions such as buying fire insurance or lottery may, on the other hand, have a long-run negative return. The latter decisions are usually made based on utility, which includes other types of gains such as peace of mind, fun, excitement, joy, and dreams (e.g., lottery).

Note that when making a decision, we may not actually compute probabilities, but most of our decisions rely on our intuition about probability. This intuition has been cultivated over many years based on a wide range of experiences. So it might come as a surprise that this intuition can be remarkably unreliable. To clarify, consider the following classical decisions:

Decision 1. Which would you choose?

Option A: 100 percent chance of gaining $250
Option B: 25 percent chance of gaining $1,000, 75 percent chance of gaining nothing

The expected win (gain) is 250 x 1 = $250 for option A and 1000 x 0.25 + 0 x 0.75 = $250 for option B. So the expected gain is the same for both options. However, in a careful study, 84 percent of respondents chose option A, preferring to take a sure gain (a bird in the hand) rather than to risk no gain at all.

Decision 2. Which would you choose?

Option A: 100 percent chance of losing $250
Option B: 25 percent chance of losing $1000, 75 percent chance of losing nothing

Expected loss of $250 for both cases. However, in a careful study, 80 percent of respondents chose option B.

Another related study puts these results in a different light when small probabilities are involved.

Decision 3: Which would you choose?

Option A: 100 percent chance of winning $5
Option B: One-in-one-thousand chance of winning $5,000

Decision 4: Which would you choose?

Option A: 100 percent chance of losing $5
Option B: One-in-one-thousand chance of losing $5,000

Options have the same expected return (a gain of $5 in decision 3 and a loss of $5 in decision 4). However, with seventy-five respondents, 75 percent chose option B in decision 3 and 80 percent chose option A in decision 4.

Now, here is what is interesting. With small probability (one in one thousand), we see different patterns. In decision 3, the respondents believed the one-in-one-thousand chance of a gain was large enough to be worth taking; in decision 4, they believed the one-in-one-thousand chance of a loss was large enough to be avoided.

Combining this finding with the previous example, we have a general rule that, at least in a gambling context, people tend to be risk-seeking to avoid a loss and risk-averse to protect a gain. This is true at least when the probabilities are not extremely small. And all these reveal facts about our intuition about risk.

Media Effect

Studies have found that people's attitudes toward risk are often formed based on the coverage of these risks in their media. To support the idea, one study examined the coverage by newspapers in New Bedford, Massachusetts, and Eugene, Oregon, and found the following patterns:

Although diseases take about sixteen times as many lives as

accidents, the newspapers contained more than three times as many articles about accidents. Although diseases claim almost one hundred times as many lives as do homicides, there were about three times as many articles about homicides as about disease deaths. Furthermore, homicide articles tended to be more than twice as long as articles reporting disease and accident deaths.

The people who read these newspapers assessed the risk of death by homicide as much greater than the risk of death by accident, which, in turn, was much greater than the risk of death by disease. Although this assessment was incorrect, nevertheless, it was a correct interpretation of the deaths about which they read. In other words, the readers' number sense was working correctly on bad information.

Another study investigated the ordering of the perceived risk by two groups of ordinary citizens and the experts regarding certain risky activities and technologies. Out of twenty such activities, ordinary citizens ranked nuclear power as the riskiest; whereas, experts ranked it the least risky. The deviation from experts' ordering is believed to be partly because of the news coverage and the group membership.

Risk Perceptions

1. In a study to demonstrate our perception about the risk, students at the University of Oregon, recruited by an ad in the student newspaper, were randomly given form I or form II of a questionnaire about vaccination.

Form I described a disease expected to affect 20 percent of the population and asked people whether they would volunteer to receive a vaccine that protects half of the people receiving it.

Form II described two mutually exclusive and equally probable strains of the disease, each likely to affect 10 percent of the population. The vaccine would give complete protection against one strain of the disease and no protection against the other.

Even though both vaccination forms proposed no risk to the

volunteers and offered a reduction of the probability of the disease from 20 percent to 10 percent, the wording or "framing" of the situation led to a "yes" response of 57 percent for form II but only 40 percent for form I.

A different study published in *Scientific American* explored at length the role that the framing of a choice plays in the resulting decision. Compare, for example, the following pairs of choices:

You have choice between

    A.   collecting $50 for sure, and

    B.   gambling with a 25 percent chance of winning $200 and a 75 percent chance of winning nothing.

You have choice between

    C.   losing $150 for sure, and

    D.   gambling with a 75 percent chance of losing $200 and a 25 percent chance of losing nothing.

Most people prefer A to B and D to C, even though the expected return of A and C are identical and the expected outcomes to B and D are identical.

    2.   In a similar study, the participants were offered an option of paying $1 for a one-in-one-million chance of winning $1 million. Nearly half of the respondents accepted the offer, including a large number of risk-averse individuals. The study also showed that replacing the offer with $1 for, for example, one-in-ten-million chance of winning $10 million led to a significant increase in number of participants in general and risk-averse participants in particular. This explains why the recent lottery attracted so many people whose real reward was, in fact, sweet dreams of having that much money.

3. Finally, here is an interesting example. The winning ticket in a lottery was 865304. Three individuals compare the ticket they hold to the winning number.

John holds 361204, Mary holds 965304, and Bill holds 865305. All three lost, since the chance of winning was only one in a million. Why, then, was Bill very upset about losing, even more so than Mary, and John was only mildly disappointed?

A Working Definition of "Risk"

We purchase fire insurance to protect ourselves against the consequences of a severe loss; in deciding to insure against the risk of fire, many of us consider only the magnitude of the possible loss and pay little attention to its probability. In fact, we may have no estimate of the probability that a fire loss will occur. In some other situations, we may equate "risk" with "probability," as in the following statements:

- The risk of a broken leg is greater for an inexperienced skier than for an experienced skier.
- The risk of death or injury in automobile accidents is greater for those not wearing seat belts than for those who are.

Lack of a fixed definition for risk may not be troublesome in our individual decision-making, but it can be a hidden source of disagreement when groups are comparing the magnitudes of various risks.

Those scholars and scientists who practice systematic risk assessment stipulate the following: Risk deals with the uncertainty of a possible loss or damage, but the risk is not the same as the uncertainty, nor is it the same as the loss. A correct characterization of risk involves both. The *risk* of a particular undesired event is the *expected loss* associated with the occurrence of the undesired event. In mathematical terminology, this characterizes risk as an expectation (i.e., a product of a probability and a numerical consequence, if numerical values are available)[1]. A formula for calculating the risk associated with a particular undesirable event is

$$\text{Risk} = \begin{array}{c} \text{probability} \\ \text{of event} \end{array} \quad \text{x} \quad \begin{array}{c} \text{probable cost} \\ \text{if event occurs} \end{array}$$

For example, if the probability that our $80,000 home is destroyed by fire in a given year is 1/2000 or 0.05 percent, then

$$\text{Risk due to fire } (0.0005) \times (\$80,000) = \$40.$$

This calculation shows how the probability and the outcome are combined to give a measure of the "risk."

Risk and Investment

Financial advisors often try to evaluate investor's tolerance for risk. But, what do they mean by risk? Here are the main issues:

1. How do we define risk?
2. How do we measure (quantify) risk?
3. Whose views of risk should we consider?
   a. individuals? public? experts?
   b. Are individuals rational when assessing risks?
   c. Are experts always right?
   d. Is perception of public reliable?

Other issues that require attention include risk estimation, risk management, and risk sharing (insurance).

Numerous factors affect markets and investing and require basic understanding of the risks involved. The risk of a security has two main components: unique risk and market risk. Investors with a low tolerance for risk try to reduce or avoid it by actions such as *diversification*. Unique risk can be *reduced* or even *eliminated* by holding a well-diversified portfolio. Market risk associated with market's wide variations *cannot be removed* by individual action. As such, the risk of a fully diversified portfolio is the market risk or its sensitivity to market changes. This sensitivity is generally measured by a quantity known as beta (β).

A security with $\beta = 1.0$ has average market risk; a well-diversified portfolio of such securities has the same risk as the market index. A security with $\beta = 0.5$ has below-average market risk—a well-diversified portfolio of these securities tend to move half as far as the market moves and has half the market's risk. For the security with $\beta = 1.0$, if the market index changes 20 percent, its return (or price) would change 20 percent too. For the security with $\beta = 0.5$, if the market index changes 20 percent, its return (or price) would change 10 percent. For the security with $\beta = 2.0$, if the market index changes 20 percent, its return (or price) would change 40 percent. The regression analysis provides an effective approach for finding β.

History and Perception

Risk can be associated with many facets of life from health risks to business risks to risks related to natural disasters. As risk theory is examined, it becomes evident that often people have different perceptions of risk; what may seem catastrophic to some may be of only minor concern to others. With these discrepancies, it is necessary to see how risk is actually defined and then question if an uniform definition even exists. This discussion will then lend itself to four different methods for handling risks. Throughout this report, examples relating to the risks associated with natural disasters and business ventures will be used. Finally, some of the mathematics behind risk assessment will be reviewed.

In discussing risk, there is often difficulty because, as mentioned, individuals have different perceptions of risk. For example, one individual might consider hurricanes to be the most risky natural disaster, while another person might consider earthquakes to be the riskiest. Surely, a Californian would be likely to follow the latter line of thinking. A variety of classifications for risk could be given to help reveal how one labels something to be "risky."[1] The combination of some of the following factors may play a role in risk classification. Some of the factors are whether something is catastrophic or minor, controllable or uncontrollable, and direct or indirect. Other characteristics that are often evaluated are whether the event is familiar or new, equitable or inequitable, internal or external, and temporary or permanent. Still, other items that are considered in classification are whether something is expected or unexpected, fatal or nonfatal, general or specific, interactive or independent, instantaneous or gradual, and, finally, whether the individual is affected voluntarily or involuntarily. Some suggest that the knowledge theory can explain some perceptions. Restated, one person might consider something to be risky because they "know" it to be risky. For example, those who have been in motorcycle accidents might consider motorcycles to be more risky than automobiles, while experts would tend to differ.

A survey was done to see why people have different perceptions of risk. The population sample included college students, a League of Women Voters, and experts. These people were asked to rank a list of technologies and activities according to the risk associated with each one. The survey included items such as nuclear power, motor vehicles, swimming, firefighting, large construction, motorcycles, and commercial aviation. Interestingly, the majority of college students and League of Women Voters ranked nuclear power as most risky. In contrast, motor vehicles were at the top of the experts' list as most risky. Eighty-four percent of college students and League of Women Voters answered similarly, while only 35 percent of the League of Women Voters and the experts had matching answers.

The media also plays its role in shaping perceptions of risk. Society often makes the association that what happens in the news is the reality of a situation. So, when airplane accidents are given more news coverage than automobile accidents, conclusions are often drawn. Individuals may tend to conclude that airplane accidents occur more frequently than car accidents or, better yet, that airplane travel (commercial) is more risky than travel in motor vehicles. In reality, more automobile accidents occur annually than airplane crashes. The above mentioned survey also provided the expert ranking that motor vehicles are indeed more risky. In another case, it is found that newspapers typically print more articles about death by homicide than death by disease. As a result, readers might assess the risk of death by homicide as greater than the risk from disease, when in fact the opposite is true.

When trying to uncover what people mean when they say that something is risky, we begin to wonder, "What exactly is risk?" The following are some of the common textbook definitions identified: Risk is the chance of loss. Risk is the possibility of loss. Risk is uncertainty. Risk is the dispersion of actual from expected results. Risk is the probability of any outcome that is different from the expected one. Risk is the condition in which a possibility of loss exists. Risk is the potential for the realization of unwanted negative consequences of an event. Risk is the probability of an undesired event times a projected cost of the negative consequence if the undesired event would occur. Risk for business ventures is the beta value for a particular security.

These definitions have varying degrees of usefulness. In fact, most individuals have difficulty providing a single definition for risk. Some of the underlying ideas behind most of these definitions are the existence of some uncertainty about an event and the potential for loss, whether it is financial, material, or otherwise. Without the existence of a commonly accepted definition of risk, often researchers of risk have different frames of reference. This is not to say that we necessarily need a singular definition, but it must be acknowledged that there is some disagreement. Some of the definitions given seem rather vague using words such as "uncertainty" and "chance," while other definitions lend themselves rather well to mathematical interpretation. For example, risk as the dispersion of actual from expectation is better known to mathematicians as variance. One might be familiar with expected return, which is the mathematical name for the definition that begins with "risk is the probability of an undesired event times . . ." The mathematical assessment of risk will be addressed in more detail later in this paper.

When people think that they are "at risk" or "facing a risk," there are different ways to handle the situation regardless of your definition of risk. Below are four methods for handling risk: avoidance, retention, reduction, and transfer (insurance). Avoidance is when an individual realizes that a risk is present and tries to avoid it. The individual does not want to suffer the negative consequences of an event, and, therefore, he or she does not participate. The person tries to remove himself of herself from the risky situation. For example, to avoid losing money in a business venture, simply do not get involved at all. To avoid losing money on the stock market, simply do not buy stock. With the case of natural disasters, avoidance is slightly more difficult. The avoidance of one type of disaster may be possible by moving to a different location, but in doing so, he or she may be exposing himself or herself to other disasters. For example, people can avoid the risk of a flood by relocating to a nonflood plain region. Unfortunately, the new region may be prone to hurricanes or earthquakes. With Mother Nature ultimately in control, avoidance is not always the best solution, particularly when it is done by means of evacuation. This procedure is costly and relies heavily on prediction techniques that are done with a degree of uncertainty.

Another option for handling a risk is retention. People can voluntarily retain the risk in some instances; but in other cases, it is involuntary. Regardless of whether it is voluntary or involuntary, risk retention means that the individual is responsible for loss if it occurs. When people agree to retain the risk, they are accepting to bear the consequences if the undesired outcome results. For example, in business ventures where stock is purchased, the risk is losing money. If a person willingly purchases stock, then he or she has practiced risk retention. That means that the person must bear the losses that may be suffered by undesired drops in stock prices. With regard to natural disasters, risk retention is practiced by most people. Without a form of insurance to protect people from natural disasters, most people simply hope that they will not suffer loss because of a natural disaster; but if a loss does occur, they are responsible. In some cases, government relief money has been allocated to help bear the loss of some disasters such as flooding, but too often that is the exception, not the norm.

A third method for handling particular risks is risk reduction, which is often associated with preventative measures, which will reduce the chances of the negative consequence from occurring and reduce the losses if the event does occur. Preventative measures can be easily described using the example of health risks. People have begun to eat foods lower in fat and cholesterol to reduce their chances of suffering a heart attack. In the business world, investors can help to reduce the risk of their stocks by purchasing a portfolio of stocks, which is a collection of stocks that have varying measures of riskiness. Buying several stocks rather than a single stock provides a balance; in case one stock may drop dramatically, we can only hope that one of the other stocks would increase considerably.

Risk reduction for natural disasters has been the topic of great interest since the beginning of this decade because the United Nations declared the 1990s as the International Decade for Natural Disaster Reduction. Considering that a natural hazard in and of itself is not problematic, we realize that problems do arise when humans enter the picture. If the natural phenomenon such as the earthquakes, volcanoes, landslides, and hurricanes occurred in uninhabited areas, they would be of little societal consequences. With a quadrupling of the world's population during the twentieth century, hazards are striking heavily populated areas and disasters are becoming more frequent and more severe. Recall, the 1954 flood in China, which resulted in approximately forty thousand deaths, and similarly, in 1988, an earthquake rumbled in what used to be USSR, and the death toll was well above forty thousand.

Risk reduction for natural disasters can be achieved through engineering technologies and improvements in the structures of buildings and infrastructures. In California, there are certain building codes that must be met before a building will pass inspection, and some of these pertain to making buildings "earthquake-proof." This measure actually relates to reducing the consequences of a disaster. The decade has certain goals; "it aims to reduce through appropriate action, the loss of life and property damage due to natural disasters." This will be a global effort and encourages people to work together. Such a project will test our ability to accept that people have different perceptions of risk, particularly when funds are allocated or research is done for specific hazards that some (but not all) people face. For example, consider the problems that may arise when the people of Bangladesh want the problem of flooding addressed while Californians want earthquake issues addressed. To achieve the goals of the decade, we must all be sensitive to one another's motives and beliefs, which may have a strong impact on our attitudes toward risk and risk reduction.

Looking back, we can try to reduce the risk or reduce the loss, or both. More often than not, with regard to natural disasters, reducing the loss is the best we can hope to do, since, once again, Mother Nature is ultimately in control.

The final method for handling risk that will be addressed is risk sharing, which is often referred to as insurance. The basic concept behind insurance is that many people face a situation where there is the potential for an undesired outcome. Ironically, only a select few will actually suffer the undesired consequence or loss. To help share the loss and make it easier for the few to bear, the entire group concerned pools their money together (collected by and insurance company), and then it is redistributed. The money is handed out as needed to those who actually suffered a loss. Today, there are many different types of insurance available. Everything from health insurance to life insurance to fire insurance to car insurance to earthquake insurance can be purchased in some areas. For example, in California, on September 21, 1990, Governor Deukmejian signed into law a bill that mandates earthquake insurance for all residential property owners. In flood plain regions such as Bloomsburg, Pennsylvania, flood insurance can be bought from an insurance company through the National Flood Insurance program here in specific parts of the United States. Unfortunately, natural disasters are not covered under traditional homeowners' insurance. Although it seems that for additional charges, some other disasters can be insured.

To help share the losses resulting from natural disasters, proposals have been made suggesting a type of international insurance for disasters. This is a hopeful idea that I do not expect to see materialize for many, many years. Part of the problem is a lack of information about past disasters. The historical records are incomplete to make accurate estimations about the predicted amount of disasters in a given time frame. With the help of mathematics, some of this information is becoming available as models are being formulated to fit the data of the past.

Considering risky business ventures, risk sharing can be accomplished by having several investors involved in an investment. As a result, any loss will be divided among all the investors. Unfortunately, just as all the losses would be shared, so too would all the profits be shared.

The handling of risk often goes hand in hand with risk assessment. Intuitively, we can agree to retain those risks that are not too risky. The description that one option is more risky than another could better be accepted if these ideas could be quantifiable. In fact, there are several ways for quantifying risk. The simplest measure for risk is expected value (also called expected return). This defines risk as

RISK = (probability of undesired event) x (cost of loss).

If the expected value for two situations is calculated, then a conclusion can be made about the comparison of risk between the two. The condition that yields the larger expected return is less risky than one with a smaller expected return. Consider the following two lottery options:

Lottery A: 50 percent gain $100; 50 percent lose $30
Lottery B: 60 percent gain $150; 40 percent lose $80

To evaluate which lottery option is more risky, calculate the expected return.

$$E(A) = (0.5)(100) + (0.5)(-30) = \$35$$
$$E(B) = (0.6)(150) + (0.4)(-80) = \$58$$

After comparing the expected returns, we can assess that the first lottery is more risky than the second one; therefore, the second one is less risky than the first one. Looking at another example, we can see the weakness of using the expected value. Consider there is a 50 percent chance of losing $1,000 and a 50 percent chance of gaining $1,000. In another situation, 50 percent, lose $1; and 50 percent, gain $1. Looking at the expected returns, they are the same.

$$\#1: (0.5)(1,000)+(0.5)(-1,000)=0$$
$$\#2: (0.5)(1)+(0.5)(-1)=0$$

In a real-life situation, most people would agree that there is more risk in possibly losing $1,000 as opposed to $1.

Realizing the shortcomings of this method, there is another way to calculate risk using what is known as variance. Variance is defined as the dispersion of actual from expected results. This can be more easily explained as how much the data varies from a given point. Looking at business dealings involving stocks, the variance of a stock describes how much the stock fluctuates. If a stock has a large variance compared to another stock, then the initial stock is said to be more risky than the second. Having a large variance implies that the stock has a past record of increasing in price and then decreasing considerably, and this pattern of change continues to persist again and again.

Calculations can also be made to assess risks apart from the business world. Particularly, the risks associated with earthquakes can be examined. When trying to evaluate earthquake occurrences, it is useful to predict the number of earthquakes that will occur in a particular area during a specified period based on the historical records of the fault. This can be done using the Poisson process because the occurrence of earthquakes follow the Poisson distribution.

Overall, risk assessment can be paramount in evaluating different situations and decision-making. It does not dictate or prescribe a particular behavior but helps to more clearly illustrate a circumstance to those studying the risk of a particular situation.

Interestingly, people continue to use personal preferences to make decisions even when situations are deemed to be very risky. This may seem ironic, but some people are attracted to risks more than other people. This is not to say that one is better or worse than the other. Risk assessment simply allows for better educated and informed decisions.

In conclusion, risks are everywhere and cannot be entirely ignored. We must try to expand our ideas in an attempt to understand other people's perspectives and work toward addressing those risks affect not only our lives but also the lives of other people. We must make a conscious effort to handle risks appropriately and try to reduce when the situation calls for such action. The actions of those in science and technology have led to advances in prediction and risk assessment that have helped to save lives. I hope that people will become more aware of the risks that they take or the risks that they face.

Public Views of Risk

In this part, different expressions of risk together with policies for adapting to risk, quantifying risk, and public views of risk are briefly discussed. Some examples involving medical risks are also included for demonstration. Topics often include policies for adapting to risks, including perceptions of risk; quantifying risk, including defining risk, mathematical expressions of risk, risk assessment, and assessment by words; and public views of risk, and medical risks, including medical risks (some examples), a clinical decision, and life expectancy criterion.

When specifying acceptable levels of risk, the public and their political representatives often have a different perception of the relative risks from what is supported by the facts. For example, nuclear power is considered a major risk by members of the public; whereas, according to experts, the risks are much less than those associated with swimming, cycling, or traveling by train. The ordering is based on the geometric mean risk ratings within each group. Rank 1 represents the most risky activity or technology. Note the difficulty here of comparing like with like.

One implicit distinction made in the public assessment of risk is between *voluntary risk* and *involuntary risk*. We all accept voluntary risk when crossing the road or driving a car. The advantages are thought to justify the risk. However, the risks and the consequences suffered by the people of Bhopal in India when 2,500 people died as a result of leaks from the Union Carbide factory are *universally condemned*.

Some risk is inevitable. Remember that not taking a risk may imply taking a risk of a different kind. In the field of medicine, whooping cough inoculations have saved many lives, but there is a finite risk in having the vaccination. With many new drugs that save lives, there is also some risk of damaging side effects. Many people choose to undergo major surgery, aware that there is a one-in-a-hundred chance of something going seriously wrong because they judge or are advised that without it their risks are much greater. The insect killer DDT is now generally deplored, but when banned in Sri Lanka in the early 1960s, when there was a raging and virulent epidemic of malaria, many people died unnecessarily because the means of controlling mosquitoes was prohibited.

Role of statistician. As analysis, statisticians should try to make available to decision makers the best possible estimates of risks. They should make them aware of the limitations of the estimates and of the need for better monitoring procedures. They should be able to expose and criticize bad emotionally based quasi-scientific estimates of risk. But they should not expect decisions to be made on the basis of their estimates alone; there are other legitimate factors.

Risk Analysis of Investment

In today's business, both managers and individual investors face risk and uncertainty. This article discusses the risk of the security by considering its two main components: the unique risk, which can be diversified by means of portfolio; and market risk, which cannot be removed by individual action. The measure and the method used are respectively the sensitivity of the stock change to market change and regression analysis. Data from Delta Airline and Disney company are used for illustration. A brief discussion concerning the recent developments on the subject is also included.

Introduction

Risk has a variety of definition. It usually refers to the probability of loss or the dispersion of actual from expected results. In the world of business, risk has a profound implication to the investors. So knowing how to deal with risk is very important. There are two kinds of risk for a security. One is the unique risk; the other is the market risk. In this article, these two kinds of risk, their measurement, diversification, and estimation, are discussed.

Unique Risk and Its Diversification

Unique risk stems from the fact that many of the perils that surround an individual company are peculiar to that company and perhaps its immediate competitors. Stock market is considered risky because there is a spread of possible outcomes. An usual measure of this spread is the standard deviation of the return, Note that the bigger the standard deviation is, the riskier is the security. Fortunately, the prices of different securities do not move exactly together. So we can select some kinds of securities to set up a portfolio to diversify the unique risk. In the following part, we will discuss how the portfolio works.

A portfolio is a group of securities. Suppose we have several types of stocks. We define an index involving each security's standard deviation and the correlations between that pair of securities, weighted by the proportions invested in each securities.

Here, correlation coefficient measures the degree to which the two stocks "covary." So if stocks in the portfolio are relatively independent, correlation would be zero. For a very large number of stocks, the portfolio correlation would tend to zero. Of course, this is only a theoretical assumption. In fact, for a portfolio with large numbers of stocks, correlation of any pairs of stocks cannot be zero.

Second, if we choose the stocks carefully, the correlation will be small and even a negative figure. In this case, the correlation will give us a deductible part from the average standard deviation, and, hence, the risk is diversified. Thus, we can conclude that the investors can choose some kinds of securities to set up a portfolio, and the portfolio can diversify the unique risk of the security.

Market Risk and Its Estimation

As discussed, the investors can eliminate unique risk by holding a well-diversified portfolio, so what is left is the market risk associated with market's wide variations. Thus, the risk of a fully diversified portfolio is the market risk.

A stock's contribution to the risk of a fully diversified portfolio depends on its sensitivity to market changes. This sensitivity is generally known as beta, denoted by -. A security with = 1.0 has average market risk; a well-diversified portfolio of such securities has the same standard deviation as the market index. A security with = 0.5 has below-average market risk; a well-diversified portfolio of these securities tend to move half as far as the market moves and has half the market's standard deviation. In other words, for the security with = 1.0, if the market index changes 20 percent, its return (or price) would change 20 percent too. While for the security with = 0.5, if the market index changes 20 percent, its return (or price) would change 10 percent; and for the security with = 2.0, if the market index changes 20 percent, its return (or price) would change 40 percent. This relations can be shown by the slope of the line. So, to the investors, what really matters is the security's beta value and the way it can be estimated. Fortunately, the regression analysis provides an effective approach to get the answer. And the development of the computer software makes this task much easier.

Discussion

In this part, we have discussed the risk of the security. For the unique risk, investors can diversify it by means of portfolio. What really matters is the market risk, which cannot be removed. To measure this risk, we should use the sensitivity of the stock change to the market change—that is, - .To estimate, we have introduced an approach based on the regression analysis using MINITAB. What the method does is estimation of beta value using the available data— that is, the past information. Note that although the beta value does appear to be stable, one should remember that it is just an estimation. However, despite the difficulties, it is still possible to calculate the risk if we desire to invest in the capital market.

Risk: A Motivating Theme

Teaching introductory probability and statistics has always been a challenge to instructors. Statisticians have attempted numerous innovations to impart statistical knowledge to students. Here, the challenge lies on how to take basic concepts and tie them to everyday problems so that the subject matter becomes meaningful to students.

The purpose of the proposed research is to introduce and apply a new idea for motivating students in an introductory probability and statistics course. The motivating theme is risk, and the process begins with a first-day-of-class questionnaire that samples attitudes of students and involves them in analysis of events and decisions from their daily lives.

Questionnaire responses serve as a context for the instructor to develop the technical concepts of probability and statistics and provide possibilities to increase student motivation and involvement in the course.

Use Risk as a Unifying Theme

I propose an approach different from each of the motivational strategies that we have found in popular textbooks. We suggest that motivation of students to study probability and statistics can be accomplished by linking its study to an idea that they have dealt with all their lives—namely, risk. With this focus, the instructor can help students to build their study of the science of statistics on a foundation—an understanding of risk—already possessed by all students. Throughout the course, students and instructor can work together to enlarge and clarify student understanding of this concept using the basic ideas of probability and statistics. Moreover, this approach (which we are using with college students, primarily at the sophomore level) is adaptable at all levels at which statistics is taught from junior high to professional schools.

This idea is one that has been germinating in me for some time, and during the fall 1991 and spring 1992 semesters, risk has finally grown into a full-grown theme. This proposal describes how it worked and requests support for completing the investigation.

Method

The following is a description of the method used during the fall 1991 and spring 1992. Our risk-centered approach began with a questionnaire, shown below. On the first day of class, the professor gave each student the assignment of completing this questionnaire (spread over four pages with space for thoughtful answers) for the next class period. To ensure that students would carefully participate in this activity, he promised to count it as a quiz. When handing out the questionnaire, he asked students to read it carefully and to make sure that they understood each question; he spoke with students individually and offered clarifications.

We make no special claims for the questionnaire itself. It was constructed quickly, and it turned out that a couple of the questions were misunderstood by some students. In addition, several questions elicited almost identical answers. Despite these factors, responses from about 150 students gave us a rather clear picture of the degree of their naiveté about probability and statistics and, more importantly, provided us with information from students' day-to-day lives to which we can help them attach the concepts from probability and statistics that we seek to teach.

Throughout the semester, the professor referred to particular groups of answers and examples provided by students on their questionnaires. This led to a high level of interaction among students and between students and the professor; sometimes the discussions were heated as students defended differing points of view. These differences provided the perfect opportunity for the professor to point out that in probability and statistics, there are no single unified agreed-upon definitions of either "probability" or "risk."

At semester's end, as the professor compared each class with which the questionnaire was used with previous nonquestionnaire classes, the results were remarkable. Not only had student involvement and motivation been high, but also a significantly higher percentage of students in questionnaire-using classes elected to take a second statistics course. Needless to say, we are pleased with these results and would like to do more research and investigation.

The instructor will be responsible for revising the questionnaire and making it more consistent and for collecting the data and improving the ways student responses may be used for introducing the key statistical concepts. She will evaluate the findings and prepare the final report.

β and Investment Risk

Numerous factors affect markets, and investing requires basic understanding of the risks involved. For example, risk of a security has two main components: unique risk and market risk.

Diversification versus Risk

Portfolio Standard Deviations

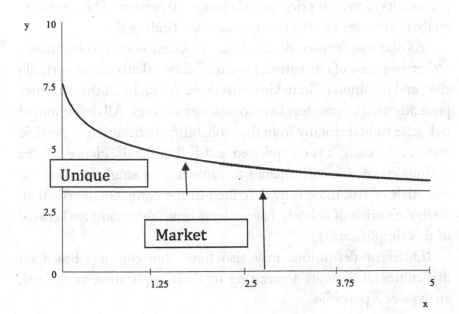

Number of shares

Investors with a low tolerance for risk try to reduce or avoid it by actions such as *diversification*. Unique risk can be *reduce* or even *eliminate* by holding a well-diversified portfolio. Market risk associated with market's wide variations *cannot be removed* by individual action.

As such, the risk of a fully diversified portfolio is the market risk or its sensitivity to market changes. This sensitivity is generally measured by a quantity known as beta (β).

A security with $\beta = 1.0$ has average market risk; a well-diversified portfolio of such securities has the same risk as the market index.

A security with $\beta = 0.5$ has below-average market risk; a well-diversified portfolio of these securities tend to move half as far as the market moves and has half the market's risk.

For the security with $\beta = 1.0$, if the market index changes 20 percent, its return (or price) would change 20 percent too.

For the security with $\beta = 0.5$, if the market index changes 20 percent, its return (or price) would change 10 percent.

For the security with $\beta = 2.0$, if the market index changes 20 percent, its return (or price) would change 40 percent. The regression analysis provides an effective approach for finding β.

As discussed earlier, deciding on a risk definition involves *trade-off* between ease of estimation, forecast ability, calculation of portfolio risk, and intuition of individual investors. As such, all the measures presently used by investors have some shortcomings. All definitions of risk arise fundamentally from the probability distribution of possible returns. As such, it is complicated and full of detail. Hence, all the definitions of risk will attempt to capture in a single number the essentials of risk more fully described in the complete distribution. Each definition of risk will have at least some shortcomings because of this simplification.

Different definitions may also have shortcomings based on difficulties of accurate forecasting or their application to the risk analysis of a portfolio.

Proposed Risk Definitions

The standard deviation measures the spread of the distribution about its mean. Investors commonly refer to the standard deviation as the volatility.

If these returns were normally distributed, then two-thirds of the monthly returns would have fallen within 6.3 percent of the mean—that is, in the band between -4.7 percent and 7.9 percent. These statements are based on empirical rule. Here, we can also make statements using Chebyshev's rule in place of empirical rule. For example, at least three-fourths of the years fund's annual returns were in a band – 24.4 percent and 62.8 percent.

Because of certain considerations the standard deviation was Harry Markowitz choice who started analysis that included definition of risk. This has been the standard in the institutional investment community ever since. It is a very well-understood and unambiguous statistic. Standard deviations tend to be relatively stable over time (especially compared to mean returns and other moments of the distribution). Econometricians have developed very powerful tools for accurately forecasting standard deviations.

Critics of the standard deviation point out that it measures the possibility of return both above and below the mean. Most investors would define risk based on small or negative returns (though short sellers have the opposite view). This has generated an alternative risk measure: semi-variance of downside risk. Semi-variance is defined in analogy to variance based on deviations from the mean but using only returns below the mean. If the returns are symmetric—that is, the return is equally likely to be x percent above of x percent below the mean—then the semi-variance is just exactly one-half the variance. Authors differ in defining downside risk as the square root of the semi-variance in analogy to the relation between standard deviation and variance. Investors with different preferences may choose different definitions of downside risk.

Downside risk clearly answers the critics of standard deviation by focusing entirely on the undesirable returns. However, there are several problems with downside risk. First, its definition is not as unambiguous as standard deviation or variance, nor are its statistical properties as well known, so it isn't an ideal choice for a universal risk definition. Second, it is computationally challenging for large portfolio construction problems. Third, to the extent that investment returns are reasonably symmetric, most definitions of downside risk are simply proportional to standard deviation or variance and so contain no additional information. To the extent that investment returns may not be symmetric, there are problems forecasting downside risk. Return asymmetries are not stable over time and so are very difficult to forecast. Realized downside risk may not be a good forecast of future downside risk. Moreover, we estimate downside risk with only half of the data, losing statistical accuracy.

Shortfall probability is another risk definition and perhaps one closely related to intuition of what risk is. The shortfall probabilities are the probability that the return will lie below some target amount. Shortfall probability has the advantage of closely corresponding to an intuitive definition of risk. However, it faces the same problems as downside risk: ambiguity, poor statistical understanding, difficulty of forecasting, and dependence on individual investor preferences for shortfall targets. Forecasting is a particularly thorny problem, and it's accentuated for lower shortfall targets. At the extreme, probability forecasts for very large shortfalls are influenced by perhaps only one or two observations.

Value at risk is similar to shortfall probability. Where shortfall probability takes a target return and calculates the probability of returns falling below that, value at risk takes a target probability (e.g., 1 percent or 5 percent lowest returns) and converts that probability to an associated return. Value at risk is closely related to shortfall probability and shares the same advantages and disadvantages. Where does the normal distribution fit into this discussion or risk statistics? The normal distribution is a standard assumption in academic investment research and is a standard distribution throughout statistics.

It is completely defined by its mean and standard deviation. Much research has shown that investment returns do not exactly follow normal distributions but instead have wider distributions—that is, the probability of extreme events is larger for real investments than a normal distribution would imply. Thus, we either need to apply other rules such as the Chebyshev's rule or seek a broader definition and measurement of risk.

I and a colleague (Dr. Smith) have developed such a measure based on tail thickness and have discussed its advantages over the measurements presented above.

The new measure using the parameters k and σ can compute measures of tail thickness for a given value of the threshold u that incorporate the characteristics of tail behavior.

One very appealing measure is the conditional mean exceedance (or cme) function,

$$M(u) \equiv E(Z\text{-}u \mid Z>u) = E(Y \mid Y>0),$$

which differentiates among different types of upper tail behavior. The cme is the average amount by which the random variable Z exceeds the threshold u given that it is larger than u. Formally, Davison (1984) observes that for a GPD,

$$E(Y \mid Y> 0) = (\sigma - ku) / (1\text{-}k)$$

where, as before, σ denotes the scale parameter for the GPD of the excess Y ($\equiv Z\text{-}u$). This simple formal result provides a convenient and meaningful measure of tail thickness for a given value assigned to u. As one can readily observe, the right side (RHS) of this equation is "large" for a "long-tailed" distribution (tail shape parameter $k<0$) and "small" for a "short-tailed" distribution ($k>0$).

Of particular interest is the result that for a "medium-tailed" or exponential distribution (k=0), the RHS reduces to σ, here denoting the standard deviation of an exponential distribution. The above five risk definitions all attempt to capture the risk inherent in the "true" return distribution. An alternative approach could assume that returns are normally distributed. Then the mean and standard deviation immediately fix the other statistics: downside risk, semivariance, shortfall probability, and value at risk.

Portfolio Risk

Suppose that risk is measured by the standard deviation of the returns. When there are only *two* shares in a portfolio, the variance of the expected portfolio return is given by the formula

$$x_1^2 \, \sigma_1^2 + x_2^2 \, \sigma_2^2 + 2x_1 \, x_2 \, \sigma_1 \, \sigma_2 \, \rho_{12} \tag{1}.$$

Here, x_1 and x_2 are the proportions of the portfolio in shares 1 and 2, σ_1 and σ_2 are the standard deviations of the rate of return for each share, and ρ_{12} is the coefficient of correlation between the returns for share 1 and those for share 2.

This is because the returns of each share, r_1 and r_2 as well as portfolio return, r, are all random variable with the following relationship:

$$r = r_1 x_1 + r_2 x_2$$

$$Var(r) = Var(r_1 x_1 + r_2 x_2) = Var(r_1 x_1) + Var(r_2 x_2) + 2 \, cov(r_1 x_1 + r_2 x_2)$$

$$= x2/1 \, Var(r_1) + x2/2 \, Var(r_2)$$

$$+ 2x_{1x2} \, cov(r_1, r_2)$$

This formula can be generalized for a portfolio of n shares in the form

$$\sum_{i=1}^{n} \sum_{j=1}^{n} x_i \, x_j \, \sigma_i \, \sigma_j \, \rho_{ij} \tag{2}$$

where $\rho_{ii} = 1$ As more share are added to the portfolio, the covariances become more important, since the number of terms in expression (2) rises with the square of n (in fact, n^2-n), although the terms not involving ρ_{ij} only rise linearly (in fact, n).

Thus, the variability of a highly diversified portfolio reflects mainly the covariances. If the covariance (the term used to describe σ_i $\sigma_j \rho_{ij}$) average were zero, it would be possible to eliminate all the risk by holding sufficient shares. Unfortunately, shares move together, not independently. Thus, most of the shares that the investor can actually buy are tied together in a web of positive covariances that set the limits to diversification. It is the average covariance that constitutes the bedrock of risk remaining after diversification has done its work. Note that the risk in minimum whten $\rho_{12} = _{-1}$ when there are two shares in a portfolio. When there are three, then negative correlation between share 1 and 2 and also between 1 and 3 implies positive correlation between 2 and 3. If a portfolio consists of n different securities, the proportionate contribution of the jth security to the overall risk of the portfolio is given by the expression

$$x_j \sigma_{jm} / \sigma^2_m \qquad (3)$$

where σ_{jm} is the correlation of the jth security with the market portfolio, σ^2_m is the variance of the market, and x_j is the proportion by value of the portfolio in share j.

While diversification makes sense for individual investors, it is not automatically true that it is the best course of action for a commercial firm to follow.

Portfolio Diversification and Its Limitations

A first step in examining the diversification effect in portfolios is to consider the characteristics of hypothetical portfolios formed with varying numbers of shares.

For this purpose, *three* simplifying assumptions are made:

1. All holdings in the portfolio are assumed to be of equal (monetary) size.
2. All holdings are assumed to be equally risky in the sense described in the previous section.
3. The risks of each pair of holdings in the portfolio are assumed to be mutually independent.

An examination of price changes in recent years shows that, on average, about *30 percent* of the price movement of a share has been contingent on what has been happening to the market as a whole. Of course, any two shares selected from the same industry group of shares would have had considerably more in common than just this 30 percent. However, since the object of the hypothetical portfolio is to measure the maximum effect that diversification can reasonably be expected to have, it will be assumed that such duplication is never necessary in a portfolio and that the shares in the portfolio have only the market influence in common.

The maximum theoretical benefits from diversification are secured with a portfolio composed of an infinitely large number of holdings.

If the only relationship between any two holdings in a portfolio lies in the fact that 30 percent of each share's prospects is contingent on the behavior of the market, then calculations on the assumptions given above show that no amount of diversification can reduce the risk or volatility of possible returns below *74 percent* of that of a one-share portfolio.

Not only is the potential total benefit from diversification limited, but also a large part of this potential can be realized with a portfolio of relatively few shares.

A portfolio of *ten* shares provides *88.5 percent* of the possible advantages of infinite diversification; one of *twenty* shares provides *94.2 percent* of these advantages, etc.

Risk and Education

Although I have been teaching introductory probability and statistics for more than forty years, I still find the task challenging. To help this, I try to take basic concepts and tie them to everyday problems so that the subject matter becomes meaningful to students. Recently, I designed and implemented an experiment using an idea for motivating students in an introductory probability and statistics course. The selected theme was risk, and the process started with a first-day-of-class questionnaire that sampled attitudes of students toward risk and involved them in analysis of events and decisions from their daily lives. Questionnaire responses served as a context for developing the technical concepts of probability and statistics and for increasing students' motivation and involvement in the course. I selected this theme, since my earlier research indicated an increased student involvement and interest in statistics.

The idea required an extensive research concerning statistics education and concepts related to risk and risk analysis. I did spend lots of time to revise the questionnaire, collect data, and perform statistical analysis. The findings were used in introductory probability and statistics courses for several semesters. I think risk is an excellent motivating theme for such courses.

Significance

Why? Well, reluctant students ask why they should study probability and statistics. Our answers each time have been a variation on the theme "to learn how to make sensible decisions in the face of uncertainty." Many textbooks try to motivate students by introduction of varied applications. This technique addresses students' apparent desire to see the relevance of their studies to the outside world and their skepticism about whether statistics has value. In recent years, several good textbooks have been published that are based on the belief that applications—to students' everyday experiences or to their chosen career fields—are the key to motivation. For our introductory students, this approach often fails to motivate. It may work with students who beforehand are deeply committed to a particular academic or career field, for they can select and investigate research questions that interest them. However, even for these committed students, applied examples and exercises may fail to motivate because they are introduced after the presentation of complex technical concepts; by then, some students have already turned against statistics. Moreover, these examples are not usually of immediate concern to the majority of students, since they lie too far away into a student's future instead of relating to their present experiences. For example, sophomores who plan to be teachers are not highly motivated by statistical experiments that compare different teaching styles; this lack of motivation occurs even among those who are highly committed to a career in education. As sophomores, they are simply too far away from their future teaching experience to be excited by examples of this sort.

Several investigations have discussed the importance of using real data to improve the teaching of applied statistics. I found that most of the students who enroll in our introductory statistics courses are not fascinated by real data and its careful analysis. Perhaps this occurs because the data typically comes from discipline such as economics, education, psychology, or yet another field in which immature students do not yet have a deep interest.

Another frequently made suggestion is to involve students in data collection and its analysis using a computer. Most of our students take only one statistics courses, and one semester does not provide enough time for proper implementation of this idea. I found that computer instruction steals time from statistical topics, and students lack enthusiasm for data collection because of problems related to sampling. Additionally, students find that statistics applied to someone else's problems seems dead and dull rather than useful and interesting.

However, the prospects for motivating students are not entirely bleak. Despite many voids in students' interests, despite many ideas that fail to motivate them, they do have common interests that we can build on as we try to teach them statistics. People of all ages perform experiments, evaluate uncertainties, and analyze information as an integral part of everyday life. This occurs whether or not they have studied statistics. As their teachers, we could relate statistics to their student experiences and help them to use the subject to become more effective decision makers. Based on our experience, linking their studies to a familiar subject with which they have immediate concern almost always works better.

Goals and Objectives

Using risk is an approach different from each of the motivational strategies that may have been found in popular textbooks. It suggests that motivation of students to study probability and statistics can be accomplished by linking its study to an idea that they have dealt with all their lives—namely, risk. With this focus, the instructor can help students to build their study of statistics on a foundation—an understanding of risk—already possessed by all students. Throughout the course, students and instructor can work together to enlarge and clarify student understanding of this concept using the basic ideas of probability and statistics. Moreover, this approach (which I experimented using college students, primarily at the sophomore level) is adaptable at all levels at which statistics is taught from junior high to professional schools.

This idea is one that has been germinating in me for some time, and during the last few semesters, risk has finally grown into a full-grown theme. Since my study produced results that, despite its weakness and inconsistencies, led to a publication in a high-class journal, I am very hopeful that a further research on statistics education as well as risk theory will produce results useful to professionals in general and to the instructors teaching statistics in particular.

Methodology

The following is a description of the methodology that I implemented. The risk-centered approach will begin with a questionnaire like the one shown below. On the first day of class, the professor will give each student the assignment of completing this questionnaire (spread over several pages with space for thoughtful answers) for the next class period. To ensure that students will participate in this activity, he or she will promise to count it as a quiz. When handing out the questionnaire, he will ask students to read it carefully and make sure that they understand each question; he will speak with students individually and offer clarifications.

Throughout the semester, the professor will refer to particular groups of answers and examples provided by students on their questionnaires. In my pilot study, this led to a high level of interaction among students and between students and the professor; sometimes the discussions were heated as students defended differing points of view.

At the end of the semester, the professor will utilize methods developed in this project to compare each class with which the questionnaire is used with previous nonquestionnaire classes. The results of the study showed that not only had student involvement and motivation been high but also a significantly higher percentage of students in questionnaire-using classes elected to take a second statistics course, and some even decided to minor in statistics.

Statistics, Opening-Day Questionnaire

1. Risk is an everyday problem. What do you know about it? Explain.
2. Which of the following is more risky? Explain.
    (a) car   train   plane
    (b) smoking   X-ray   nuclear power plant

3. How do you deal with risk? (try to avoid? buy insurance? enjoy it? . . .)

4. Do you think that the following statement is correct? Explain.
   Not taking a risk is taking a risk of another kind.

5. Have you ever been involved in any risky activities? Describe one case.

6. Have you ever made a decision that involved risk? Describe one case.

7. For your decision in question 6, how much risk was involved?

8. How did you arrive at your risk estimate in question 6?

   a. If your answer to question 7 is a word like "high" or "low" rather than a number, can you express it numerically using percentage? Explain.

   b. Order the items in question 2(a) from the least risky to the most risky. If the risk of the least risky item is r, then what should be the risks of the other items (e.g., 2r? 3r? . . .)?

9. What kinds of background information does a person need to make reasonable numerical estimates of the risks that you mentioned in questions 6 and 7?

10. What branches of science are involved in the process of estimating risk? What role do you see for "probability and statistics" in this task?

11. Look back and summarize your answers: overall, how do you measure risk?

12. Which of the following definitions of risk do you prefer? Explain.

    (a) Risk is the chance of loss.

    (b) Risk is the possibility of loss.

    (c) Risk is uncertainty.

    d) Risk is the dispersion of actual from expected results.

(e) Risk is the probability of any outcome different from the one expected.

(f) Risk is the probability of an event times the probable cost (or loss) if the event occurs.

13. What do all the definitions in question 12 have in common? What role does "probability and statistics" play in each? Explain.

14. Do you agree with each of the following statements? Explain why or why not.

    a. Risk may be objective or subjective.

    b. Quantifying risk means determining all the possible outcomes of a risky activity and determining the relative likelihood of each outcome.

15. Suppose that you are thinking about flipping a fair coin. To quantify this, which of the following approaches do you prefer?

    a. Repeat the flip a large number of times until you have established the result that half of the time it comes up tails and half of the time heads.

    b. Assign fifty-fifty likelihoods, since there are only two possible outcomes and you have no reason to say that they are not equally likely.

16. Have any of the later questions changed your answer to early questions?

17. Can you see ways that learning probability and statistics will enable you to come with more accurate answers to these questions? Explain.

Can I Be Sensible about Risk?

This section examines some current attitudes toward risk. The objective is to convey some of the complexity that is involved in assessing risk and to provide background that will help the decision maker in future risk assessment.

In day-to-day conversation, we make statements like "Mary is a risk taker" or "John is not a risk taker" to describe the attitudes of people we know toward actions that may lead to loss or damage. We take a risk when we cannot predict or control fully the consequences of an action.

Acceptance of risk by individuals is often very different from the acceptance of the same type of risk by groups. For risks that involve ourselves alone, we may feel comfortable with subjective decision-making, relying on hunches or gut feelings; however, groups often must justify their decisions to others and, thus, may need to make decisions on objective bases. Individually, we take risks when we decide whether to get married or to smoke or when to cross the road. Communities also take risks, such as when they build roads or protection against flooding or when they fluoridate the water supply. Often these decisions are based on some quantitative analysis. They may decide not to protect against flooding if the costs are too great; they may decide to fluoridate the water supply because they believe that the good effects outweigh the bad ones.

Risk has two factors: the uncertainty of the occurrence of an undesirable event, often expressed as a probability; and the variable magnitude of its consequence, often measured in dollars. Both components are difficult to estimate. In what follows, we examine some current attitudes toward risk with the objective of providing a background that will help the reader in future risk-assessment.

Television, reflecting great public interest and concern, provides extensive coverage of natural disasters with strong visual impact. Yet television is often criticized for the way it presents risks. To date, very few attempts have been undertaken to assess the role of the broadcast media in elaborating and reinforcing dominant public perceptions of risks during disasters. This research proposes to study the effects of broadcast coverage of natural disasters on the audiences' perception of the risks involved in the occurrence and aftermath of a natural disaster. Frame analysis (Goffman 1974; Edelman 1993; Entman and Rojecki 1993; Fiske and Taylor 1991) will be applied to identify expectations projected by the broadcast news media that are used to make sense of the potential impact of a natural disaster report at a given point in time. Frames can make bits of information more salient by placement or repetition, or by associating them with culturally familiar symbols (Pan and Kosicki 1993). We propose to undertake a content analysis of televised news reports of natural disasters to analyze the latent messages imparted that affect the audiences' perception of risks involved. The study will investigate the role of framing in the broadcasters' use of production elements in their reporting of the risks and hazards of natural disasters.

Experts versus Public

In today's world, there is much disagreement about which risks are of greatest concern. Nuclear power, for example, is considered a major risk by members of the public; experts, on the other hand, consider it more risky to travel by train. This disagreement is clearly seen in table 1, which gives an ordering of the risks that people see associated with twenty different activities and technologies.

Here, we mean "mathematical expectation." If we face an outcome that has a numerical value—such as winning $400 in a raffle—and if we multiply the probability of the outcome by the value of the outcome, that value is our expectation. If we have one chance in one thousand of winning $400 in a raffle, then our expectation is

$$(0.001) \times \$400 = \$.40.$$

Thus, the expectation of anyone holding a lottery ticket is forty cents. Another way of looking at this value is as the "average" or mean amount won by all of 1,000 ticket holders, one of whom wins $400 and 999 of whom win nothing.

Consider the following table. We observe that even groups of educated, relatively well-informed individuals (college students and members of the League of Women Voters) differ from the "experts."

	College Students	League of Women Voters	Experts
nuclear power	1	1	20
handguns	2	3	4
smoking	3	4	2
pesticides	4	9	8
motor vehicles	5	2	1
motorcycles	6	5	6
alcoholic beverages	7	6	3
police work	8	8	17
contraceptives	9	17	11
firefighting	10	11	18
surgery	11	10	5
food preservatives	12	20	14
large construction	13	12	13
general (private) aviation	14	7	12
commercial aviation	15	14	16
X-rays	16	18	7
electric power (non-nuclear)	17	15	9
railroads	18	19	19
bicycles	19	13	15
swimming	20	16	10

Table 1. Ordering of Perceived Risk for
Twenty Activities and Technologies.

"Experts" and the public often disagree in risk assessment. The table from *The Business of Risk* (1983) by Peter G. Moore illustrates how groups of people rank the risks associated with certain activities and technologies. Even groups of educated, well-informed individuals differ from those with expertise in risk assessment.

Lack of agreement, shown in table 1, between the "experts" and the others is supported by our calculation of the correlations for the ratings for the different groups; these are displayed in table 2.

	League of Women Voters	College Students
College Students	84%	
Experts	35%	34%

Table 2. Rank Correlations for Ratings.

Who are the "experts" of table 1? Primarily, they are persons with expertise in risk assessment—that is, people who treat risk systematically as an expectation and who have had experience in estimating and working with probabilities. Although they gather extensive information to use in their assessment, the experts are not, as a rule, scientists or practitioners in a field of knowledge—such as nuclear power or environmental science in which major risks have been identified. Their expertise lies in the process of using available information to assess risks.

Are the experts right? On the one hand, there is plenty of evidence on which to base an assessment of the risks of train travel, but there has never been a total nuclear disaster. Just what are the chances that it will occur? Experts might use information like that presented in table 3, which is based on data collected in the United States and shows the annual death rates for people killed in various ways within twenty-five miles of a nuclear reactor (Moore 1983).

Type of accident	Death rate
Car accident	1 in 4,000
Accidental fall	1 in 10,000
Fire	1 in 27,000
Reactor accident	1 in 750,000

Table 3. Annual Death Rates in the
Neighborhood of a Power Plant.

Estimation of Probabilities

How does an informed member of the general public estimate probabilities of various risks? According to Paulos (1988), Americans are innumerate and, in particular, woefully poor at using numbers meaningfully to make estimates about the objects and events in the world around them. However, Combs and Slovic (1979) suggest a different view—namely, that errors in probability estimates are based on biased coverage of events by the media.

Combs and Slavic examined the coverage by newspapers in New Bedford, Massachusetts, and Eugene, Oregon, and also surveyed readers of these papers. The people whom they surveyed had attitudes toward causes of death that were directly related to the amount of coverage of various types of events in their local newspapers. In fact, these patterns emerged:

Although diseases take about sixteen times as many lives as accidents, the newspapers contained more than three times as many articles about accidents as about diseases, noting about seven times as many accidental deaths.

Although diseases claim almost one hundred times as many lives as do homicides, there were about three times as many articles about homicides than disease-related deaths. Furthermore, homicide articles tended to be more than twice as long as claim almost 100 times as many lives as do about 3 times as many articles about homicides.

The people who read these newspapers assessed the risk of death by homicide as much greater than the risk of death by accident, which, in turn, was thought to be much greater than the risk of death by disease. Although their assessment had an incorrect result, nevertheless, these readers made interpretations consistent with the numbers of the deaths about which they read. In other words, the readers' estimation processes were correct, but the results were incorrect because they relied on incomplete information.

In a *Scientific American* article that contains numerous thought-provoking examples, Daniel Kahneman and Amos Tversky (1982) explore at length the role that the framing of a choice plays in the resulting decision. Compare, for example, the following pairs of choices:

CHOICE 1

On the table in front of you is $200; you will get it and possibly something more. To determine how much more, you must choose between the following:

> A: avoid a gamble and collect $50 more for sure;
> B: gamble with a 25 percent chance of winning $200 more and a 75 percent chance of winning nothing more.

Do you prefer A or B?

CHOICE 2

On the table in front of you is $400; you will get it and possibly something more. To determine how much more, you must choose between the following:

> C: avoid a gamble and lose $150 for sure;
> D: gamble with a 75 percent chance of losing $200 and a 25 percent chance of losing nothing.

Do you prefer C or D?

According to Kahneman and Tversky, most people prefer A to B and D to C, even though the expected gains are identical for A and C and also for B and D.

Choice B has the expectation

$$(0.25) \times \$200 = \$50,$$

to be added to the initial \$200, yielding \$250;
and choice D has the expectation

$$(0.75) \times \$200 = \$150,$$

to be subtracted from the initial \$400, yielding \$250. Thus, both choices lead to the same expectation of \$250.

Consider also the following example in which we see several different ways of looking at the risk of not winning in a lottery (Kahneman and Tversky 1982).

> The winning ticket in a lottery was 865304. Three individuals compare the ticket they hold to the winning number.

> John holds 361204;
> Mary holds 965304; and
> Peter holds 865305.

If one million different tickets have been sold, then every ticket holder is very likely not to win, and each just as likely as the others. The chance of winning is only one in a million. Why, then, is Peter very upset about losing, even more so than Mary, and John is only mildly disappointed?

We may suppose that the ticket holders' disappointments are related not only to winning or losing, since all have lost, but also to the chance of matching the winning ticket. Only fifty-four of the one million tickets match the winner as closely as do Peter's and Mary's tickets, differing only in a single digit. John's ticket differs from the winner in half of its six digits, and, thus, he has not "come close" to winning. A question that still remains is, why should Peter be more disappointed than Mary? Is he thinking not of a single winning ticket but of a list of digits drawn in order? Reading the digits from left to right, one sees at once that Mary is not a winner but does not know that the same is true for Peter until the end.

Paradoxes Involving Probabilities

To reason correctly when we use probabilities is difficult and requires thinking that seems contrary to intuition. For example, consider the following hypothetical situation concerning a diagnostic test for cancer. Suppose that in the over-thirty-year-old population in the United States, 0.5 percent or one out of two hundred people actually have cancer. Suppose further there is a test for cancer that is 98 percent accurate: if the test is performed on persons who have cancer, the test results will be positive 98 percent of the time; and if the test is performed on persons who don't have it, the test results will be negative 98 percent of the time. Table 4 shows the expected results if this test is performed on ten thousand people (of whom about fifty actually will have cancer).

	Number of positive test results	Number of negative test results
People free of cancer (9950)	199	9751
People having cancer (50)	49	1
TOTAL	248	9752

Table 4. Hypothetical Results of 98 Percent
Accurate Cancer Detection Test.

Suppose you have had the test and the results come back positive. Do you have cancer? Probably you would reason that since the test is 98 percent accurate that it is almost certain that you do.

However, a positive test result is wrong for most people and you probably would not have cancer. Look back at table 4. What portion of the persons with positive test results actually have cancer? The answer is 49/248, or slightly less than 20 percent. Although the test is 98 percent accurate, since most people are healthy, most of the test's errors are on healthy people, and a positive test is likely to be in error.

Here is a second counterintuitive probability example. This one deals with the unlikely event that two randomly chosen people will share the same birthday. Surprisingly, the likelihood of coincident birthdays becomes high in groups that are much smaller than 365.

A particular professor with an announced policy of occasional surprise quizzes comes in to meet her class of thirty students and asks each student to write down his or her birthday month and day, ignoring the year, on a slip of paper. She collects the slips and then announces that there will be a surprise quiz today if two of the slips name the same birthday. As preparations are being made to tally the dates, do you suspect that the students are already beginning to relax, confident that with only thirty students and 365 days to choose from, there will not be a match?

Although we cannot say for certain what will happen in this particular class, elementary concepts of probability theory and a straightforward calculation can be used to discover that the probability of a repeated birthday in a group of thirty people (with no twins and no February 29 birthdays) exceeds 70 percent. In fact, the probability of a repeated birthday is over 50 percent for a group as small as twenty-three people (Hogg et al. 1974).

There is perhaps no more appropriate way to end a consideration of the pitfalls inherent in working with probabilities than with a quote from *The Bending of the Bough* by George Moore,

"There is always a right way and a wrong way, and the wrong way always seems the more reasonable."

Eliminating Risks

Almost any risk can be eliminated if we are willing to pay the price. If you pay $300 per year to insure your $80,000 home against loss by fire, and if there is only one chance in two thousand that this loss will occur, then you might evaluate this situation using the following procedure:

Imagine two thousand people like yourself paying $300 per year to insure their $80,000 homes against loss by fire. They pay a total of $600,000 in premiums, but only one collects $80,000. The insurance company, thus, is collecting $520,000 ($260 per policyholder) more than it is paying out for this loss. (Of course, the costs of operating an insurance company and small claims will reduce this profit somewhat.) Looking at it from the insured's point of view, most people lose money each year on fire insurance. Is purchase of fire insurance, then, an unwise decision?

The decision to purchase insurance can be evaluated using mathematical expectation. Using positive numbers to denote gains and negative numbers to denote losses, we multiply each consequence by its probability and add them:

$$(0.0005) \times (\$80,0000 - \$300) + (0.9995) \times (-\$300) = -\$260.$$

Thus, the mathematical expectation from purchasing this fire insurance policy is a loss of $260. This figure agrees with our observation above that the insurance company is collecting $260 more per policyholder than it is paying out. Rather than being foolish when they purchase insurance, people are making a trade, accepting a loss of money in exchange for the "peace of mind" that comes from protection against the very unlikely but very devastating loss of a home.

We may also be paying for an intangible item, the pleasure of dreaming, when we buy lottery tickets. Even those of us who know how slim our chances of winning are and who know that our mathematical expectation from the purchase of a fifty-cent or one-dollar ticket is only a few cents still occasionally buy tickets. Although we know that the purchase of a ticket is almost equivalent to throwing the money away, the attractiveness of the prize makes us dream, and we trade the price of a ticket for the pleasure of holding the dream for a while.

What Needs to Be Done?

In the paragraphs above, we have tried to convey some of the complexity that is involved in assessing risk. What, then, you may wonder, is an intelligent, sensible person to do to increase his or her understanding of risk and to evaluate it properly?

First, we must accept the discipline of a uniform definition of risk. One might compare risky situations sensibly by considering the probabilities of each of the risks; likewise, one might compare them by considering the consequences of each. However, to compare a variety of risks—including ones with enormous consequences but low probabilities and ones with mild consequences but high probabilities—experts evaluate risk as a product of a probability and a consequence.

We cannot make sensible decisions about risks if we do not have accurate information. Media reporting often spotlights spectacular events and ignores commonplace ones. Readily available references such as the annual *World Almanac and Book of Facts* may be used as a source of accurate statistics about death and accident rates, for example.

The birthday problem and the cancer detection test remind us that correct thinking about probabilities can seem paradoxical, running counter to our intuition. In the eighteenth century, the mathematician James Bernoulli began a campaign advocating basic knowledge of probability for all. Perhaps, eventually, our concern about risk assessment will motivate us to reach Bernoulli's goal. Acceptance of the idea of risk as quantifiable, accompanied by a belief that the risks of disasters can be reduced and perhaps eliminated, is a modern approach to risk that is gradually gaining acceptance, replacing the belief that disasters are caused by the wrath of the gods, who must be appeased by sacrifices. Current scientific and medical practices even attempt to alter the risks associated with the survival selection processes of nature.

One successful risk-elimination project was conducted by the World Health Organization; their campaign to eradicate smallpox costs about $300 million and is saving an estimated two million lives per year or twenty million lives in ten years. Whatever cash value is assigned to a human life, it surely is more than $15 (the quotient of $300 million and twenty million). Often risk elimination is not so easily costed out. For some environmental risks such as global warming, by the time we know what the costs will be—either to eliminate the risk or to live with it—it may be too late to do anything. Choices to eliminate risks are very difficult to make, but as a minimum requirement, the following cost-benefit relationship should hold:

$$\text{Cost of avoiding risk} < \text{probability of risk} \times \text{cost of consequence.}$$

Other reasonable quantitative principles also are commonly used in decision-making; however, it would be incorrect to suppose that all decisions are, or even should be, made on a quantitative basis. There are often personal factors, such as "peace of mind," and other pressures, such as satisfying the demands of an electorate, on decision makers.

As we learn more about risk assessment, we also learn that the assessment itself is a risky matter. Because both probabilities and costs may be difficult to estimate, our assessment is likely to be wrong. The "experts" persist, believing—contrary to popular wisdom—that a little knowledge is more valuable than dangerous.

Choices and Risks

Some people take big risks—a gambler rolling the dice, a wildcatter drilling for oil, a tightrope walker venturing a tentative first step. But experts argue that all of us take risks all the time, that risk is inherent in almost all ordinary activities. As we eat our buttered toast, we ingest cholesterol. As we walk through the autumn woods, we breathe polluted air. We take a risk each time we do something for which we cannot predict or fully control the consequences.

Experts—people who treat risk systematically, people with experience in estimating probabilities—understand that a degree of risk may or may not be easily measured; it may be objective or subjective. They consider it an objective risk to call out "heads" or "tails" as someone flips a quarter, because mathematicians have established standard probabilities for flipped coins. While even experts cannot predict the outcome of any particular flip of the coin, they understand the odds they are facing in making a call.

Experts disagree in their judgment of complex environmental risks, such as whether burned waste will produce damaging amounts of toxic substances. From the same data, one researcher might estimate the probability of damage at 1 percent, another at 3 percent. Both would describe any assessment of that risk as subjective. Each takes the available information and provides a number that describes a degree of belief that an event will happen. Although they gather extensive information to use in their assessments, these experts in risk assessment often are not scientists or specialists in environmental matters. Their expertise lies in their skill in working with available information to assess risks.

Many risks, they note, are not so much personally chosen (diet drinks, oral contraceptives) as socially imposed (nuclear power, hazardous wastes). And some of the social risks—those to public health or safety—have become major political issues, provoking widespread uneasiness about scientific progress. We value the products of modern chemistry yet are preoccupied with chemical contamination. We use electricity from nuclear power plants yet fear the prospects of a nuclear accident. We have trouble telling at what point a hazard (something that may cause harm) could turn into a disaster (the actual harm caused by a hazard). We have trouble judging the relative benefits and dangers of the risks we face.

Is Air Travel Safer than Car Travel?

Some people think that traveling by plane is inherently more dangerous than driving a car. According to the National Safety Council, during the life of a randomly selected person, the odds of dying in a motor vehicle accident and in air and space transport are 1 in 98 and 1 in 7,178, respectively. This indicates that flying is far safer than driving. However, for some, flying may feel more dangerous because our perception about risk is usually formed based on factors beyond mere facts. For example, most people think that they are good drivers and, as such, feel safer because driving affords personal control. Plane crashes are often catastrophic. It kills many at once and grabs the attention of major media, which makes people more sensitive to them.

Are You an Average American?

In general, there is a lot more to calculating and comparing risk than one might think. According to the experts, for an average American, the annual risk of being killed in a plane crash and a motor vehicle are about one in eleven million and about one in five thousand, respectively. But, is this all? First, most of us are not the average American. Some people fly more others, and some do not fly at all. So, if we take the total number of people killed in commercial plane crashes and divide that into the total population, the result gives a good general guide, but it is not specific to our personal risk.

Here are other useful numbers:

1. Dividing the number of people who die into the total number of people gives us the risk per person.
2. Dividing the number of victims into the total number of flights passengers took gives the risk per flight.
3. Dividing the number of victims into the total number of miles all of them flew gives you the risk per mile.

Example

In 1995, out of every one hundred million, about 16,300 and 111 people were killed in automobile and commercial flights accidents, respectively. The numbers of deaths per one hundred miles were respectively 3 and 100 for one hundred million miles traveled, and 30 and 20 for every one hundred million trips made.

This shows that the risk of death per mile is 33 times higher for car. However, the risk of death per trip is about 1.5 times higher for airplanes.

Discussion

All the above calculations produce useful and accurate numbers. However, which one is most relevant to us depends on our personal flying patterns. Some fliers take many short flights. Some take less but longer flights, for example. Since the overwhelming majority of the plane crashes take place in connection with takeoffs and landings, the risk is less a matter of how far you fly and more a matter of how often. If you are a frequent flier, then the risk per flight means more to you. For occasional long-distance fliers, the risk per mile means more. A frequent long-distance flier would want to consider both.

The number of plane crash fatalities in the United States varies widely from year to year. So calculation of risk based on one year and average of five years or ten or twenty may be different significantly. In some years, no plane crashes occur, or at least very few do. This makes the value of the risk per year misleading. If we average things over, say, five years or ten, some other factors will muddy the waters. In the last five years, safety factors have changed. A ten-year average might be misleading too.

Despite all these caveats, numbers are a great way to put risk in general perspective. Without question, most metrics shows that flying is less risky than traveling by cars. But wait: Just when you thought it was safe to use numbers to put risk in perspective . . .

Numbers are not the only way—not even the most important way—we judge what to be afraid of. Risk perception is not just a matter of the facts. For example, consider the risk awareness factor. The more aware of a risk we are, the more concerned about it we become. This explains why, when there is a plane crash in the news, flying seems scarier to many of us, even though that one crash has not changed the overall statistical risk significantly.

CHAPTER 6

The Mathematics of Uncertainty

We may not actually compute probabilities in our daily lives, but many of our daily decisions rely on our intuition about probability. This intuition has been cultivated over many years, based on a wide range of experiences. So it might come as a surprise that this intuition can be remarkably unreliable.

Probability

Probability theory investigates (systematically) the laws concerning phenomena influenced by chance. It is the branch of science that studies methods for making inference for situations that involve uncertainty. The theory starts with introducing a measure for degree of certainty (uncertainty) about outcomes. In recent decades, probability theory has become exceedingly important. One reason for this stems from the fact that almost everything in real life involves some degree of uncertainty. This is even true for science itself, since, except for mathematics, which is manmade, other disciplines are empirical-based and as such there is a margin of error. Well-known examples are evolution, big bang, tectonic theories, etc.

Probability theory also provides tools for making quantitative statements about uncertainty and allows one to draw conclusions from such statements using certain rules. It is (as opposed to the word "probability") a mathematical theory (a model of reality) that enables us to calculate the likelihood of outcomes. Areas of applications include reliability and risk analysis, quality control, business, social sciences, medical sciences, insurance, etc.

In 1620, Francis Bacon argued that learning about the world could only take place by observation and induction. With the rapid expansion of experimental sciences, scientists began to make use of the probability theory. For example, in a monumental piece of experimental research between 1856 and 1863, Mendel laid the statistical laws of genetics. Without any precedent, Mendel perceived that the genetic mechanism operated like a random device.

The probability theory investigates the laws
concerning phenomena influenced by chance
It is the branch of science whose goal is to do things
like quantitative inference for instance
It is also a measure of certainties about
occurrence of an event of interest
It has applications in all real-life problems that involve
uncertainty, such as investing and insurance
In recent years it is used extensively as
our understanding is enhanced
It is a new way to look at the world, I look
at it as a significant advance.

What Is Probability?

Probability is not just a vehicle but a standalone field with a large number of unanswered questions. It is, in a way, a human response to the seeming lack of complete determinism in the world where there is uncertainty. Probability may even be seen as human reaction to fear and something that guides much of the decisions we make.

Most people use probability for the exact reason that the field was originally created: to handle the concern of the unknown. It is used to provide a way to quantify the visceral fear of unpredictability. Like any other field of mathematics, probability is a human way to shape the uncertain world into something we can understand or feel comfortably about.

For example, in the medical field, with each test, there are four possibilities: the patient has the disease and test result is positive, the patient has the disease and the test results is negative, the patient does not have the disease and the test result is positive, and the patient does not have the disease and the test result is negative. The probability of each of these events varies depending on how accurate the test is and how common the disease is. Think about diseases such as neurofibromatosis, Huntington's, Tay-Sachs, and cystic fibrosis, which are all genetic diseases that can be passed from parents to their children. A couple with a family history of these diseases may consult to figure out the probability of the child inheriting a genetic disease. Finally, probability is used in all other branches of science. For instance, probability is used in the discipline of engineering for various reasons ranging from quality control to quality assurance.

Uncertainty Makes Life Exciting

Although uncertainty may seem a problem, it could make the life interesting and exciting. Indeed, the world would be a dull place if things were completely predictable. Among many real-life examples of this, we find sports exciting because the outcomes are not completely predictable. Of course, uncertainty can also cause grief and suffering.

Let me start with an everyday example. Probability theory is also a useful tool for making inference. Suppose that as you come out of your apartment you notice a man looking inside your car through the window. You will probably infer that the man is trying to rob your car. How did you arrive at that conclusion? You considered the possible set of circumstances that might have produced the sample of event that you observed. Was the man trying to rob your car? Was the man interested in a car like yours and wanted to see how the interior of the car looks? Was the man trying to figure out if this was his friend's car? Of these and many other possibilities for the event you observed, you picked the outcome that you thought was most probable—that is, you based your inference on your assessment of a set of probabilities.

Probability Modeling

Mathematical modeling is a very fertile field in modern scientific investigations. Its goals is to analyze and translate real-life situations into scientific terms. No one can analyze the real world (by definition of analyze). One can only analyze a picture (conceptual model) of the world.

Approaches to modeling and data analysis have changed a great deal in recent decades. Probability is one important concept used in such approaches to mathematical modeling. By using probability, data is decomposed into a smooth part and a rough part or systematic part and random part (signal and noise), and models are developed for each part separately.

It is the attention given to the second component that perhaps most distinguishes probabilistic approaches to the modeling. In fact, mathematics mostly deals with the analysis and modeling of the systematic part. In the modern approaches, however, models are judged based on their signal-to-noise ratio. This ratio measures the contribution of the systematic part relative to the random part. Often a large value of this ratio occurs if the context of data is well understood. When this happens, usually only the systematic part is used as a model to describe the situation and more importantly to make predictions. The random part is then used to evaluate the model and to produce bounds on prediction errors.

In addition to this, today the ideas of randomness are central to much of the modern scientific disciplines. Clearly the real world can only be analyzed and explained scientifically through physical sciences. However, many scientists expressed their frustration when attempting to do this utilizing classical method based on determinism. Historically, it was natural for physicists to be interested at first in the macroscopic world that surrounds us. To make quantitative predictions about it, they devised deterministic models, which perform impeccably. Such are the origins of mechanics, of thermodynamics, of optics, of electromagnetism, and of relativity. These theories remain valid in the domains for which they were designed, and they continue in a state of vigorous development. But as regards fundamentals, physics today has its cutting edge on the microscopic level, where progress is achieved by means of probabilistic models, models that allow precise quantitative predictions for random phenomena. In short, chance is inherent in the basic nature of microscopic processes, reducing determinism to a mere consequence of chance regarding mean values that is on the macroscopic level.

Birthday Problem

Birthday is an important date for most people, and as such, they show interest in problems related to it. Think, for example, about birthday matches in a classroom or church. There are 365 days in a year, and so we need to assemble 366 people in a room to guarantee at least one birthday match. That is straightforward. But how many people do we need in a room to guarantee a 50 percent chance of at least one birthday match? I ask you to guess a number. When I ask this question, usually the numbers vary a great deal. The answer, to most people's surprise, is twenty-three.

Here are further facts about birthday matches that often surprise people:

- With only forty-one people, the chance of at least one common birthday is more than 90 percent.
- With eighty-eight people, the chance of at least three common birthdays is more than 50 percent.

Here are more interesting facts:

- If we want a 50 percent chance of finding two people born within one day of each other, we only need 14 people.
- If we are looking for birthdays a week apart, the magic number is 7.
- If we are looking for a 50 percent chance of finding someone having a specific person's exact birthday, we need 253 people.

I often ask students to calculate the probability that two randomly selected individuals have birthdays, for example, on May 25 and October 13. I usually choose two students in the class and use their birthdays. The answer (after providing hints) they should find or suggest is $(1/365)^2$. We then change the dates to, say, June 9 and June 9 (again, we choose someone's birthday in the class). Not very sure, they eventually produce the same answer. In both cases, the probability is less than 1/133,000. I ask them if they consider this a small probability. Most students answer yes. We ask which of these two events they find more surprising. All of them, of course, refer to the latter.

This reveals that, although these two events are equally likely, the latter creates a great deal of surprise and, therefore, may be classified as a coincidence. We ask if they are convinced that the two main components of a coincidence are probability and the degree of surprise.

Why do we classify some events as coincidences? According to John Allen Paulos, human beings are "pattern-seeking animals." It might just be part of our biology that conspires to make coincidences more meaningful than they really are. A sport-related example is hot hand in basketball. We notice certain pattern and ignore many equally probable others.

Probability Is Complex

Probability is a measure of "chances" or certainty about the occurrence of an event of interest. The measure could be objective or subjective or even a mixture of these. This means that there are more than one possible way for defining, choosing, or introducing such measures or for assigning probabilities. An important step in the study of probability was the introduction of rules or postulates that help with identification of inconsistent probability assignments. These rules provide guidelines leading to probability selections free of contradictions.

Suppose that to each event like A in the sample space S, a number "probability of occurrence of A," $P(A)$, is assigned, which obeys the following three postulates or axioms:

1. $P(A) \geq 0$
2. $P(S) = 1$
3. $P(A \cup B) = P(A) + P(B)$, if A and B are incompatible (disjoint).

Using these, one can construct a deductive theory that will include all the rules needed to study a chance experiment.

Note that the above axioms do not completely determine the assignment of probabilities to outcomes. They only serve to rule out assignments inconsistent with our intuitive notions of probability. For example, when flipping a coin, once we assign probability to occurrence of a head, the probability of occurrence of a tail is one minus that number, no more, no less. So the question that arises at this point is, how do we assign probabilities to different outcomes of a chance experiment? In what follows, some frequently used definitions and methods are presented that could provide a partial answer to this question.

The Monty Hall Problem

Imagine a *Let's Make a Deal*-type scenario. You are given the option to pick one of the three doors. Behind two of the doors is nothing, and behind the third door is a brand-new car. The goal of this game is, not surprisingly, to win the car. After you select your door, the host of the game will choose one of the other doors that has nothing behind it to open. After he does this, you are given the choice to keep your current door or switch to the remaining door. What do you do? But, more importantly, what should you do? What does probability tell you to do? The answer to this is somewhat counterintuitive. In fact, until computers were programmed to simulate this event, even world-class mathematicians refused to believe the result. The answer is that you should switch doors. At the beginning of the game, you have three choices and no information about any of them. The probability that you pick the car is 1/3 or 33 percent. You make your selection, and the host opens one of the other doors. Now, two doors remain. It is tempting to think here that your chance of winning has increased to 50 percent, but this is not the case. Your probability remains unchanged. However, the probability of winning if you switch doors is 2/3, or 67 percent. If you do not believe me, suppose you switch. Then the only possible way you can lose is if you initially pick the door with the car behind it. The probability that you pick the door with the car from three choices is 1/3. So the probability that you lose if you keep your door is 67 percent, and since only one other door remains when you are offered the choice, the probability that you win if you switch is 67 percent.

Classical (Equally Likely or Theoretical) Definition

If n has only finitely many elements (outcomes), say n, and if by symmetry each of these elements (outcomes) has the same chance to occur—that is, if elements (outcomes) are equally likely—then an appropriate assignment for each possible outcome is $1/n$. If A has m elements (outcomes), then

$$P(A) = m/n = \text{Number of outcomes in } A$$
$$/ \text{ Number of possible outcomes.}$$

In practice, it is not always easy to establish whether simple events (outcomes) are equally likely or not. To resolve this, if we have no reason to say that simple events (outcomes) are not equally likely; we assume that they are. This is sometimes referred to as *principle of insufficient reason*. For example, *if* we assume a given die is a fair die (or have no reason to say that it is not), then the probability of event $A=\{2,4,6\}$ is $3/6 = 2/2$.

It should be pointed out that the classical definition of probability has a logical problem in that it uses the probability (chance) in the definition of probability. Recall that equally likely means that outcomes have the same chance or probability of occurrence. Another problem relates to the fact that the only way to produce convincing evidence that a given coin is fair is by experimentation—that is, by flipping it. For fair coin, the relative frequency of the occurrence of event H is expected to tend to $P(H) = \frac{1}{2}$ if we perform the experiment (in this case, flipping the coin) over and over (law of large numbers). Based on this, we may use the relative frequency as the probability of occurrence of an event, *since* it is based on experiment. Calculation of probability based on this approach is discussed below and is demonstrated in example 2.3.2.

Objective (Empirical) and Subjective (Judgmental) Definitions

Definition of probability based on relative frequency is usually referred to as "empirical probability or statistical probability." Formally, for an event A, it is defined as

Number of times event A has occurred / Total number of times experiment has been performed = Relative Frequency.

This definition of probability has wider applications compared to the classical definition.

This interpretation of probability is also referred to as objective interpretation because it rests on results of the experiments rather than any particular individual concerned with the experiment. In practice, this interpretation is not as objective as it might seem, since the limiting relative frequency (when an experiment is performed infinitely many times) of an event will not be known. Moreover, it refers to the experiments and events that are repeatable. Thus, we will have to assign probabilities based on our beliefs about the limiting relative frequencies. Note that there are many real-life situations where we have no control on occurrence of the event of interest. For example, we can create natural disasters such as an earthquake. In fact, in many cases, repetitions may not be possible. For instance, no frequency interpretation can be given to the event that New York Jets will win the Super Bowl next year. Other situations where this definition may not be applicable is when no or very few data regarding the event of interest is available. An example of the former is the probability calculation of a nuclear failure. For the latter case, if after six rolls of a die four is not yet observed, we cannot assign a probability $0/6 = 0$ to it, since observing four has certainly a nonzero probability. For cases such as occurrence of an earthquake or failure of a nuclear power plant, we could only use experts' judgment and subjectively passing probabilities.

Summarizing these, probability is a number between zero and

one (inclusive) that provides an indication of how likely is an event to occur. Here, zero refers to impossible and one refers to definite and everything else is, of course, between. Mathematics works mostly with items with probability of one. That is why you see theorems in mathematics books. Physics works with items with probability very close to one. That is why you see laws in physics books. Then there are subjects such as biology, geology, etc., where you see theories or hypotheses where probabilities are not very close to one.

Which Definition of the Probability Should We Use?

So far, we discussed different definitions of probability. But when, for example, should we use equally likely definition? If we have no reason to doubt that outcomes are not equally likely, we will assume that they are. This, for example, applies to problems such as flipping a coin, rolling a die, drawing cards from a deck of cards, or choosing a student from a group of students randomly. But it cannot be applied to, for example, weather prediction or being dead or alive tomorrow (we hope they are not equally likely), since otherwise we should expect half of the people dead by tomorrow. For these cases, relative frequency (statistics) is more appropriate and may be used. As a different example, think about buying insurance for your car. During a period, you may make a claim or no claim. But can an insurance company assume that these are equally likely? Of course not. So insurance companies rely on statistics to find the probability that you will make a claim. The statistics they use includes things such as your age, type of the car, etc.

Now, if no past statistics are available or if an experiment is not repeatable (e.g., natural disasters), then one may consider a subjective assessment, such as an expert opinion, to assign probabilities to different possible outcomes. For example, what is the probability of complete failure of a nuclear power plant close to your home? For this, there is no past data (statistics), so experts may examine the components of the system and subjectively assign or estimate likelihoods. Then combine these likelihoods to arrive at a quantity presenting or estimating the risk (probability) of such events. In summary, subjective probabilities result from intuition, educated guesses, and estimates. Here, probabilities are considered to be measures of personal belief (or knowledge about the subject) and, hence, are usually different for two different people. As a sport-related example, given an athlete's extent of injuries, a doctor may think that he or she has 80 percent chance of full recovery. For the same athlete, a trainer with a different set of experiences may suggest a different probability estimate.

Finally, some words of caution about probability. First, probability is an undefined term, or it has many different meanings, to put it differently. The two most usual interpretations are (1) it is an intrinsic property of a physical system, and (2) it is a measure of belief in the truth of some statement. Second, there are many examples demonstrating the fact that assigning likelihoods to events, especially rare events, may lead to serious problems. Following one's intuition often results in numbers very different from the true likelihoods. For example, if you have thirty students in your class, what is the probability that at least two of them would have the same birthday? To anybody's surprise, more than 70 percent (why?). As a sport-related example, consider baseball and football hall of famers who share birthdays. Is this highly likely, or does it have a small chance? George Halas and Reel Schoedienst were both born on February 2; Fran Tarkenton and Lang MacPhail, February 3; and Hank Aaron and Roger Stanback, February 5. Noting that these are just examples from early February, one can see that this is a very likely event.

Type	Definition	Description
Classical (Theoretical)	Number of Outcomes in the Event Number of Possible Outcomes	Finite Number of Equally Likely Outcomes
Objective (Empirical)	Frequency of the Event Total Frequency = Relative Frequency	Based on Available Statistics or Relative Frequency From an Experiment

Subjective (Judgmental)	No Definition	Based on Intuition, Available Knowledge or Educated Guesses
Relationships: As the number of experiments is increased empirical probability will approach the theoretical (actual) probability (The Law of Large Numbers). As our knowledge about the matter under investigation is increased, our guess gets closer to the actual probability		

Different Definitions of Probability.

Probability Assisting Mathematics

The ideas of randomness are central to much of modern physics and have overthrown the "clockwork universe" conceptions of earlier centuries. The laws of probability and statistics were developed by such mathematicians as Fermat, Pascal, and Gauss, and received their first major application in physics in the kinetic theory of gases developed by Maxwell and Boltzmann.

Here, the use of probability is necessary because the number of particles involved is too great for a deterministic/mathematical calculation. With the advent of quantum theory, physics seemed to be based on an essential randomness, whose reality was debated by Bohr and Einstein until the end of their lives. Only later, in the experiments of Alain Aspect, has a convincing demonstration been given that the inescapable randomness of quantum theory is a fact of nature.

Since the molecules and their collisions are so numerous, and the velocities so varied, and since our ignorance of the initial conditions is almost total, Maxwell postulated that positions and velocities are distributed at random; and he was confident that this assumption would describe the gas adequately and would allow one to calculate the mean values of the macroscopic variables. His breathtaking intuition was confirmed half a century later by the work of Albert Einstein (1905) and of Jean Perrin (1908) on Brownian motion.

To clarify, think about one mole of gas. One mole of any molecular substance such as O2 contains 6.02 x 1023 molecules. To construct the theory governing a deterministic system of 1,023 molecules, physicists exploited probabilities through ignorance, and with complete success. The reason for this success deserves discussion. The Soviet physicist Lev Landau has shown that a classical system requiring infinitely many parameters would behave in a totally random fashion; in other words, it would be random unavoidably and not merely by reason of our ignorance.

Boltzmann bases his analysis on the fallowing observation: the molecules are so fast and their collisions so frequent that the system rapidly loses or at least appears to lose track of the initial conditions. Typically, this leads us into the realm of probabilities through ignorance. His based his theory on the following postulates:

1. Every molecule has equal a priori probability of being in region A.
2. The system evolves spontaneously from the less toward the most probable state.

This type of studies led to model for phenomenon such as Brownian motion. The phenomenon was discovered and studied first by the botanist Robert Brown in connection with the erratic motion of pollen grain suspended in fluids. Einstein first presented a quantitative theory of the Brownian motion in 1905 based on kinematic theory and statistical mechanics. He showed that the motion could be explained by assuming that the immersed particle was continually being subjected to bombardment by the molecules of the surrounding medium. Wiener developed rigorous mathematical explanation in 1918 based on stochastic process called Wiener-Levy Process.

Examples

*   A dust particle suspended in water moves around randomly, executing what is called Brownian motion. This stems from molecular agitation through the impacts of water molecules

on the dust particle. Every molecule is a direct or indirect cause of the motion, and we can say that the Brownian motion of the dust particle is governed by very many variables. In such cases, one speaks of a random process; to treat it mathematically, we use the calculus of probabilities described.

- A compass needle acted on simultaneously by a fixed and by a rotating field constitutes a very simple physical system depending on only three variables. However, we shall see that one can choose experimental conditions under which the motion of the magnetized needle is so unsystematic that prediction seems totally impossible. In such very simple cases whose evolution is nevertheless unpredictable, one speaks of chaos and of chaotic processes; these are the terms we use whenever the variables characterizing the system are few.

Method allowing one to predict the future exactly from initial conditions that are likewise exact are called deterministic. The best example of a deterministic theory is classical mechanics. However, this definition of determinism is based on the tacit assumption that the deviations on arrival diminish roughly in proportion to the deviations at departure. In that case, the idealized limit can be envisaged quite clearly; and in fact, gunners can realize excellent approximations to it. But what would happen if initial and final deviations were connected by a relation more complex than simple proportionality?

Misleading Use of Probability

Mark Twain, in his book published in 1924, mentions a famous line attributed to Benjamin Disraeli, "There are three kinds of lies; lies, damned lies, and statistics." Also, we frequently hear the sayings "That is just statistics" and "You can prove what you wish to prove with statistics." In fact, both statistics and statisticians have a poor image in the minds of many people. This is because most data can easily be manipulated in an unethical and unscientific fashion to draw desired conclusions. In other words, it is easy to distort the truth. This is because most people are unfamiliar with concepts and the language of statistics.

An interesting comparison with medical science and doctors is as follows. If you do not find your doctor's instructions helpful, most probably you will blame him or her, not the medical science. But when it comes to statistics, most people blame the science or the methodology used, not the user. There are good and bad engineers, and good and bad lawyers, despite the fact that both professions require a license to practice.

One problem that arises when using probability is that most of us in our daily decisions rely on our intuition about probability, the intuition that has been cultivated over many years based on a wide range of experiences. Unfortunately, our intuition can be completely unreliable. For example, why do people tend to think that parents with four baby girls in a row are due for a baby boy? Why do we often get more concerned about low risks like nuclear power plants than high risks like driving? Why do people tend to think a lottery ticket with the numbers 17, 15, 18, 19, 44, 51 has a better chance of winning than one that has the numbers 9, 8, 7, 6, 5, 4 when both actually have the same chance? Researchers have studied questions like these for many years, and their discoveries are illuminating.

Fortunately in sports and the related, contexts people rely on statistics more than anything else. But as is discussed in several sections of this book, the way statistics are presented may be misleading. A good example of this is Simpson's paradox, discussed in section 2.7 related to what is called hot hand. Here, we may illustrate the point by presenting a different example.

When writing this section, it was middle of December 2001. Looking at NBA records, I noticed that almost all the teams in the Eastern division had a double-digit number of wins and losses. So it was hard to pick a team as the best. I thought this can be related to the discussion of this section as follows.

Suppose that in the area you live, people like to bet on outcomes of the games played between the teams in a particular tournament or conference (in this case, Eastern division). You can take, for example, sixty-four of these people and send them a letter with a forecast for the next six games. Since there are a total of sixty-four possible outcomes such as WLLWWL, LLWWWL, etc., you can send each of these outcomes to one individual. By doing so, one person will get all the forecasts correct, and six people will get five correct forecasts. These people may then start believing that you must have some special information about the games. After all, the probability of correctly guessing outcomes of six successive tosses of a fair coin is only $(0.5)6 = 0.031$. If each week you do this by sending forecasts to several different groups of sixty-four people and start new groups each week, you may be able to create a reputation and generate clients to make significant profit.

One real story regarding the prediction is related to the price of an average family home in the year 2000. In mid-1980s, housing prices in a large part of United States rose by more than 20 percent per year. Based on this, some "experts" suggested that the price of an average family home would exceed $1 million. However, when year 2000 arrived, prices were nowhere near their prediction. This again led some people to doubt and blame statistical forecasting, not the so-called experts. A similar thing happened to stock market where experts were predicting fifteen thousand for Dow Jones Industrial Average and six thousand for Nasdaq for the year 2002, while in fact the opposite happened.

Summarizing these, statistics do not lie, but people do. This science, together with the methodology used, like a paring knife, can be a very valuable tool when it is properly used. When it is misused, it can lead to some very bad results. Some of the ways to intentionally or inadvertently distort the truth with statistics are described in the book *How to Lie with Statistics*, by D. Huff (Norton 1954, New York).

Perception

Television, reflecting great public interest and concern, provides extensive coverage of natural disasters with strong visual impact. Yet television is often criticized for the way it presents risks. To date, very few attempts have been undertaken to assess the role of the broadcast media in elaborating and reinforcing dominant public perceptions of risks during disasters. This research proposes to study the effects of broadcast coverage of natural disasters on the audiences' perception of the risks involved in the occurrence and aftermath of a natural disaster. Frame analysis (Goffman 1974; Edelman 1993; Entman and Rojecki 1993; Fiske and Taylor 1991) will be applied to identify expectations projected by the broadcast news media that are used to make sense of the potential impact of a natural disaster report at a given point in time. Frames can make bits of information more salient by placement or repetition, or by associating them with culturally familiar symbols (Pan and Kosicki 1993). We propose to undertake a content analysis of televised news reports of natural disasters to analyze the latent messages imparted that affect the audiences' perception of risks involved. The study will investigate the role of framing in the broadcasters' use of production elements in their reporting of the risks and hazards of natural disasters.

Flying versus Driving

Some people think that traveling by plane is inherently more dangerous than driving a car. According to the National Safety Council, during the life of a randomly selected person, the odds of dying in a motor vehicle accident and in air and space transport are 1 in 98 and 1 in 7,178, respectively. This indicates that flying is far safer than driving. However, for some, flying may feel more dangerous because our perception about risk is usually formed based on factors beyond mere facts. For example, most people think that they are good drivers and, as such, feel safer because driving affords personal control. Plane crashes are often catastrophic. It kills many at once and grabs the attention of major media, which make people more sensitive to them.

Are You an Average Person?

In general, there is a lot more to calculating and comparing risk than one might think. According to the experts, for an average American, the annual risk of being killed in a plane crash and a motor vehicle are about one in eleven million and about one in five thousand, respectively. But, is this all? First, most of us are not the average American. Some people fly more others, and some do not fly at all. So, if we take the total number of people killed in commercial plane crashes and divide that into the total population, the result gives a good general guide, but it is not specific to our personal risk.

Here are other useful numbers:

1. Dividing the number of people who die into the total number of people gives us the risk per person.
2. Dividing the number of victims into the total number of flights passengers took gives the risk per flight.
3. Dividing the number of victims into the total number of miles all of them flew gives you the risk per mile.

1995

In 1995, out of every one hundred million, about 16,300 and 111 people were killed in automobile and commercial flights accidents, respectively. The numbers of deaths per one hundred miles were respectively 3 and 100 for one hundred million miles traveled, and 30 and 20 for every one hundred million trips made.

This shows that the risk of death per mile is 33 times higher for car. However, the risk of death per trip is about 1.5 times higher for airplanes.

Discussion

All the above calculations produce useful and accurate numbers. However, which one is most relevant to us depends on our personal flying patterns. Some fliers take many short flights. Some take less but longer flights, for example. Since the overwhelming majority of the plane crashes take place in connection with takeoffs and landings, the risk is less a matter of how far you fly and more a matter of how often. If you are a frequent flier, then the risk per flight means more to you. For occasional long-distance fliers, the risk per mile means more. A frequent long-distance flier would want to consider both.

The number of plane crash fatalities in the United States varies widely from year to year. So calculation of risk based on one year and average of five years or ten or twenty may be different significantly. In some years, no plane crashes occur, or at least very few do. This makes the value of the risk per year misleading. If we average things over, say, five years or ten, some other factors will muddy the waters. In the last five years, safety factors have changed. A ten-year average might be misleading too.

Despite all these caveats, numbers are a great way to put risk in general perspective. Without question, most metrics shows that flying is less risky than traveling by cars. But wait: Just when you thought it was safe to use numbers to put risk in perspective . . . Numbers are not the only way—not even the most important way— we judge what to be afraid of. Risk perception is not just a matter of the facts. For example, consider the risk awareness factor. The more aware of a risk we are, the more concerned about it we become. This explains why, when there is a plane crash in the news, flying seems scarier to many of us, even though that one crash has not changed the overall statistical risk significantly.

Rare Events

This part aims to explore the theory regarding the low probability events such as exceedances and extreme values with a particular focus on its application. It deviates from material taught in introductory statistics courses in that it concentrates on frequency and values of extremes and the situations where such values are of greater concern than averages. The theory of exceedances, together with a brief account of extreme value theory, threshold theory, and theory of records, is presented. It is hoped the instructors of statistics will find this article of some value for teaching statistics and for its focus on a popular motivating theme.

Traditionally, statistics has focused on the study of averages and typical values with high frequencies. This is evident from examination of course descriptions and textbooks for introductory statistics courses offered in universities and colleges. Most classical approaches treat the data as a message and seek to decompose it into a systematic part (signal or trend) and a random part (noise). The techniques such as smoothing are then used to recover the systematic part. Smoothing usually is achieved by some sort of averaging (low-pass filtering). The difference between the original message and the systematic part (i.e., the random part) is then used to provide a reliability statement regarding the systematic part.

In wider applications, however, it is becoming increasingly impractical to focus on averages alone. It has become important to understand the distributions of extreme values and rare events, since such events are usually accompanied by severe consequences. Consider, for example, a natural disaster such as an earthquake. Clearly, a moderate earthquake is of less concern than a severe one. In the reliability study of systems and structures, it is the weakest link that is worrisome. In risk management and insurance industry, the highly complex situation and possible large claims are what one needs to prepare for. The celebrated *central limit theorem* has given statistics its focus on averages—for we do what we know how to do. The theories of extremes and record values are less simple, less unified, and more recent—but not less important.

Extreme values are usually analyzed using one of the three major theories. These theories are concerned with the actual values of the extremes. The following is a brief description of each:

1. The *extreme value theory* usually deals with annual maxima or minima. This theory is limited to the absolute largest or absolute smallest data value in a specific period (e.g., a year).
2. The *threshold theory* deals with values above or below a specified threshold.
3. The *theory of records* deals with values larger or smaller than all previous values.

As pointed out, these theories deal with the values of the extremes. The frequencies of extremes are usually analyzed using a theory known as the *theory of exceedances*. This theory deals with number of times a chosen threshold is exceeded. In this article, the theory of exceedances and some of its applications are discussed. The approach follows that of Castillo (1987) but with emphasis on applications of the theory in the analysis of sports data. We start by presenting some examples:

- In track and field, athletes may be required to meet or exceed prespecified performance levels to qualify for major competitions. We may be concerned with the number of athletes that succeed, or, alternatively, we may be interested in a level that ensures a certain number of qualifiers.
- In sports such as basketball, we may be interested in the number of players who score more than a specified number of points per game.

In most applications, the number of exceedances and values of excess over a threshold are combined to yield a more detailed analysis. For example, insurance companies analyze both the number of times a "large" claim is made and the amount by which the claims exceed a large threshold. Reliability engineers study both the number of large earthquakes and their sizes. We apply the same methodology to sports.

Exceedances

Assuming independent and identically distributed (i.i.d.) trials, determine the probability of r exceedances in the next n trials. For example, in a season of n games, how many times will a basketball player or a team score more than x points?

This situation can be described by a Bernoulli experiment with two possible outcomes: "exceedance" or "not exceedance." Since the experiment is repeated n times, the number of exceedances has a binomial distribution with parameters n and $p(x)$, where $p(x)$ is the probability of exceedance of the level x of the variable of interest. Note that $p(x) = P(X > x) = 1 - F(x)$, where $F(x)$ is the cumulative distribution function of X.

How Often, Return Periods

Consider an event (breaking a record, winning a championship, etc.) whose probability of occurrence in a unit of time (normally one year) is p. Assume that occurrences of such events in different periods are independent. Then, as time passes, we have a sequence of Bernoulli experiments with two possible outcomes: (a) occurrence or (b) nonoccurrence. Thus, the time (measured in units of a period of interest) to the first occurrence is a geometric random variable with parameter p and a mean value of $1/p$. This motivates the following definition.

Definition. Let E be an event and T the random time between consecutive occurrences of E. The mean value of the random variable T is called the return period of the event E.

Note that if $F(x)$ is the distribution function of the yearly maximum of a random variable, the return period of that random variable to exceed the value x is $1/[1-F(x)]$ years. The return period of the variable to go below the value x is $1/F(x)$ years.

Example. Consider the return period for a basketball player like Michael Jordan. Data from the National Basketball Association (NBA) show that the distribution of points per minute (PPM) for the guards who played in 1992–1993 season can be approximated by a normal distribution with mean 0.4236 and standard deviation 0.1159.

Using this and the fact that Jordan's PPM was 0.8291, the probability of exceeding Jordan's PPM is

$$P(PPM>0.8291)=1/5000.$$

Thus, an estimate of the return period for a player who would perform like or better than Jordan is five thousand players. If we assume that one hundred guards are added to the NBA each year, it takes, on average, about fifty years to find a player who will perform like or better than Jordan.

Example. In an interesting paper, Greenhough and colleagues analyzed soccer games from 169 countries between 1999 and 2001 (Greenhough et. al 2002). Specifically, they investigated the distributions of the number of goals scored by the home team, the away team, and the total. The results revealed that the total number of goals scored in a game exceeds ten once in every ten thousand English top-division matches or cup games, or about once in every thirty years. Worldwide, though, such a score is seen once in every three hundred top-division games, or about once a day.

Applications to Sports

The above results are applicable when observations form a sequence of independent and identically distributed random variables. This assumption is not reasonable for most sporting events such as the one-hundred-meter dash in which records are broken more frequently than what theory predicts. One way to deal with this problem is to assume that every year more athletes compete or more attempts are made, resulting in a higher chance of setting a record. Yang (1975) proposed a model that assumes that the observed records are generated by a geometrically increasing population. His model differs from the classical model in the limit distribution of the interrecord times. A generalization of the Yang model includes consideration of distributions such as $P(x)$ and is presented in Arnold (1998).

To clarify, suppose that a competition starts with one hundred participants and the rate of population or participant growth is 4 percent. This means that with each passing year, more participants would compete, thereby increasing the chance of setting a new record. The number will increase to $100(1.04)^{10} = 148$ after ten years and to $100(1.04)^{20} = 219$ after twenty years. If we assume that each participant has an equal chance of breaking a record, the chances of breaking the original record after ten and twenty years are 1.48 and 2.19 times of that for starting year, respectively. Note that using the past history, we can estimate both the number of first-year participants and the geometric rate of growth. Alternatively, we can choose a *(t)* such that as time passes, the chance of setting a record increases.

Uncertainty and Diagnostic Tests

Diagnostic tests are usually assessed by two numbers known as *sensitivity* and *specificity*. Sensitivity is the probability (percentage) of a positive test result when the disease is present, and specificity is the probability (percentage) of a negative test result when disease is absent. The higher the sensitivity and specificity, the better (and often more expensive) the test.

Despite their importance, sensitivity and specificity are not of main concern to most patients. Instead, informed patients want to know what percentage of people who test positive actually have the disease. In fact, when the disease is rare, the probability that a patient with positive test results actually has the disease can be surprisingly low even for tests with a high sensitivity and specificity.

Consider, for example, breast cancer in women. It has been estimated that of women who get mammograms at any given time, only 1 percent truly have breast cancer. (Note: this number should not be confused with likelihood of breast cancer in women's lifetime.) For mammograms, typical values reported are 0.86 for sensitivity and 0.88 for specificity. With these specifications, for a positive test result from a mammogram, the probability that the woman truly has breast cancer is less than 8 percent.

How can this probability be so low? Consider a random sample of one hundred women selected for screening. Here, one woman is expected to have breast cancer, 1 percent of the sample. For the woman with breast cancer, the chance of detecting it is 0.86. Therefore, we would expect the one woman with breast cancer to have a positive test result. For a woman without breast cancer, the chance of detecting it is 0.88. Therefore, for this group of ninety-nine women, we would expect about $0.88 \times 99 = 87$ negative results and $0.12 \times 99 = 12$ positive (false positive) results. This shows that of the thirteen women with a positive test result, $1/13 = 0.08$ actually is expected to have breast cancer.

The issues are not limited to infectious disease. For example, a study done at a Lyme disease clinic in Boston in 1992 found that of eight hundred patients who had been referred to the clinic because of positive results on blood tests, 45 percent were false positives—people who did not, in fact, had Lyme disease. Other examples of this type of situation are prostate specific antigen (PSA) blood test for prostate cancer in men and Down syndrome.

Covid

If you get a positive result on a Covid test that only gives a false positive one time in every one thousand, what is the probability that you actually have Covid?

You would not know because you do not have enough information. Without knowing what percentage of population have, in fact, Covid, you do not know how likely it is that a result is false or true.

To see this, imagine you undergo a test for a rare disease. The test is amazingly accurate: if you have the disease, it will correctly say so 99 percent of the time; if you do not have the disease, it will correctly say so 99 percent of the time. Assume that the disease in question is very rare; just one person in every one thousand has it. This is your "prior probability."

Suppose you test one million people. With one in one thousand, there are one thousand people who actually have the disease, and your test correctly identifies $(1,000 \times 0.99) = 990$ of them. There are 999,000 people who do not have the disease; your test correctly identifies $999,000 \times 0.99 = 989,010$ of them. But that means that your test, despite giving the right answer in 99 percent of cases, has told $999,000 \times 0.01 = 9,990$ people that they have the disease, when in fact they do not. Therefore, if you get a positive result, in this case, your chance of actually having the disease is 990 in 10,980 (9,990 + 990), or just around 9 percent.

Without knowing the prior probability, you do not know how likely it is that a result is false or true. If the disease were not so rare—if, say, 1 percent of people had it—your results would be very different. Then you would have not only 9,900 false positives but also 9,990 true positives. So, if you had a positive result, it would be more than 50 percent likely to be true.

So what should we do when facing uncertainties of this kind?

Be your own advocate in your care.

Trust science and professionals.

Talk with friends and family to learn about their experiences.

Bayesian Approach

The Bayesian theorem is used in probability studies to help predict future events based on an event that already occurred, which is known as a prior probability. When data is not available, a prior distribution is used to calculate knowledge about the parameter. When data is available, we can update the prior knowledge using the conditional distribution of parameters. The Bayesian theorem makes it possible to transition from the prior to the posterior (Bayesian).

Some of the uses for the Bayesian theorem include medical diagnosing for predicting the outcome of the disease process and predicting how fast an infectious disease can spread. The theorem can also be used in sports to predict the outcome of a game and how many points a player might score during a game. Bayes' theorem can also be used to help predict odds in gambling. There are many other uses for Bayes' theorem; it can be applied whenever you need to predict future outcomes given previous data.

The Bayesian model is

$$P(A|B) = \frac{P(A\cap B)}{P(B)}.$$

By breaking down the model part by part, hopefully this will help you interpret why the Bayesian model helps predict future outcomes. I'm going to start breaking the model down by starting on the left side of the equal sign, which is $P(A|B)$. This part means the probability of A given (|) B; meaning, you want to find the probability of A given that you already have the information (data) for B. The second part is the numerator of the fraction $P(A\cap B)$. In this part, it makes it easier to set up your data in a table, tree diagram, or Venn diagram so you can see what you need to multiply, $P(A\cap B) = P(A) * P(B)$ (for independent events). The third part is the denominator of the fraction $P(B)$, which is the probability that you already know. Hopefully this helps you understand why the Bayesian model helps predict outcomes; but if not, let's use an example.

When betting that the second card drawn out of a deck will be a king given that the first card is a king, what will be your probability of winning being that the dealer is not replacing the cards as they are drawn?

$$A=\{2^{nd} \text{ card is a king}\} \qquad B=\{1^{st} \text{ card is a king}\}$$

$$P(A|B)=\frac{P(A \cap B)}{P(B)}=\frac{P(1st \text{ card is a king and } 2nd \text{ card is a king})}{P(1st \text{ card is a king})}$$

$$=\frac{\frac{4}{52}*\frac{3}{51}}{\frac{4}{52}}=\frac{3}{51}=0.059$$

This means that you have about a 6 percent chance of winning.

Probability and Odds

Recall that if outcomes of an experiment are equally likely, then we may assign probability to an event A as

$$P(A)=\frac{number \ of \ times \ A \ occurred}{number \ of \ times \ trial \ is \ repeated} \quad \text{or}$$

$$P(A)=\frac{number \ of \ outcomes \ in \ A}{total \ number \ of \ outcomes}.$$

For such cases, odds are viewed as ratios of the number of outcomes in A and the number of outcomes in not A. We could also look at it as the number of successes to the number of failures, where a success is occurrence of event A. Formally,

$$\text{Odds}(A) = \text{Odds in favor of A} =\frac{P(A)}{1-P(A)}=\frac{P(A)}{P(A')}.$$

For example, rolling a fair die, the odds of rolling a three are one to five, or 1:5. as there is one way to roll a three (success), while there are five ways to not roll a three (failures). Note that since the probability of rolling a three is 1/6,

$$\text{Odds(Observing 3)} =\frac{P(3)}{1-P(3)}=\frac{1/6}{5/6}=\frac{1}{5}.$$

Similarly, rolling two fair dice, the possible outcomes are

$$(1,1), (1,2), (1,3), (1,4), (1,5), (1,6)$$
$$(2,1), (2,2), (2,3), (2,4), (2,5), (2,6)$$
$$(3,1), (3,2), (3,3), (3,4), (3,5), (3,6)$$
$$(4,1), (4,2), (4,3), (4,4), (4,5), (4,6)$$
$$(5,1), (5,2), (5,3), (5,4), (5,5), (5,6)$$
$$(6,1), (6,2), (6,3), (6,4), (6,5), (6,6)$$

Since there are three ways to roll sum ten, and thirty-three ways to not roll sum ten, the odds of rolling sum ten is then 3:33. Note that when the probability of success is low, the odds and the probability are very close.

A more interesting example is the ratio of odds for winning a home-versus-away game,

$$\frac{P(\text{home win})/P(\text{home loss})}{P(\text{away win})/P(\text{away loss})}$$

which is expected to be larger than one. In fact, some recent investigations revealed that this ratio varies between 1.2 (for major league baseball) and almost 2.5 (for college basketball).

The odds of a lightning strike are 1:135,000. The odds of dying as a result of a meteor strike anywhere in the world, like the kind of rare but catastrophic geologic event that shapes an eon, are 1:75,000. The odds of winning the Powerball lottery are 1:195,249,054.

Even though the odds of dying from a meteor strike are better than winning the Powerball, many play the lottery and think that they have a good chance of winning but hardly think that a meteor strike will affect them.

Odd and Gambling

Gambling has grown rapidly in the last few decades especially in relation to the sports competitions. People who gamble on outcomes of games and tournaments may be divided into three possibly overlapping categories:

1. The casual gamblers who gamble for possible enjoyment and may not be aware of odds or strategic subtleties and may not even be very knowledgeable about players and the teams.
2. The compulsive gamblers who are happy primarily when in the act of gambling and do not care about the sport itself or even the outcome of the competitions.
3. The professional gamblers who look to gambling as another profession. Such individuals understand odds, the games involved, and also know players, teams, and their potential well, although they may not have a good background in mathematics and the calculus of the odds.

One interesting fact is the following: many individual gamblers who understand the odds and their implications may not fully understand the difference between the true and house odds. To clarify this, consider the following table comparing the true and house odds for the game of roulette in a well-known place such as Las Vegas, Monte Carlo, and Atlantic City. For example, the true and house odds in Las Vegas roulette are given in table below.

Type of bet	True odds	House odds
Color(Red or Black)	20 : 18	1 : 1
Parity(Even or Odd)	20 : 18	1 : 1
18#'s (1-18 or 19-36)	20 : 18	1 : 1
12#'s(columns or dozens)	26 : 12	2 : 1
6#'s(any 2 rows)	32 : 6	5 : 1
4 #'s (any 4 number square)	34 : 4	8 : 1

3 #'s (any row)	35 : 3	11 : 1
2 #'s (adjacent)	36 : 2	17 : 1
Single #'s	37 : 1	35 : 1

If a one-dollar bet is made on red, then the chance of winning a dollar is 18/38; whereas, the chance of losing is 20/38. So in the long run (or on average), one will lose two dollars for every thirty-eight games he or she plays. With so many players and so many tables to play, it is not hard to imagine what could happen in the long run.

Turning to a sport-related example, betting works as follows: The bookmaker sets odds on each of the possible outcomes. Suppose that an odds of thirty-nine to one is set for a player or a team. If this player wins, the person who has bet on this player winning will receive amount given in the odds, plus his or her money. This means a gain of thirty-nine dollars. If the chance of winning for this player is 1/40, then the bet is called a fair bet. As in gambling clubs, if we convert the odds to the probabilities, we find that total probability assigned to the players or teams is less than one. The difference benefits the bookmakers and in the long run makes them the winners. Remember that bookmakers have various ways to guarantee making money. People who know probability theory usually see no merit whatever in betting at racetracks or casinos, or in buying lottery tickets. After all, multimillion-dollar gambling places are built using gambling proceeds. Why would somebody bet with people who make a lot of money gambling?

Rare Events in Statistics Curriculum

This part aims to emphasize the need for inclusion of statistical theories of extreme values in introductory statistics courses. It points out situations where extremes values are of greater concern or importance than averages. The theory of exceedances, together with a brief account of extreme value theory, threshold theory, and theory of records, is discussed. It is hoped the instructors of statistics find this article of some value for teaching topics related to nontypical and nonaverage values.

Traditionally, statistics has focused on the study of values with high frequencies and on averages. This is evident from examination of course descriptions and textbooks for introductory statistics courses offered in universities and colleges. In the modern world, however, it is becoming increasingly impractical to focus on averages alone; it has become important to pay attention to rare events and extremes, since they are newsmakers and are often accompanied by severe consequences.

The celebrated *central limit theorem* has given statistics its focus on averages—for we do what we know how to do. The statistical theories of extreme values are less simple, less unified, and more recent. However, they are not less important. In fact, there is a need for inclusion of extreme value statistics in introductory courses.

Why Averages?

As pointed out, most textbooks focus on values with high frequencies and on averages. This is because most classical approaches treat the data as a message and seek to decompose it into a systematic part and a random part. Another representation is

Message = Signal (systematic part) + Noise (random part).

For data with time index, known as times series, the Wald decomposition theorem states the following:

Time Series = Deterministic (Trend) + Stochastic (Noise).

When the form of a signal is known or is assumed, it is easy to separate the two parts. In the absence of information concerning the systematic or random parts, smoothing is used to recover each part. Smoothing is an exploratory operation, a means of gaining insight into the nature of data without precisely-formulated models or hypotheses. This is usually carried out assuming that the systematic part is smooth and the random part is rough—that is,

Message = Smooth part (trend) + Rough part (noise).

Smoothing is often achieved by some sort of averaging (low-pass filtering). Several techniques (low-pass filters) are available for doing this. Once the smooth part is determined, the difference between the original message and the smooth part is used to provide an estimate for the rough part. The rough part is usually used to make reliability statement regarding the systematic part.

For time series, for example, one popular smoothing technique is known as moving averages. The idea is to average the neighboring values and to move it. To clarify, suppose that Y_t, $t = 1, 2, ..., n$ is an observed time series of length n. The moving average of order $2p+1$ is defined as, where $w_j > 0$ are weights such that

$$\sum_{j=-p}^{p} wj = 1.$$

The weights are usually selected based on prior information. One classical approach referred to as exponential smoothing chooses the weights

$$S_t = \sum_{j=-p}^{p} w_j \, Y_{t-j} \quad , \quad t = p + 1, ..., n - p$$

according to a geometric progression. It is useful when assigning bigger weight to more recent values makes sense.

Why Extreme Values?

The average wealth of me and Bill Gates is around forty billion.

As noted earlier, in many applications, it is not appropriate to focus on averages. In fact, there are many instances where extreme values and values with low frequencies are of more concern than average values or values with high frequencies. Examples include a large natural disaster compared to a moderate or an average one, a weak component or link compared to their average counterpart, and a large insurance claim compared to an average claim. In sports, of course, fans only remember out-of-ordinary or exceptional performances or games.

Let us now see how extreme values are analyzed. Extreme values are usually analyzed using one of the three major theories:

1. the *extreme value theory*, which deals with "annual" maxima or minima;
2. the *threshold theory*, which deals with values above or below a specified threshold;
3. the *theory of records*, which deals with values larger or smaller than all the previous values.

These theories deal with the actual values of the extremes. The frequencies of extremes are analyzed using the *theory of exceedances*. This theory deals exclusively with number of times a chosen threshold is exceeded. For example, in track and field, athletes may be required to meet or exceed a prespecified performance levels to qualify for major competitions. Here, we may be concerned with the number of athletes that succeed. Alternatively, we may be interested in a level that ensures a certain number of qualifiers. Note that, in most applications, the number of exceedances and values of excesses over a threshold are combined to yield a more detailed analysis. For example, insurance companies analyze both the number of times a "large" claim is made and the amount by which these claims exceed a specific large threshold. Reliability engineers study both the number of extreme loads, such as large earthquakes, and their magnitudes. In what follows a brief account of each theory is given.

Extreme Value Distributions

Extreme value theory generally deals with the annual (or any other period) maxima or minima. Specifically, the theory is based on dividing the sample into a subsamples and fitting a distribution to maxima or minima of the subsamples. For example, data may consist of largest earthquakes in California for each of the last one hundred years.

Here, like most statistical theories, first, distribution of the largest or the smallest values were derived for a finite sample. Then by letting sample size tend to infinity, the limiting distribution of extreme values was obtained. For this, a typical maxima Y_n is reduced with a location parameter β_n and a scale parameter α_n (assumed to be positive) such that the distribution of standardized extremes ($Y_n - \beta_n$)/α_n is nondegenerate. The forms of the limiting distributions are specified by the extreme value theorem. This theorem states that there are three possible types of limiting distributions (denoted by F_y (y)) for maxima:

1. the Gumbel distribution (type I) for which

$$F_y(y) - \exp(-y)) \text{ for } -8 < y < 8$$

2. the Fréchet distribution (type II) for which

$$Fy(y) = \begin{cases} 0 & \text{For } y \leq 0 \\ \exp(-y^{-k}) & \text{For } y > 0 \quad (k > 0) \end{cases}$$

3. the Weibull distribution (type III) for which

$$Fy(y) = \begin{cases} exp(-(-y)^k) & \text{For } y \leq 0 \quad (k > 0) \\ 1 & \text{For } y > 0 \end{cases}$$

These three forms can be combined to yield the generalized extreme value distribution taking the form

$$P(\max(Y_1, Y_2, ...,y_N) \quad y) = \exp\{-\lambda(1\text{-}ky/\sigma)^{1/k}\}$$

where, depending on whether parameter k is positive, zero, or negative, we get type I, type II, or type III extreme value distribution, respectively.

Most classical distributions fall in the domain of attraction of one of these three types. For example, distribution of maxima of samples from a normal distribution tends to type I (Gumbel) distribution. It shows the necessary and sufficient conditions for a particular distribution to fall in domain of attraction of one of the three types. For type I, it is

$$\lim_{t \to +\infty} n[1\text{-}f_Y)a_nt + b_n)] = e^{-y}.$$

For type II, it is

$$\lim_{t \to +\infty} \frac{1-f_Y(t \cdot y)}{1-f_Y(t)} = y^k$$

where $t > 0$ and $k > 0$ and $f(.)$ denotes the density function. Finally, for type III, it is

$$\lim_{t \to +\infty} \frac{1-f_Y(t \cdot y + u)}{1-f_Y(t + u)} = y^k$$

where u is the endpoint of the distribution for $Y(F_Y(u) = 1)$, $t > 0$, and $k < 0$.

Basic results obtained are the following:

- Only distributions unbounded to the right can have a Fréchet distribution as a limit.
- Only distributions with finite right end point ($u < \infty$) can have a Weibull as a limit.
- The Gumbel distribution can be the limit of bounded or unbounded distributions.

How do we decide which of the three limiting distributions fit to the data? Theoretically, we can use the fact that each of the classical distributions falls in the "domain of attraction" of one of the limiting distributions above. This works if distribution of the original data is known. Unfortunately, in practice, such information is not usually available, and decisions should be based on the area of application or on expert opinion. For example, in the case of sports, one may think of a possible ultimate record, in which case, the distribution bounded above (below) is more appropriate. When information about the appropriate limiting distribution is absent, statistical goodness of fit may be used.

When fitting extreme value distributions to annual maxima or minima, it is possible to discard some relevant data related to the years with several large observed values and retain less informative data from the years with no real large values. The threshold theory discussed next avoids this problem.

Summary

Assessing the probability of rare and extreme events is an important issue in the risk management of financial portfolios. Extreme value theory provides the solid fundamentals needed for the statistical modelling of such events and the computation of extreme risk measures.

Extreme value theory (EVT) is a branch of statistics dealing with the extreme deviations from the median of probability distributions. There exists a well-elaborated statistical theory for extreme values.

It applies to (almost) all (univariate) extremal problems. From EVT, extremes from a very large domain of stochastic processes follow one of the three distribution types: Gumbel, Fréchet/Pareto, or Weibull.

The *generalized extreme value (GEV) distribution* is a family of continuous probability distributions developed within EVT. The GEV combines three distributions into a single framework. The distributions are

Type I: Gumbel   Type II: Fréchet   Type III: Weibull

The GEV allows for a continuous range of possible shapes. The shape parameter, S, governs the tail behavior of the distribution. The subfamilies defined by $S \sim 0$, $S > 0$, and $S < 0$ correspond respectively to the Gumbel, Fréchet, and Weibull families. Note the differences in the ranges of interest for the three extreme value distributions: Gumbel is unlimited, Fréchet has a lower limit, while the reversed Weibull has an upper limit.

The GEV facilitates making decisions on which distribution is appropriate. The GEV distribution is often used as an approximation to model the minima or maxima of long (finite) sequences of random variables. In general, the GEV distribution provides better fit than the individual Gumbel, Fréchet, and Weibull models. For example, in most hydrological applications, the distribution fitting is via the GEV, as this avoids imposing the assumption that the distribution does not have a lower bound (as required by the Fréchet distribution).

Generalized Pareto Distribution

The threshold theory allows one to make inference about the values above or below a threshold—that is, the upper or the lower tails of a distribution. It considers the excesses, the differences between the observations over the threshold, and the threshold itself. Like extreme value distributions, there are three models for tails:

1. long-tail Pareto
2. medium-tail exponential
3. short-tail distribution with an end point

Again, most classical distributions fall in domain of attraction of one these tail models. It has been shown that the natural parametric family of distributions to consider for excesses is the generalized Pareto distribution (GPD),

$$H(y; \sigma, k) = 1 - (1 - \frac{ky}{\sigma})^{1/k}.$$

Here, $\sigma > 0$, $-\infty < k < \infty$, and the range of y is $0 < y < \infty$ ($k \leq 0$), $0 < y < \sigma/k$ ($k > 0$).

This is motivated by the following considerations:

- The GPD arises as a class of limit distributions for the excess over a threshold, as the threshold is increased toward the right-hand end of the distribution (i.e., the tail).
- If Y has the distribution $H(y, \sigma)$ and $y' > 0$, $\sigma - ky' > 0$, then the conditional distribution function of $Y - y'$ given $Y > y'$ is $H(y; \sigma - ky', k)$. This is a "threshold stability" property; if the threshold is increased by an arbitrary amount y', then the GPD form of the distribution is unchanged.
- If N is a Poisson random variable with mean λ and $Y_1, Y_2, ...,$ Y_N are independent excesses with distribution function $H(y; \sigma, k)$, then

$$P(max\,(Y_1, Y_2, ..., Y_N) \leq y) = \exp\{- \lambda\,(1-ky/\sigma)^{1/k}\}$$

which is the generalized extreme value distribution. Thus, if N denotes the number of excesses in, say, a year and $Y_1, Y_2,..., Y_N$ denote the excesses, then the annual maximum has one of the classical extreme value distributions.

- The limit $k \to 0$ of the GPD is the exponential distribution.

In practice, the proposed method is to treat the excesses as independent random variables and to fit the GPD to them. The choice of threshold is, to a large extent, a matter of judgment depending on what is considered large or small.

The theory is very useful when modeling large values based on observed large values, which is of main concern. Clearly, the modeling and prediction of large earthquakes should be based on past large earthquakes, not on past medium or small earthquakes. The same is true for sports records and performances, as moderate values do not carry information about the exceptional future performances.

Summary

The relation between records and their frequencies for sports events has been considered by many investigators. Here, like many other areas of statistical application, we are primarily interested in drawing inference about the extreme values of a population of records or performances. The most commonly used method, developed at length by Gumbel, is to divide the data into subsamples and to fit one of the extreme data points in each subsample. In many cases such as environmental series, the natural subsample is one year's data; in fact, Gumbel's method is often referred to as the method of annual maxima. This also makes sense for sports, since here, too, usually annual or seasonal records are of interest. However, in other areas of application such as earthquake engineering, there is no natural seasonality in the data, and the subsample method appears artificial and wasteful. This may also be the case for some sports because of the expansion of sports-related activities and the fact that seasons do not happen at the same time all over the world. An example of this is tennis for which there are tournaments all year round. When applying to a given sport, subsample method presents a further difficulty. To clarify this, consider a sport such as men's two-hundred-meter race and take the period to be one year. Gumbel's method fits one of the extreme value distributions to the data presenting the best time for each year. Now, suppose that in a given year (an exceptional year), more than one record was set. Gumbel's method drops these records and uses only the best for that year together with the best records for other years, which could not be as good as the second-best record of the year in which more than one record was set. This means that a great deal of relevant information regarding the records (extremes) may be ignored by this method. With this and other drawbacks of the methods used for prediction of the extreme values, consideration of an alternative approach that avoids some or all of these difficulties is much needed. Here, we consider a theory concerning the tail behavior of statistical distribution introduced by Pickands (1975). This approach is based on fitting a suitable parametric model to a few of the largest order statistics corresponding to the best records regardless of when they were set. Since this approach has several advantages over the traditional subsample method, in recent years, a great deal of effort has been put in by various investigators to develop the theory and the methods based on that to its fullest potential. These have led to introduction of new classes of alternative procedures commonly referred to as "the threshold method." In fact, so far, two main procedures have been made available for practical applications: one by hydrologists called POT (peaks over threshold) method, and one by statisticians based on the GPD (generalized Pareto distribution) and the use of extreme order statistics. These two areas of work may be regarded as contributions to the same general problem—that is, the modeling of the extreme characteristic of a series of observations in terms of its exceedances (in fact excesses) over a high (low) threshold level.

Because of the importance of the results concerning the tail distribution and also power and flexibility of the methods based on them, this chapter is devoted to threshold method and discussion of its application to sport data.

Analysis of Records

Records in general and sports records in particular are of great interest, and their occurrence usually results in a great deal of media attention. Examples include the chase of the single season home run record by baseball players and breaking of the men's one-hundred-meter record.

Records occur everywhere from sports to the stock market. *The Guinness Book of World Records* is popular reading around the world. We plan to discuss the mathematics behind the theory of records and present some applications.

Theory of records deals with values that are strictly greater than or less than all previous values. Suppose that data consists of the real numbers Y_1, Y_2, ... Y_n with Y_n representing the most recent measurement. Usually, Y_1 is counted a record, as it is the largest value at the starting point. Y_i is a record (upper record or record high) if it is bigger than all previous values or measurements—that is, if

$$Y_i > \max(Y_1,...,Y_{i-1}) \text{ for i} \geq 2.$$

The study of such values, their frequency, times of their occurrence, their distances from each other, etc., constitutes the theory of records. Formally, the theory of records deals with four main random variables:

1. the number of records in a sequence of n observations
2. the record times
3. the waiting time between the records
4. the record values

It is interesting to note that the first three can be investigated using nonparametric methods; whereas, the last one requires parametric methods.

Example: Think about total annual snowfall in New York in the next ninety or one hundred years.

Q: What is the chance that a baby born this year would experience, for example, six personal record snowfalls? ten personal record snowfalls?

Q: How many personal record snowfalls should she expect to experience?

Q: How many record snowfalls is the most likely case?

The first year is a record. No history. What about the second (next) year?

The chance that the second year is a record is 50 percent, or 1/2. What about the third year?

The chance that the third year is a record is 1/3. The same applies to fourth, fifth, and so on.

Let R_n denote the number of records in n years. Then we have

Term Years	Expected Number of Records
1	1
2	$1/2$
3	$1/3$
4	$1/4$
.	.
.	.
.	.
n	$1/n$

$$E(R_n)=1+\frac{1}{2}+\frac{1}{3}+\frac{1}{4}+\dots+\frac{1}{n}$$

For $n = 90$

$$E(R_n)=1+\frac{1}{2}+\frac{1}{3}+\frac{1}{4}+\dots+\frac{1}{90}=5.08\approx5$$

that is, a ninety-year-old person is expected to see five record snowfalls in her lifetime.

Note: Five is also the most likely scenario (probability = 0.21).

What are the values of this series for $n = 1000$?, $n = 1000,000$?, $n = 1000,0000,0000$?

The answers may surprise you. For the first two, it is 7.5 and 14.4.

The following is an example of mathematics of records in traffic regulations.

Car Caravan in a One-Lane Tunnel

When traffic moving in one direction is confined to a single lane, a slow car is likely to be followed closely by a queue of vehicles whose drivers wish to go faster but who cannot pass. If there is no exit from this lane, then more and more following vehicles will catch up and be added to the slow-moving "platoon" or "caravan" . . . until there happens to be following vehicle traveling at a lower speed. This vehicle will not catch up but will accumulate its own caravan. Thus, cars whose drivers all desire different speeds in fact will travel in caravans at actual speeds determined by record lows in the sequence of desired speeds.

Applying the simple probability model to a random sequence of drivers, the frequency of record lows corresponds to the number of caravans formed by n drivers. In addition, the number of trials between successive record-breaking low values correspond to the lengths of caravans.

Since caravans will be successively slower, separations between caravans will increase as time passes. This explains why cars near the exit of long tunnel tend to travel faster and in smaller bunches more widely separated than cars in the tunnel near the entrance. This model of traffic flow also has been mentioned by several authors and has been used in places such as Bush Gardens near the BBC headquarters to help the flow of traffic.

Note that five is also the most likely scenario (probability = 0.21). Every record will eventually be beaten. In fact, surprisingly, the intuitive idea that every record will be beaten also leads to mathematical proof that the harmonic sum

$$1+1/2+1/3+\cdots$$

grows without bound, becoming bigger than any finite number.

An Application

Values of n such that $E(R_n) \geq N$ for the first time is

N	2	3	4	5	6	7	8	9	10
n	4	11	31	83	227	616	1674	4550	12367

Minimum number of years one needs to wait to see N records (theoretically).

For example, $1/1+1/2+1/3+1/4 > 2$. From this table, in a random series, on average,

- an eleven-year-old child has seen *three* personal records;
- a mother, thirty-one years old, *four* personal records;
- a grandmother, eighty-three years young, *five* personal records.

Isn't this a key to the fact that in our youth, winters were colder with more snow? summers were warmer? generally things were better?

Questions

1. How long would the present record stand (survival time)?
2. What would be the value of the next record?
3. How long should we expect to wait for the rth record to be set?
4. What can we say about the time of the rth record?
5. What can we say about the value of the rth record?
6. Is there an ultimate record?

As n tends to infinity, what does $E(R_n)$ tend to? Well, using the approximation $\int(1/x)\,dx$, it tends to $Ln\,(n) + \gamma(=0.5772)$, where γ is also an Euler's number.

Note that $Ln(n)$ tends to infinity as n tends to infinity. It is known among people who follow records that every record will eventually be beaten. In fact, surprisingly, the intuitive idea that every record will be beaten also leads to mathematical proof that the harmonic sum $1+1/2+1/3+\cdots$ grows without bound, becoming bigger than any finite number.

W_r: waiting time between the $(r-1)^{th}$ and r^{th} records

The expected value of W_r is infinite even for $r = 2$.

Record Number r	2	3	4	5	6	7	8
Median W_r	4	10	26	69	183	490	1316
Med W_r/Med W_{r-1}		2.50	2.60	2.65	2.65	2.68	2.69

Med W_r/Med W_{r-1} tends to e. The median number of attempts required to arrive at a new record is $e=(2.718...)$ times the median number of attempts that was required to arrive at the previous record.

This suggests a geometric increase with rate $e = 2.718...$ If we assume that one unit of attempts was needed to arrive at the second record, the total number of attempts to arrive at record number twenty may be calculated as

$$1+ e + e^2 +\cdots + e^{18} = 103{,}872{,}541.$$

This is a slight overestimation, as for early records, the ratios are less than e. This leads to probability estimates for a new record for one-hundred-meter dash as

0.152461 for the next year 0.562681 for the next 5 years 0.808753 for the next 10 years

Recall that after seeing the second record, the median wait time to the third record ten observations (attempts). Other results regarding

W_r include a law of large numbers, $\log W_r/r \to 1$, and a result indicating that $\log W_r$ is approximately equivalent to the arrival time sequence of a Poisson process. Since sports records are more frequent than records generated by independent and identically distributed sequences, it is possible to model $\log W_r$ as a nonhomogeneous Poisson process.

Let us look at the data for long jump. For long jump, the fifth record was set in 1991. Using the theory of records, seventy-three attempts are needed to produce five records, and these should have occurred during the period 1962–1991 (thirty years). This leads to geometric increase with rate $i = 1.055$. Noting that the waiting time to the sixth record is 183 attempts, it takes (in median sense) forty-nine years for a new record to be set. This means waiting till the year 2040. The return period of the present record (8.95) was found to be 64.5 years based on the tail model I obtained.

We end this part by noting that, rather than records and waiting times between them, one could consider improvements of equal size and analyze the corresponding waiting times. This seems a reasonable approach, since, as records improve, increase in number of attempts could offset the decrease in number of record-breaking performances. For example, consider the rise in pole vault records and their waiting times shown in the following table.

Data for Pole Vault

Improvement (feet)	Number of Years
14 to 15	13
15 to 16	22
16 to 17	1.5
17 to 18	7
18 to 19	10
19 to 20	10

Here, one can consider smaller improvements and apply some of the classical statistical methods. In the case of pole vault, for example, the goal of such analysis should be to predict the number of years it would take to go from twenty to twenty-one.

Examples of Applications of Records

Structural engineering. Modern building codes and standards provide information on extreme winds and other maximum forces expected to be acting on a structure in its lifetime.

Ocean engineering. The designs of offshore platforms, breakwaters, and dikes rely on knowledge of the probability distribution of the largest waves and the periods associated with the largest waves.

Pollution studies. Codes to control pollution require that pollutant concentration (expressed as the amount of pollutant per unit volume of, say, air or water) remains below a given critical level. Here, the largest value plays a fundamental role.

Meteorology. Extreme meteorological conditions influence many aspects of human life as well as the behaviors of some machines, the lifetimes of certain materials, and so on. In these cases, engineers are concerned with accurate prediction of rare events rather than mean values.

Insurance. The future solvency of an insurance company depends on its ability to predict, with some degree of accuracy, the magnitude and frequency of enormous claims. If a terrorist bombing destroys a client's skyscraper in Manhattan or if hurricanes, floods, or earthquakes hit large numbers of a company's clients, the company must be prepared.

Materials strength. An important application of extreme value theory to materials strength is the analysis of size effect. In many engineering problems, the strength of actual structures must be inferred from the strengths of small elements or reduced-size prototypes or models, which are tested under laboratory conditions. Extreme value theory is used to make reliable extrapolation possible. In general, the minimum strength of the weakest subpiece determines the strength of a piece.

Exceedances and Excesses

In sports, extraordinary individuals or teams are those who pass a threshold classified as average and ordinary. For example, in one-hundred-meter dash for men, only athletes who run the distance in less than ten seconds may be considered exceptional. In basketball, players with a point per minute or assist per minute greater than 2/3 may be referred to as exceptional. In tennis, players who win several grand slam or have more than forty-five aces in one match may be considered exceptional. If we examine the distribution of times for one-hundred-meter race for the runners or points per minute (PPM) for basketball players, then we find that the exceptional performances constitute the tail of the corresponding distribution (for one-hundred-meter times, the lower tail; and for PPM, the upper tail). Thus, to study such performances, rather than entire distribution covering the entire range of possible values, we may just study the upper (lower) tail of the corresponding distribution. We can do this either by seeking a model for values above (below) a chosen threshold or by considering, for example, the third-best performance in the history of that sport as threshold and study the number of times it was surpassed or exceeded. This type of analysis allows us to find how many basketball players would it take to produce a player that will perform like or better than, for example, Michael Jordan.

Exceedances in Sports

The most memorable moments in life are those that are out of the ordinary, those that are not average. Similarly, the most critical values in data are often the extreme values. Examples of this include large earthquakes, high-speed winds, large claims in insurance, weakest link, star players, and fastest runners. In sports, no average performances ever make the news. However, the classical statistical theories deal mostly with problems related to averages as a result of the celebrated central limit theorem. Clearly, alternative methods must be used for analysis of extreme values.

Extreme values are usually studied using three main available theories. The extreme value theory deals with annual maxima or minima. This theory is limited to the absolute largest or absolute smallest data value in a given period (e.g., a year). The threshold theory deals with values above or below a specified threshold. But choosing a threshold is often a very difficult choice. Finally, the theory of records deals with values larger or smaller than all past values. These theories are concerned with the actual value of the extreme observations.

A theory known as the theory of exceedances, on the other hand, deals with number of times a threshold is exceeded. As a result, it is a counting process. In this thesis, we follow the presentation of E. Castillo (1987), in *Extreme Value Theory in Engineering* (Academic Press, London), and this theory is applied in a new way to the analysis of sports. Some examples include the following:

1. In track and field, athletes need to exceed specified qualifying levels (e.g., time, distance, or height) to be selected for certain competitions. We are concerned with the number that succeed.

   For instance, in the long jump, a distance of eighteen feet, nine inches will qualify for districts.

2. In sports such as basketball, we may be interested in the number of players who average more than m points per game.

In most applications, exceedances and threshold theory are combined to yield more detailed analysis. For example, insurance companies worry not only about the number of times a "large" claim is made but also about the amount by which the claim exceeds the threshold of "large."

Olympic Data

In 2016 Olympics, in Brazil, two thousand athletes from more than two hundred nations were set to compete for forty-seven titles. And, according to the experts, some of these athletes and countries were on the brink of rewriting Olympic history.

Looking back to some extraordinary events of previous Olympics, other than Michael Phelps's eight gold medals in Beijing Olympics, a great moment belonged to Usain Bolt, who won the men's one-hundred-meter sprint in a world record time of 9.69. Since then, Bolt has put on another amazing performance at the world championships, shattering his own record in the one hundred meters by 0.11 seconds to an almost inhuman 9.58 seconds. Additionally, he has lowered the two hundred meters record by the same amount to 19.19 seconds. Bolt became the first man to win back-to-back Olympic sprint doubles when he successfully defended his one-hundred-meter and two-hundred-meter titles in the London Olympics. In Rio, the Jamaican sprinter could have won an unprecedented third sprint double. And if he also could have won the 4x100m, as he did so in Beijing and London, Bolt's Olympic gold medal tally would stand at nine; only one man—USA's Ray Ewry—has ever won more Olympic gold medals in athletics.

The one-hundred-meter run defines the fastest man on earth. In well over one hundred years, there have been only twenty-five men who have been named "the fastest man on earth." Bolt is the last in the list, and his performance has since forced experts to alter their predictions related to the one-hundred-meter run.

For the majority of experts and fans, the most interesting prediction is that for the ultimate record, as it shed light on human strength and limits. For teachers like me, such problems are even more interesting because of their educational value.

My personal prediction for the ultimate record is 9.40 seconds, with 90 percent confidence. This prediction has received a great deal of attention by the sport community and media during the London Olympics. I obtained this number by applying several newly developed statistical methods and their modifications. Surprisingly, independent of my calculations, Bolt thinks that world records will stop at 9.40.

Of course, it is hard for anybody to predict the magnitude of the athletic talent at the extreme margins of humanity with certainty. Bolt, as it turns out, is a perfect example. He combines the mechanical advantages of taller men's bodies with the fast-twitch fibers of smaller men.

Is there any other runner who could set a new record? I personally think that this is less likely but still possible. Other than experienced runners like Gatlin, there are several young talents in the top-ten fastest men that could do it on a good day and with plenty of luck.

Bloomsburg Floods

I applied the theory to the floods in Bloomsburg, the city I live. The probability of a record flood during the next ten years is 0.06. Also, with a 90 percent confidence, the largest possible flood in Bloomsburg will not exceed 33.00.

"Best" Applicant

Suppose that there are one hundred applicants. Probability of one record (first applicant is the best) is $p_{1,100} = 0.01$; two records is $p_{2,100} = 0.05$. Occurrence of five records has the largest probability, $p_{5,100} = 0.21$. So, if the manager decides to hire the applicant who creates the *fifth record*, then the probability of hiring the best applicant will be 0.21.

One-Hundred-Meter Dash

During the 2009 world track and field competitions, Bolt put on an unbelievable performance shattering his own record in the one-hundred-meter, lowering it by 0.11 seconds to an amazing 9.58 seconds.

Question: How long his record will survive?

In most sports, records have occurred more frequently than what the theory predicts.

Men's one-hundred-meter data of 1912–2012 includes twenty records.

To produce twenty records, more than one hundred million independent and identically distributed attempts are needed.

Also, unlike the theoretical expectation, the waiting times have decreased significantly with time, especially in the recent past.

To account for these and other contributing factors, we made up for the increase in probability and frequency of the records by inflating the number of attempts.

Bloomsburg Flood Data

Recall that the median number of attempts required to arrive at a new record is $e=(2.718...)$ times the median number of attempts that was required to arrive at the previous record (the one before that).

This suggests a *geometric increase* with rate of $e=2.718...$ Using this the probability estimates for a new record can be calculated as:

0.152461 next year   0.562681 next 5 years   0.808753 next 10 years

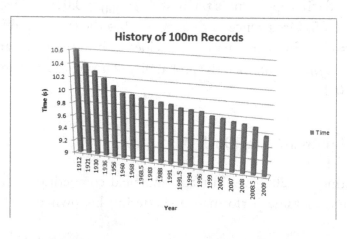

Other Applications of Records

1) Test of randomness. Observing many record highs or lows suggests that the data is not a simple random sample—that is, an alternative hypothesis should be sought to fit the data better.

Of course, it is possible for one hundred random observations to be ordered so that the sequence has as many as one hundred record highs.

But detailed calculation shows that the probability of ten or more record highs in a one-hundred-long random sequence is less than 5 percent.

Formal procedures are available to test the hypothesis of randomness based on the sum or the difference of record-high and record-low frequencies.

2) Sequential strategy for destructive testing. We usually teach mathematical methods for finding the minimum of a function. However, we never discuss any strategy for this when dealing with the real-world problems.

Many products fail under stress. For example, a wood beam breaks when sufficient perpendicular force is applied to it, electronic component ceases to function in an environment of too high temperature, and a battery dies under the stress of time.

But the precise breaking stress or failure point varies even among "identical" items.

Suppose that I can observe an item's exact failure point in a laboratory by gradually increasing stress (force, temperature, time, etc.). From such destructive testing of one hundred items, I could find all their failure points, say, $X_1, X_2, ...X_{100}$. But now suppose that I only need to find the weakest item in my sample: I only want the minimum value among failure stress, $X_1, X_2, ...X_{100}$. Then I need not stress most of the items to their failure points.

The minimum failure stress among any batch of items can be determined sequentially. Test the first item until it fails and record its failure stress X_1, Stop the next test (short of failure) if the second specimen survives this amount: so the second specimen's failure stress X_2 is determined exactly if $X_2 < X_1$; otherwise, obtain only the "censored" information that $X_2 > X_1$, and hence $X_1 = \max(X_1, X_2)$. In either case, proceed to the third specimen and stop the test if this item survives a stress equal to $\min(X_1, X_2)$: so X_3 is observed only if $X_3 < \min(X_1, X_2)$; but $\min(X_1, X_2, X_3)$ is always determined. In general, the ith item survives its stress test if $X_i > \min(X_1, \ldots X_i) < \min(X_1, \ldots X_{i-1})$ or the test concludes with stress-to-failure if $X_i = \min(X_1, \ldots X_i) < \min(X_1, \ldots X_{i-1})$. In either case, the value $\min(X_1, \ldots X_i)$ is known after the i trial.

The items destroyed in this sequential procedure are those with "record low" failure points. The frequency of such record lows fits the same probability model as the low in a sequence of weather records. For a sample of n items, the expected number of items destroyed is $1 + (1/2) + (1/3) + \ldots + (1/n)$. This harmonic sum grows very slowly compared to sample size n. For example, the sum is only 5.19 when $n = 100$ and is only 7.49 when $n = 1000$.

The sequential strategy to find the minimum value generalizes easily to find the 2, 3, . . ., or j smallest values among $(X_1, X_2, \ldots X_n)$. To begin, test j items until they fail at stresses $(X_1, X_2, \ldots X_1)$. Thereafter, stop the ith trial if the item survives the j lowest failure stresses among all $i - 1$ previous specimens. The probability of stress-to-failure for the ith item $(i > j)$ is the probability that it is among the j smallest of i independent observations from the same continuous distribution.

All ranks are equally likely for X_i, so the desired probability is j/i. The expected number of items destroyed is the sum of failure probabilities over all trials:

$$1 + 1 + \ldots + 1 + j/(j+1) + j/(j+2) + \ldots + j/n \text{ j terms}$$
$$j[1 + 1/(j+1) + 1/(j+2) + \ldots + 1/n] \text{ j}[1 + 1/2 + 1/3 + \ldots + 1/n]$$

If j is much less than the sample size n, then so is the expected number of failures. For example, to find the weakest four items in a sample of one thousand, I expect to destroy only about twenty-six; and to find the weakest eight items, I expect to destroy less than fifty.

Hiring the Best

A company is looking for a typist. An advertisement was answered by n people, and the company decided to hire one who was the fastest (or any other measure of performance). Each applicant wants to know immediately whether he or she is hired. Applicants who are refused are no longer interested in this job, leave, and will not return. The problem is to find a strategy that ensures the highest probability that the best applicant is hired.

Suggested Solution 1

Let X_i be the typing speed of ith applicant. Assume that applications appear in random order and that all the variables X_i are different. One managerial strategy can be based on the theory of records.

Consider $n = 100$, for example. Table indicates that in random order of applicant, the series $X_1, X_2, ..., X_n$, $n = 100$ has one record with probability $p_{1,100} = 0.01$, two records with probability $p_{2,100} = 0.05$, and so on.

The occurrence of five records has the largest probability $p_{5,100} = 0.21$. If the manager decides to hire the applicant who creates the fifth record, then the probability of hiring the best applicant will be 0.21.

Suggested Solution 2

Another manager's strategy is to examine first $s-1$ applicants but to hire none of them.

The manager hires the first of the remaining applicants (if that person is present) who has shown better performances than the previous applicants. If such an applicant is not present, the manager must hire the last one. For a given s, we calculate the probability $p(s,n)$ that the best applicant is chosen. Finally, we choose s such that $p(s,n)$ is maximal. With this strategy, the manager examines the first thirty-seven applicants and hires the next one whose test score exceeds that of the preceding applicants. The probability that the best applicant will be chosen is 0.371. If n is large, we can use the Euler formula. Then

$$p(s,n) \approx \frac{s-1}{n} \ln \frac{n-1}{n-2}$$

For $n = 10$, he should examine three and hire the next one whose test score exceeds that of preceding applicants. For $n = 5$, examine 2. For $n = 15$, examine 5.

Questions

1. How long would the present record stand?
2. How many records do we expect to observe in n attempts (n years)?
3. How long can we expect to wait for a record to be broken?
4. What would be the value of the new record?
5. How long should we expect to wait for the rth record to be set?
6. What can we say about the value of the rth record?
7. Is there an ultimate record?

A nice feature of the theory of records is that several of these questions can be answered with a nonparametric or distribution free analysis—that is, we do not have to know what the probability distribution of observations is.

Prediction of Records

To predict future records, the following approach can be used.

Short Term

For independent and identically distributed observations,

$$1 + 1/2 + ... + 1/83 = 5.$$

Long jump data has five records in forty-three years. Assuming that i is the geometric rate of improvement (increase in number of attempts)—that is, every year, the number of attempts has been i time the previous year—we can find its value by solving the equation

$$1 + i + i^2 + ... + i^{42} = 83.$$

Solving for i, we obtain a value of 1.027, which is a rate of improvement 2.7 percent (2.7 percent more attempts per year). For the long jump data, this means that attempts are increased as

1962	1963	1964	...	2000	...	2004
1	1.027	$(1.027)^2$		$(1.027)^{38}$		$(1.027)^{42}$

Now, to predict records for the next ten years, we set

$$n_1 = 83, \ n_2 = (1.027)^{43} + ... + (1.027)^{52} = 36.51.$$

For this example, the probability of a new record occurring before 2014 is

$$n_1/ (n_1 + n_2) = 0.31.$$

Using the maximum likelihood method, a better estimate is $n_1 = 73$. This will lead to a smaller value of i.

Ultimate Record

Let $Y_1 > Y_2 > ... > Y_n$ be the order of statistics of distances in the men's long jump.

Then it can be shown that a level $1 - p$ confidence interval for R_u (ultimate record) is given by

$$\{Y_1 - (Y_2 - Y_1) / [(1-p)^{-\alpha} - 1], Y_1\}$$
$$\alpha = \log_{k(n)} / \log [(Y_{k(n)} - Y_3)/ (Y_3-Y_2)]$$

$k(n)$ is any sequence such that $k(n) \to \infty$ and $k(n)/n \to 0$. One natural choice is $k(n) = \sqrt{n}$.

It can be shown that a better choice is

$$k(n) = (eT_r)^{1/2} + (t_r)^{1/2}$$

T_r: time between the last and previous ultimate record
t_r: time the last record has held to date
To demonstrate, suppose that we are in the year 2004. For long jump,

$$Y_1 = 8.95 \ (1991) \qquad t_r = 2004 - 1991 = 10$$
$$Y_2 = 8.90 \ (1968) \qquad T_r = 1991 - 1968 = 23$$
$$Y_3 = 8.86 \ (\text{year } 1987) \qquad Y_{11} = 8.63 \ (\text{year } 1997)$$

$k(n) = (eT_r)^{1/2} + (t_r)^{1/2} = 11$
$\alpha = \ln(k(n)) / \ln [(Y_{k(n)} - Y_3)/ (Y_3-Y_2)] = \ln 11 / \ln [(8.63 - 8.86)/ (8.86-8.90)] = 1.37085$
$Y_1 - (Y_2 - Y_1) / [(1-p)^{-\alpha} - 1] = 8.95 - (8.90-8.95) / [(.95)^{-1.37085} -1] = 9.64$

With 95 percent confidence, the ultimate record for men's long jump will lie within the interval [8.95, 9.64].

Alternative Short-Term Prediction

In a recent study, Noubary (2005) has developed a method utilizing the following three results of the theory of records for independent and identically distributed sequence of observations.

a. If there is an initial sequence of n_1 observations and a batch of n_2 future observations, then the probability for this additional batch to contain a new record is $n_2/(n_1+n_2)$.

b. As sample size n→∞, the frequency of the records among observations indexed by $an < i < bn$ tends to a Poisson count with mean $ln(b/a)$.

c. If $F(y)=1-e^y$, $y >0$ and Y_N denote the record values and if $D_1=Y_{N_1}$, $D_r =Y_{N_R} - Y_{N_{r-1}}$, $r \geq 2$, then the improvements D_1, D_2 ... are independent and identically distributed (i.i.d.) random variables with common distribution function $F(y)$.

Clearly, the results of theory of records for independent and identically distributed sequences are not suitable for sports, since sport records are usually more frequent than what the theory predicts. To account for this, some adjustment is therefore necessary. Noubary has treated the problem as if either participation has increased with time or more competitions have taken place so that the chance of setting a new record was increased.

For Boston Marathon, the participation has steadily increased during the years. Usually, 1970 is selected as the starting year, because during this year, a qualifying time was introduced for participation. One simple approach would be to model the increase and use that together with result (a) above for prediction. For example, using regression, the model given below was found for the number of participants as a function of the year ($R^2 = 0.938$).

Number of participants in year $t =$ $-1294 +$ $1088t - 57.5t^2 + 1.25t^3$

In this model, the data for the year 1996 was replaced by the average of the two neighboring values, since this was the one-hundredth running of the Boston Marathon and more than thirty-eight thousand runners were allowed to participate. Now, this approach seems reasonable, as it could account for other factors such as advanced training programs, better equipment, diet, coaching, and even use of steroids, all of which increase the chance of setting new records.

To clarify, suppose that it is now year 1924, and the best record for men's one-hundred-meter run is b and the probability of breaking this record is p. Suppose further that in year 2000, the population of word has tripled. If we divide this population to three groups, then each group could break the record with probability p.

Thus, the probability that at least one group breaks the record is $1- (1 - p)^3$. For example, this probability is 0.143 for $p = 0.05$.

Berry (2002) has discussed the effect of population increase on the breaking of sports records and has introduced the following exponential model for the growth of the world's male population since 1900.

$$\text{Population in year } t = 1.6 \exp [.0088(t - 1900)]$$

Note that this means a geometric increase of $\exp (0.0088)- 1 = .00884$ per year since year 1900.

We can use this assuming that the number of participants or attempts is proportional to the population size at time t. However, this approach does not use information from the sports itself and the way records were set. In other words, it is the same for all sports regardless. We can also apply result *(b)* assuming a geometric increase.

According to this result, the frequency of the records among observations 84 and 137 (sum of 84 and 53) has approximately a Poisson distribution with mean

$$\lambda = \ln (137/84)= 0.489$$

Using this, the probabilities of zero and one record in the period 2000–2009 are respectively 0.613 and 0.300. Here, $1 - 0.613 = 0.387$ is an estimate for probability of at least one record in the next ten years. Finally, let us demonstrate application of the result *(c)*.

The probability for occurrence of a record larger than m_2, say, in the next n_2 years can be calculated using the following formula obtained by combining results *(a)* and *(b)*.

$$P(m > m_0) = \frac{n_2}{n_1+n_2} \exp(-(m_0 - 8.95)/0.195)$$

Note that, here, 8.95 is the value of the last (fifth) record.

As an example for $m_0 = 9$, $P(m > m_0)$ is respectively equal to 0.0329 and 0.3024 for the future one and ten years, assuming a geometric increase.

To account for these and other contributing factors such as diet, shoes, and track type, etc., we treat the problem as an independent and identically distributed one but make up for the increase in probability and frequency of records by inflating the number of attempts. We start by stating results of the theory of records for independent and identically distributed observations that we plan to use:

a. If there is an initial sequence of n_1 observations and a batch of n_2 future observations, then the probability that the additional batch contains a new record is $n_2/(n_1+n_2)$.
b. For large n, $P_{r,n}$, the probability that a series of length n contains exactly r records is given by

$$P_{r,n} \sim \frac{1}{(r-1)!n} (\ln (n)+y)^{r-1}$$

where $y = 0.5772$ is Euler's constant (see appendix).
c. As sample size $n \to \infty$, the frequency of the records among observations indexed by $an < i < bn$ tends to a Poisson count with mean $ln(b/a)$.
d. The median of W_r, the waiting time between the $(r-1)^{th}$ and r^{th} records, are

Record Number r		2	3	4	5	6	7	8
Median (W_r)		4	10	26	69	183	490	1316
Med (W_r) /Med (W_{r-r})			2.50	2.60	2.65	2.65	2.68	2.69

Table 1. Medians of Waiting Times between
Successive Records and Their Ratios.

Moreover,

$$\frac{\text{Median}(W_{r+1})}{\text{Median}(W_{r0})} \approx e = 2.718...$$

Application to One-Hundred-Meter Data

Using the maximum likelihood method and maximizing $P_{r,n}$ with respect to n, we find $n = 100,212,150$. The number $100,212,150$ is an estimate for the number of i.i.d attempts that is required to produce twenty records. Next, we need to distribute these attempts over the period of one hundred years using an increasing function or pattern. The median number of attempts required to arrive at a new record is $e = (2.718...)$ times the median number of attempts that was required to arrive at the previous record. This suggests a *geometric increase* with rate $e = 2.718...$

If we assume that *one unit of attempt* was needed to arrive at the second record, the total number of attempts to arrive at record number twenty may be calculated as

$$1 + e + e^2 + ... + e^{18} = 103,872,541.$$

This is a slight overestimation, as for early records, the ratios are less than e.

Suppose i is the annual *geometric rate* of increase in number of attempts.

The number of attempts in year k is i time the number of attempts in year $(k-1)$.

Then i can be found by solving the equation

$$1 + i + i^2 + \cdots + i^{100} = 100,212,150$$

$i = 1.179888$—that is, 17.9888 percent more attempts per year. This leads to probability estimates for a new record 0.152461 next year, 0.562681 next five years, 0.808753 next ten years.

Note: one would expect even better results if the geometric increase is replaced by

increase in male population of the world. Also improvement can be achieved utilizing models such as Logistic or Gompertz or more generally a model of the form

$$y_{n+1} - y_n = H(y_n) = i^* f(y_n) (1 - g(y_n)),$$

where y_n : number of participants or number of attempts at year n Noubary (2005) has considered the following simpler model:

$$Y_{n+1} = y_n \exp[r^*(1 - y_n/h)].$$

For example, the number of attempts in the future one and ten years are 412 and 4876 using $y_0 = 100$, $r^* = 0.04$ and $h = 50$. The corresponding numbers using the logistic equation are 402 and 4742. These numbers result in smaller probability estimates compared to the geometric increase of 4 percent.

Further Analysis of Ultimate Records

Although ultimate records in any area are of great importance, here, we concentrate on sport records. Sports provide an inexhaustible source of fascinating and challenging problems in many disciplines including mathematics. According to authoritative opinion, modern sports have been getting more and more intellectualized, and more advanced scientific analysis is presented. At the same time, more attention is paid to the fact that it is possible to study many situations in sports from a mathematical perspective and that it is desirable to obtain more valid quantitative and qualitative estimates of the things happening in sports. For example, mathematical methods are applied to estimate an athlete's chances of success, to identify the best training conditions for him/her, and to measure of their effectiveness. Information theory makes it possible to estimate the amount of eyestrain in mountain skiing, table tennis, and so on. Mathematical physics is used to identify the best shape of rowboats and oars. In general, applied probability and statistics has been instrumental in analysis of sport data. For example, being profitable, baseball has long since been an object of attention for sport and business interests. A vast volume of statistics has been accumulated, enabling experts to draw conclusions about the quality of a team's performance (the average number of successful pitches, depending on the pitcher's and catcher's proficiency, the law of distribution of hits, and so on). A simulation model of baseball was constructed with the help of the probabilistic Monte Carlo method. Soon after, mathematical methods were applied to football. One paper contains the analysis of 8,373 games in fifty-six rounds, including the U.S. National Football League table. It supplies important recommendations on offensive strategy.

In sum, athletic competitions provide the researcher with a wealth of material that is registered by coaches, carefully preserved and continuously piled up. There is plenty of opportunity out there to experiment and to test mathematical models and optimal strategies in situations occurring in sports. Only a tiny part—quite possibly not the most intriguing one—of the problems arising in sports has been described in the pages of books and journals. Think how many yet unsolved problems arise in different sports.

We note that apart from intrinsic interest, there are several medical and physiological reasons why we would like to know how fast a human being could, for example, run a short distance such as one-hundred-meter dash or a medium distance such as four hundred meters. While there is a general agreement among the physiologists and physical educators about existence of an upper limit for such a speed (or a lower limit for the time needed to run this distance), the limit is not known at the present time. Because of a great interest in this question, apart from physiological research, there have been some attempts to estimate (predict) the limits via statistical modeling.

Methodology

We suggests consideration of an innovative procedure for estimating the best attainable (ultimate) records for athletic events. We can base our method on a newly developed theory that provides an approximate confidence interval for the minimum of a function. We propose using both the order values corresponding to the winning times and the time interval between the recent records. We anticipate that the procedure will require no numerical calculations. Our plan is use the theory to develop a new procedure and demonstrate its application using the annual winning times for men's one-hundred-, two-hundred-, four-hundred-, and eight-hundred-meter data.

Recent Olympic competitions provided some interesting results that will help the analysis greatly. We will also try an alternative approach by considering the distribution of ordered values corresponding to the winning times and fit a model that best describes their lower tails (extreme values). For this approach, we will apply a theory known as threshold theory.

According to this theory, the lower tail of all common statistical distributions can only take one of the possible three forms known as generalized Pareto distribution. Selection of a particular form will be made on the basis of a parameter that measures the tail thickness. Also, the best model will be selected using a statistical goodness of fit. Finally, this approach will also be applied to data from one hundred, two hundred, four hundred, and eight hundred. A comparison between the result will also be presented.

CHAPTER 7

Mathematical Modeling

MATHEMATICAL MODELING IS a description of a system using concise language with well-defined rules. The process is an attempt to simulate real-life situations with mathematical tools and rules mostly to forecast system's future behavior. This description suggests that modeling is a cognitive activity in which we make them to describe devices, objects, entities, or state of affairs. Moreover, it is representation of the essential aspects of systems in a usable form. We can also think of a *model* as an approximate representation of a physical situation.

Useful models are capable of explaining all relevant aspects of a situation. They can be used instead of experiments to answer questions regarding the specific situation, and by doing so, they allow to avoid the costs of experimentation.

In general, mathematical modeling has three potential uses, not necessarily independent.

1. A model may be *descriptive* in the sense that it synthesizes the available information on a process with no real attempt to explain the underlying mechanism.

2. A model may be also *explanatory* in the sense that it makes certain underlying assumptions about the process under study and derives the logical implications of those assumptions.

3. A model may be *predictive*—that is, it may be constructed for the purpose of predicting (e.g., the future values or the response of the system to factors that have not been observed).

There is no such thing as the best model for a given phenomenon. Models are neither right nor wrong, but in the final analysis, a model is judged using a single quite pragmatic factor, its usefulness. Models can, however, be judged as appropriate or inappropriate. Such a judgment must take into account the goals of the study and the information available. A mathematical equation has little value unless we can link its meaning to physical concepts. Of course, such linkage cannot usually be made in all generality, so we resort to simplification to gain as much understanding of the situation as we can. This is in line with the belief that it is better to understand few simple things well than to have a whole book of equations, disconnected with the physical world.

When Models Are Used?

Mathematical models are used when the observational phenomenon has measurable properties. A model consists of a set of assumptions about how a system or physical process works. These assumptions are stated in the form of mathematical relations involving the important parameters and variables describing the system. The conditions under which an experiment involving the system is carried out determine the "givens" in the mathematical relations. The solution of these relations allows us to predict the measurements that would be obtained if the experiment were performed.

Mathematical models are used extensively by, for example, engineers in guiding system design and modification decisions. Intuition and rules of thumb are not always reliable in predicting the performance of complex and novel systems, and experimentation is not possible during the initial phases of a system design. Furthermore, the cost of extensive experimentation in existing systems frequently proves to be prohibitive. The availability of adequate models for the components of a complex system combined with a knowledge of their interactions allows the scientist and engineer to develop an overall mathematical model for the system. It is then possible to quickly and inexpensively answer questions about the performance of complex systems. Indeed, computer programs for obtaining the solution of mathematical models form the basis of many computer-aided analysis and design systems.

To be useful, a model must fit the facts of a given situation. Therefore, the process of developing and validating a model necessarily consists of a series of experiments and model modifications. Each experiment investigates a certain aspect of the phenomenon under investigation and involves the taking of observations and measurements under a specified set of conditions. The model is used to predict the outcome of the experiment, and these predictions are compared with the actual observations that result when the experiment is carried out. If there is a significant discrepancy, the model is then modified to account for it. The modeling process continues until the investigator is satisfied that the behavior of all relevant aspects of the phenomenon can be predicted to within a desired accuracy.

Good Modeling

Many applied scientists think that in a great deal of stochastic modelling in the areas such as biology, ecology, seismology, etc., the emphasis has been statistics (mathematics) with little or no attempt to tie together the statistical results with biological (geophysical) observations. They view this approach of modelling essentially as a statistical (mathematical) exercise with a bit of biological (geological) justification. Some people refer to this as biostatistics (geostatistics), etc. In contrast, they consider a situation in which a biological or geological problem is investigated with statistical tools, but the statistics is considered of purely secondary interest. The objective is to derive biological (geological) conclusions that are testable, not to develop elegant statistics, though that may indeed occur. This has been referred to as mathematical biology (geology). In biostatistics (geostatistics), although the results obtained can be statistically important, too often they are either irrelevant to the problem or outside the realm of real-world testability. With regard to this, practitioners often refer to the fact that only very few models proposed by statisticians can be checked for validity of their assumptions. This is because of their selection criterion based solely on statistical facts and more importantly lack of physical interpretation for parameters in the models. Their suggestion is that, if models are constructed with some basis on biological or geological facts or observations, there is the likelihood that their analysis will lead to results that, if not testable, are at least comparable.

How?

In general, one finds two different possible attitudes toward applying statistics to biological, ecological, or the subject of interest to me, seismological problems. In one of these, a model already constructed is either analyzed further or extension is made to it, such as by considering time-dependent solutions, spatial patterns, or adding stochastic variation. The latter has been considered particularly important noting that environment varies unpredictably through time in ways that affects all the variables and also the fact that noise is present in almost any system. Thus, when modelling, a critical problem is to incorporate stochastic variation into an established deterministic model in a way that is biologically (geologically) meaningful yet statistically (mathematically) tractable.

As a result of childhood experience, I worked on earthquakes and related phenomenon for a relatively long period. I found that the best modeling approach was to incorporate random variation into the established deterministic models proposed by seismologists in a way that is physically meaningful and mathematically (statistically) tractable. I also found that depending on whether one was dealing with the modeling, there were two different approaches:

Geomathematics: A mathematical exercise with a bit of geological justification. The emphasis is mathematics (statistics).

Mathematical geology: Investigation of geological problems with mathematical or statistical tools. Mathematics is considered of purely secondary interest.

Benefits of Modeling

The benefits of a mathematically presented models are the following:

1. They are *clearly* defined and thus easily communicated so that their strengths and weaknesses may be analyzed.
2. They can be manipulated according to the clear rules of mathematics and investigated either analytically or, more experimentally, using a computer.

In fact, with the huge growth in computing power, modelling of all kinds has boomed. It is now possible to model complex systems and test them inside the computer before an actual prototype is built. Can we expect our models to be true descriptions of reality? The answer is no. Models are but approximation to certain aspect of complex reality.

In fact, a clearly stated model provides us with a constant reminder of what is real and what is modeled—that is, what is observed versus what is expected. Is this a weakness?

No. In fact, one cannot analyze the real world (by definition of analyze). One can only analyze a picture of the world (conceptual model) that is in one's head. The main requirements on such models are that they be accurate enough for the purposes at hand and that they be tractable enough for the needed accuracy.

Guideline

Mathematical modeling takes the data as a message and seeks models for them using the basic decomposition.

$$\text{Datum} = \text{Systematic Part} + \text{Random Part}$$
$$(\text{Message} = \text{Signal} + \text{Noise})$$

It is the attention given to the second component that perhaps most distinguishes statistical modeling from other kinds. In the case of a stationary process, where statistical properties of a system remains unchanged as time passes, there is a decomposition theorem:

$$\text{Stationary Process} = \text{Deterministic Part} + \text{In-deterministic Part}$$

Mathematics mainly deals with the deterministic part; whereas, probability and statistics deal with the other part also known as random or stochastic part.

Doing It Right

Models developed by mathematicians/statisticians based on criterion such as goodness of fit alone often leads to a "best" model just for the data utilized. Moreover, often the parameters in such models do not have a meaningful interpretation for their users. This section makes argument against modeling processes that disregard information from discipline related to the origin of data and presents an example of incorporating such information in modeling process.

After working for more than forty years in industry and academia, I feel that I know a little about the right ways of constructing models. Based on my experience, the models mathematicians develop do not always earn the appreciation of the practitioners. They call a model good if (a) it is general, (b) it incorporates the rules of their field of application, and (c) it is understandable to them in their own jargon.

For more than two decades, I followed progress in earthquake modeling and personally tried anything I knew or could find in books and journals. I even tried to learn some geophysics and seismology to help the matter. Many well-known earth scientists I talked to during that period expressed their concern about models mathematicians consider good. Most of them believe that in most models developed by mathematicians, the emphasis has been mathematics, and often little or no attempt was made to tie together the mathematical concepts with geophysical facts.

They viewed this approach of modeling essentially as a mathematical exercise with a bit of geological or geophysical justification. To make a distinction, we may name this geomathematics. In contrast, they expressed desire to see a geological problem being investigated with mathematical tools, where mathematics is of purely secondary interest. In other words, the objective is to derive models with physical significance (e.g., models whose parameters have physical interpretation), not to produce elegant mathematics, though that may indeed occur. Let us refer to this as mathematical geology. I learned that if we follow their advice and construct models on the bases of the geological facts, their analysis may lead to results that are mathematically meaningful and physically appropriate. Moreover, such models may be testable, and as such their validity may be verified with methods other than goodness of fit. Also, compared to other models obtained using other approaches, they may include fewer parameters. I also learned that there are two major attitudes toward mathematical modeling applied to the disciplines such as seismology. In one, modeling is carried out solely based on goodness of fit. Regression and time series modeling are examples of the tools people often use for this approach. In the second approach, either the deterministic models developed by experts in the field are analyzed further or extension is made to them by considering, for example, time-dependent solutions, spatial patterns, or by adding random variation. Reading more about it, I realized that the latter has gained a great popularity in recent decades. Again, here, the critical problem is to incorporate random variation into an established deterministic formulation in a way that is physically meaningful and mathematically/statistically tractable.

Traveling Salesman

A range of problems related to the positioning of stores and the planning of delivery routes requires information on the distances of the road, y, between different places. Where a large number of such places are involved, finding these distances by direct measurement is time-consuming. To avoid this, the usual approach is to relate the road distances to the straight line distance, denoted by x, as measured using a scale map. This relationship enables us to predict a value of y given a corresponding value of x So the question is, how do we obtain this relationship or the model? To do this, there is a whole spectrum of differing approaches. On the extreme ends of this spectrum are

conceptual approach and empirical approach.

The remainder of the spectrum usually consists of a combination of the extremes, often referred to as *eclectic approach*.

1. Conceptual approach

Derives the form of the model on the basis of our understanding of the situation. It uses logical reasoning, "known theory," to obtain the model. How? Here is the example:

a. When $x = 0$, the two points coincide, so $y = 0$.

b. If there is a straight road between the two points, then $y = 0$; but, otherwise, y will be greater than x.

c. The distance y will generally increase with x. However, the randomness in the pattern of roads will mean that even if different pairs of places have the same x, there will be random differences in the y values.

y = Part predictable from x + unpredictable random component

Thus, each y will consist of a part predictable from the x value and an unpredictable random component.

d. Provided we keep to relatively similar situations (e.g., urban roads), then the form of the relationship should not depend strongly on the distances involved. Thus, if the straight line distances are doubled, we would expect in most cases the road distances also to be approximately doubled.

With these reasonings, we can now consider a few possibilities (simplest relationships):

i. $y = x$: satisfies (a) and (d) but not (b) or (c).

ii. $y = x +$ random component [error]: this now allows (c) but not (b).

iii. $y = $ constant $+ x +$ random component: this helps with (b) but (a) now fails.

iv. $y = $ constant$^* x +$ random component: this now satisfies all four provided that the constant is greater than one and that the random component, e, is constrained so that $y \geq x$.

Putting all these together, we shall write our "conceptual" model as

$$y = Bx + e, \quad B \geq 1, \quad y \geq x.$$

Notice that this model is derived without any data being provided or used about the actual situation.

2. Empirical approach

Considers only empirical evidence and ignores the evidence we have from previous experience or similar situations. To apply this approach to the above problem, we need to begin with data. Suppose that we have data. First, we plot it to get some idea about possible pattern to relationship. Suppose that the data suggests a straight line through the origin for the main trend. If the observation (x, y) represents one of the points, the relation can be expressed as

$$y = Bx + e,$$

where e is the error in predicting y from the line. If this works, we could say that we arrived at the same model by both approaches. The conceptual approach used what we might term *prior information* (i.e., the information that we possessed prior to obtaining the data). The empirical approach ignored this information and used only the *empirical information* contained in the data. In practice, we should seek to use both sources of information, the *eclectic approach*. To see this, consider the following question: How do we obtain or calculate B?

Answer: We have to take an empirical approach and use our data to provide a numerical value for B. This value is referred to as an estimate of B. There are some methods and formulae for obtaining B.

Question: How do we test validity of our model? Answer: By testing on more data, we compare the actual measurements with predictions from the model. Alternatively, we can simulate data from a known model and develop a model using empirical approach. Finally, compare that with original model and alter our modeling approach to achieve improvements.

Note: The model as it has been defined concentrates on the

deterministic (here line) and provided no information about the random component, e. As a further stage in the model construction, we must consider how to model the properties of this random component.

For that, we examine the evidence of the observations. We make, for example, "frequency" table and use known statistical methods to analyze the errors. We, thus, can see that to discuss the random component of the model, a different language is needed from the mathematical statements of the deterministic component. This is the language of expectations and probabilities, the language of statistics.

Discrete Models of Populations

Suppose that I am trying to model the population of rabbits in a certain part of Pennsylvania or the population of Philadelphia.

Let y be the number of individuals at time n (generation n), and by and dy denote the number of individuals who were born and who died during this period. Then

y : # of individuals at time n (generation n)

by : # of individuals who were born during this time.

dy : # of individuals who died during this time

$$y_{n+1} = y_n + b\, y_n - d\, y_n = (1 + b - d)\, y_n = r\, y_n$$

For realistic models of growth, certain restrictions on r are necessary.

Suppose that the action of previous generations determines the growth of a population—that is, $r = r(y_n)$, a function of y_n. Thus, r is a density dependent growth rate.

One of the simplest density dependent models that contains a formulation that represents effects of overcrowding is when r is a linear function of population size—that is,

$$y_{n+1} - y_n = r^* \, y_n \, (1 - y_n /k) = F(y_n)$$

where r* is intrinsic rate of growth and k = environmental carrying capacity. There is a sizeable literature regarding the behavior of the logistic equation for typical values of r* and y_0 (the initial size of population). Here, an *important* question is "Do biological populations exhibit chaotic behavior?" Surprisingly, this question is unresolved at present. But it is known that if the models with *chaotic* behavior are relevant to populations, then wild oscillations of populations *need not* necessarily be the consequence of random environmental fluctuations but might be intrinsic to the population. A similar question is, can we use this model for human population (e.g., population of England)? The answer is, since population is changing instantly, continuous models are more appropriate. The continuous version of classical logistic equation is a first-order Bernoulli differential equation. Note that there is a big difference between discrete and continuous logistic models. The latter allows no oscillation.

Modeling the Stock Market

In recent years, selection of stocks in an almost random fashion has become popular among inventors on the ground that it works as well as expert advice. The individuals who are familiar with probability theory rely on the fact that random walk, Brownian motion, or models of this type provide reasonable description of the stock market. While there is a general agreement that such models are reasonable, why they work is not entirely understood or explained. Let us analyze the general characteristics of the such models in an attempt to explain their connection to the behavior of typical investors.

Paul Cootner was one of the first economists to recognize the importance of what was then largely statistical analyses of stock price movements and to begin to provide a theoretical foundation for the apparently random behavior of stock price changes. In his first paper on stock market prices, he provided an operational definition of market efficiency and its relation to random walk. The ideas later were discussed in more detail in a collection of essays that he edited, *The Random Characters of Stock Market Prices*.

Of course, as is pointed out by several investigators, it would be a negation of economic law if competitive prices truly moved in a classical random walk. This is because given enough time, prices could wander anywhere. Still, an economist would be ill advised to ignore the probabilistic jitter inherent in speculative prices. This observation has a long history and goes back to Maurice Kendall's (1953) survey of the statistics of speculative price. For spot and futures prices of commodities, for single stocks and stock indices numbers of stocks, the noneconomist Kendall found white spectra, almost zero autocorrelations. As he said, to the initial outrage of the economists in his audience at the Royal Statistical Society, it was as if the devil drew the price changes out of an urn. Kendall's observations accord with the classical Wiener definition of a random walk in terms of a process with independent increments. The random walk was discussed as early as 1900 in Sorbonne thesis of L. Bachelier and also in his popular 1914 books. Later, several corrections were made to his strict model of the arithmetic random walk. Paul Samuelson in the middle 1950s repaired the Bachelier long-run anomaly by stipulating the geometric Brownian motion based on proportionate changes. This led, as the physicist M. F. M. Osborne observed empirically at much the same time, to the lognormal Wiener process. Samuelson argued that the Bachelier anomalies could be avoided once the absolute or arithmetic Brownian motion of physics is replaced by the geometric Brownian motion of economics. This, of course, is not very surprising, since most of the modern option theory operates in terms of lognormal distribution of prices, which has a longer (fatter) tail than normal distribution. Mandelbrot (1963) has pointed out that even fat-tailed distributions of the Pareto type may occur in speculative stochastic processes. This is because meandering chance could lead the real world away from economic law even faster and further than the Wiener process had suggested. Cootner (1964) has argued that, for stock prices, the correct model is not a Gaussian process, nor even a lognormal random walk, but a martingale.

There are also some arguments in favor of Markov process, ergodic probability distribution, and, more recently, chaos.

Why Probabilistic Analysis?

Consider a single stock and the collection of individuals who are interested in it. To study the price changes, let us pick a starting point and look at the situation as a system. A priori, the choice of initial conditions presents a very considerable problem. To see this, imagine only one popular stock and examine the possibility of representing investors' behavior through some deterministic theory.

1. We would need to know the initial investment and initial direction (investment plan, motive, etc.) for every investor. This makes the initial data enormous.
2. To perform the calculation, we would need to solve a very large system of equations featuring initial conditions. The result would consist of very many numbers representing the state of the system at some given later time.
3. Finally, we would need to calculate the readily measurable macroscopic performance variables as sums over these output values.

Given our knowledge of twentieth century, we can see that attempts to implement such a deterministic program must run into several insurmountable obstacles. First, there is the practical impossibility of knowing simultaneously the investment and the direction (the plan) taken by individual investors. Next, even if we had this information, such calculations would require a very large memory and would be so lengthy as to be impracticable. Finally, there is a methodological objection that one can formulate as follows: since the aim is to calculate "macroscopically" measurable quantities from microscopic data, might one not do this more easily in some other fashion than a deterministic analysis?

The story is similar to what happened in physics when Maxwell in 1859 could not but realize that in practice, deterministic calculation related to molecules of even one mole of gas is impossible. Here, the initial investment and direction taken by an individual is similar to initial position and velocity of a molecule. It was in this context that Maxwell made the conceptual leap of introducing into the theory the notion of unpredictable. His breathtaking intuition was developed further by Boltzman and was confirmed half a century later by the work of Albert Einstein (1905) and of Jean Perrin (1908) on Brownian motion. It is important to note that Maxwell does not reject determinism as such, but he mitigates our ignorance of the initial conditions by introducing assumptions of probabilistic nature. His approach can be placed within a framework of the so-called probabilities through ignorance. The introduction of this approach in 1872 led to the birth of statistical physics.

In a similar fashion, when modeling the stock market, one can fruitfully exploit probabilities that represent our ignorance. Basically, we admit the deterministic nature of actions taken by investors but try to find a way to treat them probabilistically. In other words, to construct the economic theory governing a deterministic system of large number of investors, it is possible to exploit probabilities through ignorance and have a complete success.

The reason for this claim deserves discussion and explanation. The Soviet physicist Lev Landau has shown that a classical system requiring infinitely many parameters would behave in a totally random fashion; in other words, it would be random unavoidably and not merely by reason of our ignorance. A system of several million investors depends on very large number of parameters that, though theoretically not infinite, can be considered as infinite for practical purposes. In this way, one is led from chance by reason of ignorance to chance unavoidable, which is governed by the laws of probability. That is why we have struck so lucky with our appeal to the random walk and Brownian motion and other models, and why the use of probabilities through ignorance, far from being a mere second-best reflecting the limits of our numerical capabilities, in fact takes us very close to the laws of market.

Deterministic Chaos

In the preceding section, we saw, using Maxwell's idea, how chance could generate determinism. The present section briefly looks at a situation called chaos where determinism can generate chance and points out the way the resulting theory may be considered appropriate for stock market analysis. Chaos is a particularly unfortunate name because unlike what we understand from the word, it actually refers to a higher degree of order. To appreciate its importance, we can refer to the fact that the heart has to be largely regular or you die. But the brain has to be largely irregular; if not, you have epilepsy. This shows that irregularity, chaos, leads to complex systems. It is not all disorder. The chaos theory was pioneered by Henri Poincare near the end of the last century, but only the advent of fast computers in the early 1960s has made its development possible. Today, it is a very active field of research and has induced a far-reaching revolution in our concepts. To classify it, we make a semantic distinction between random and chaotic processes.

A dust particle suspended in water moves around randomly, executing what is called Brownian motion. This stems from molecular agitation through the impacts of water molecules on the dust particle. Every molecule, much like an investor, is a direct or indirect cause of the motion, and we can say that the Brownian motion of the dust particle is governed by very many variables. In such cases, one speaks of a random process; to treat it mathematically, we use the calculus of probabilities.

A compass needle acted on simultaneously by a fixed and by a rotating field constitutes a very simple physical system pending on only three variables. However, one can choose experimental conditions under which the motion of the magnetized needle is so unsystematic that prediction seems totally impossible. In such very simple cases whose evolution is nevertheless unpredictable, one speaks of *chaos* and of *chaotic processes;* these are the terms one uses whenever the variables characterizing the system are few.

Thus, when the system such as the one we based on behavior of individual investors is studied, one looks at a very large number of variables, and, hence, random process is more appropriate. However, when only a few variables such as interest rate, currency rate, etc., can characterize the system, the chaos may be used.

The relevance of chaos to market behavior may also be explained by the following example. A pendulum swings to and from with a regular back-and-forth motion, but if it is struck by the ball of a second pendulum before reaching its zenith, both pendulums may begin swinging in wildly erratic patters. In the financial market, a trend is enhanced or undermined by surprises in governmental announcements or economic actions by one or more influential nations.

Remarks

Based on the proceeding discussion, the possible models may be described in the following way:

1. Stochastic models use randomness as a basic concept so that market is governed, at least partially, by chance and associated laws of probability.
2. Chaos offers the fascinating possibility of describing randomness as the result of a known deterministic market. Randomness may lie on the choice of initial conditions and the subsequent mechanism of the market change. Note that the term "chaos" is usually reserved for dynamical systems whose state can be described with differential equations in continuous time or difference equations in discrete time. This means that it is possible to represent the market using these models.

The theory of chaos is fascinating if for no other reason than the blurring of the long-held distinction between random and deterministic phenomena. It is potentially capable of "explaining" very complex processes with simple, parsimonious models and essentially without error. The theory has the potential to offer traders entirely new perspectives on the movements of market.

There are many important questions about the new theory that are currently unresolved, and it is likely that some of these issues will never be resolved. One such a question in present context is the question of choosing between deterministic versus stochastic modeling of market. One reason for this is appropriateness of the random walk model. To clarify this, we note that random walk is stochastic version of first-order difference equation and structurally has similarly to logistic equation. On the other hand, it is an autoregressive model with unit root. To clarify this, we note that there is a connection between random subdivision and self-similarity, which is the basis of chaos.

Kolmogorov was probably the first to invoke the connection between random subdivision and self-similarity. His argument ran as follows: if we have some large unit (say, a large rock) that is broken up into n_i smaller pieces and then fragmented again and again so that a fragment belonging to the jth generation gives rise to n_j smaller fragments, then the number of rocks at the ith stage of fragmentation is $n_1 n_2 \ldots n_i$. Assuming that the process of fragmentation is homogeneous in space and in time (i.e., that the mechanism of breakage is always the same), the numbers n_i are identically distributed random variables.

By a straightforward reasoning, Kolmogorov found that the distribution of sizes of fragments must be lognormal. The property of lognormality has, indeed, been known for many years to describe the granulometry of sands and natural aggregates. There are many other examples of lognormality in variables resulting from processes of infinite random fragmentation.

Modeling Building

We could build models using symbols instead of things. Mathematically presented models are clearly defined and are easily communicated so that their strengths and weaknesses may be analyzed. Of course, we cannot expect our models to be true descriptions of reality. Additionally, models may give a good description of one aspect of the situation but a poor one of another. In short, models are but approximation to certain aspects of complex reality. A model can be seen as truth insofar as it reveals "unhidden" aspects of the situation being modeled that were previously hidden. Thus, the truth of the model does not depend on it being "the truth, the whole truth," but simply on whether it is adequate to reveal some aspects of reality to our insight. We, therefore, have to judge it not in terms of right or wrong, but in terms of insight and use.

Forestry

The science of ecology is becoming more and more closely involved with mathematics. Forestry is one branch of ecology that has become apparent. Math is employed in forestry research and management. Ideas in forestry thinning, inventory, and growth modeling all utilize mathematics.

The original question proposed reminded me of an operations research problem. It asked at what time a tree should be harvested to make the most money. The commercial value of a single tree depends on the quality of timber, age, and volume. The value of the tree increases with the quality and volume of usable timber. The value of a tree also increases with the age of the tree and then reaches a plateau and eventually decreases.

The simply answer to the original question would be at time t such that the value function V(t) is maximum. However, the tree could be harvested early and the money invested. To answer this problem, the interest rate must be considered. When the interest rate is higher, it is better to cut the tree earlier.

The answers to these problems are not definitive, since many variables were maintained as constant to simplify the problem. However, this study has led to future research to determine the distribution for a natural healthy forest. The initial solution is a clustered forest rather than a completely random or regularly distributed forest will lead to the best result.

Modern ecology, which deals with many important issues, has, in general, two tasks.

a. To learn about the structure of our natural world.
b. To find out about the mechanisms that have caused that structure to evolve.

The characteristics used in forestry are commonly divided into three groups describing

1. an individual tree   2. a forest stand   3. a forest region

Trees in a stand can be characterized in terms of their size, distribution, and their relative locations with respect to each other.

In forestry, math is used in two main areas:

1) forest research   2) forest management

(e.g., inventory planning, growth modeling, forest regeneration, forest thinning, etc.).

Forest Rotation

The commercial value of a single tree is determined by

(1) the volume   (2) quality of timber   (3) age of the tree

Young trees have no commercial value. The value increases with age as the volume of usable timber increases. Eventually, the tree approaches maturity, its growth ceases, and its value reaches a plateau. Ultimately, decay would set in and the value declines to zero.

Procedure

Modern ecology, which deals with many important issues, has, in general, two tasks:

a. To learn about the structure of our natural world.
b. To find out about the mechanisms that have caused that structure to evolve.

Recall that mathematical modelling has three potential uses, not necessarily independent:

1. A model may be *descriptive* in the sense that it synthesized the available information on a process with no real attempt to explain the underlying mechanism.
2. A model may be also *explanatory* in that it makes certain underlying assumption about the process under study and derives the logical implications of those assumptions.
3. A model may be *predictive*—that is, it may be constructed for the purpose of predicting (e.g., the future values or the response of the system to factors that have not been observed).

All of these uses come into play in ecological applications, though the explanatory and predictive aspects are considered to be more important. In general, there are two different possible attitudes toward applying mathematics to biological and ecological problems. In one, the emphasis is mathematics with little or attempt to tie together the mathematical results with biological observations. It is essentially a mathematical exercise with a bit of biological justification. We may refer to this as *biomathematics*. In comparison, consider the situation in which a biological problem is investigated with mathematical tools, but the mathematics is considered of purely secondary interest. The objective is to derive biological conclusions that are testable, not to develop elegant mathematics, though that may indeed occur. This may be referred to as *mathematical biology*.

Spatial Patterns of Trees

The purpose of this sort of analysis is to develop (using recent results of spatial statistics) models and appropriate methods for analyzing the relative location of trees (called spatial pattern of trees) in a given forest.

The spatial pattern of trees has an important role in many fields of forestry. It affects, for instance, (1) the sampling design of a forest inventory, (2) the growth possibilities of an individual tree and thus the timber production of the whole stand, (3) the need for silvicultural treatments in a seedling or sapling stand, and (4) the need for thinning in a young or middle-aged stand. The spatial pattern is implicitly observed in forest management—for example, in treatments of seedling or sapling stand and the thinning operations. Nevertheless, spatial patterns of trees have seldom been presented in an analytic form in forest research and even more rarely in forest management. Spatial analysis has, however, been developing already for a few decades.

The aim of the analysis of spatial point patterns is to measure, by using quantitative characteristics, how individuals are located with respect to each other and to describe with mathematical models the laws regulating the location. With mathematical models, one can obtain detailed knowledge about the underlying random mechanism that has generated the pattern. Furthermore, mathematical models allow the use of simulation procedures for artificial production of point pattern compatibility.

Analysis

The characteristics used in forestry are commonly divided into three groups describing (1) an individual tree, (2) a forest stand, and (3) a forest region. A forest stand can be assumed to consist of soil, trees, and other vegetation. Trees in a stand can be characterized in terms of their size distributions and their relative locations with respect to each other. The relative locations of trees can be illustrated by the point configuration formed by the dimensionless trees on the horizontal plane, called the spatial pattern of trees. The spatial pattern of trees can also be described by using stochastic models, called spatial point processes.

The relative spatial distribution of trees plays an important role in many areas of forest research and forest management. Examples of purposes where its estimate may be utilized are forest inventory planning, the construction of growth models of trees or stands, and problems relating to forest regeneration and thinning.

The spatial pattern affects, though the interaction of trees, the growth of an individual tree and thereby the current timber productivity of the whole stand. This fact has been recognized in forestry for a long time.

One aim of the treatments, such as thinning, applied to forests is to make the spatial pattern more regular to distribute the growth factors evenly among the trees. Further examples of circumstances under which the spatial pattern of trees might be of significance are (1) spreading of diseases and fires, (2) producing high-quality timber (the growing space usually affects the quality of timber), and (3) the simulation of artificial forests for different research purposes.

The simplest model for point configurations is the model of independently and uniformly distributed individuals, the so-called Poisson process in the plane. Therefore, a classical problem in dealing with point configurations is to determine whether the configuration can be regarded as generated by a Poisson process. The usual alternatives are the regular and clustered models. Different kinds of indices have traditionally been applied, later also graphs, such as second-order summaries. If the Poisson hypothesis is rejected, more complicated models and mathematical tools for the parameter estimation will be needed.

A natural subfamily of processes to model the interaction in point configurations is the family of Gibbs processes, originally introduced in statistical physics. The parameters of Gibbs processes can be used as a data summary.

In spatial analysis, the sampling method partly determines what kind of information we can get from the sample. The mapped data are usually the most informative and also the most expensive. So far, the parameter estimation has been based on mapped data. The measurements are distances from randomly chosen points and trees to the nearest tree in circles around random points.

Narrative

In modeling the locations of trees in a forest, the forest is generally regarded as a subset of the horizontal Euclidean plane on which the supposed origin points of trees are projected. If the area of the relevant forest is large compared with the diameters of the trees, an individual tree can be represented as a single point. The set consisting of these points is called spatial point pattern. The points of a pattern are referred to as individuals or simply trees, to distinguish them from arbitrary points of the plane. A spatial point pattern thus indicates the distribution of the horizontal space among trees within the region.

The aim of the analysis of spatial point patterns is to measure, by using quantitative characteristics, how individuals are located with respect to each other and to describe with mathematical models the laws regulating the location.

The choice of the method of analysis depends among other things on the type of spatial sampling (i.e., on the types of measurements used at data collection). The sampling methods usually fall into three classes: (1) The area in question is divided into small subareas—for example, into quadrants—and the number of individuals in each quadrant is counted. (2) The distances between an arbitrary individual or an arbitrary point of the plane and the nearest neighboring individuals are measured. (3) The locations of the individuals in the whole area or some subarea are mapped. With measurements belonging to the first and second groups, one can test whether the positions of individuals are distributed randomly, and, if not, whether the pattern can be regarded as regular or clustered. Further, different kinds of indices to measure the amount and the direction of deviation from a random pattern can be calculated.

In a more advanced analysis, mathematical laws for the relative locations of trees are searched. The most informative spatial data for this purpose are the mapped data (i.e., the data in which the coordinates of trees in some subarea are known). With mathematical models, one can obtain detailed knowledge about the underlying random mechanism that has generated the pattern. Furthermore, mathematical models allow the use of simulation procedures for artificial production of point pattern compatibility.

Applications

"A biologist is someone who can construct a real-world counter example to any statement a theoretician makes."

Returning to the main item, we note that in this approach of modeling, the emphasis is mathematical with little or no attempt to tie together the mathematical results with biological observations. It is essentially a mathematical exercise with a bit of biological justification.

In comparison, consider the situation in which a biological problem is investigated with mathematical tools, but the mathematics is considered of purely secondary interest. The objective is to derive biological conclusions that are testable, not to develop elegant mathematics, though that may indeed occur. This may be referred to as *mathematical biology*. In fact, a large number of biologists think that a great deal of mathematical work in biology, ecology, etc., has been biomathematics; and although the result can be fun as mathematics, too often they are either *irrelevant* to biological problems or completely outside the realm of *real-world testability*. In fact, very few models can be checked for validity for their assumption simply because of our lack of knowledge. If the models are constructed with a firm basis in biological fact, however, there is the likelihood that their analysis will lead to results that are testable. The same is, of course, true in other disciplines. As an example, I applied this idea successfully for modeling of seismic P-waves originated from underground nuclear explosions and natural earthquakes and presented its presented advantages over the direct mathematical modeling. Now, given that one is mathematically trained, and the objective is to do, for example, mathematical biology rather than biomathematics, how does one proceed? Experience has shown that there are two routes: either learn the necessary biology or find an expert who is willing to collaborate. Each path has its advantages, and to be truly successful, a bit of both is undoubtedly the best. Becoming an expert in a certain biological discipline can be extremely time-consuming, but it is probably the best way to ensure that any modeling efforts undertaken are firmly rooted in observation. Collaboration has the benefit of not requiring as much time spent learning the biology, but to be effective, a certain minimum effort is necessary on the mathematician's part to learn at least the basics.

As there are few who have the ability or desire to become biologists, *collaborative work is essential* even for those biological experts in one discipline. From what is gathered, a positive attitude is growing among biologists, especially ecologists, toward the utility of mathematics in aiding their understanding of biological and ecological systems. However, they view modeling as a path to a biological end, not a mathematical one. We finish this section by mentioning a definition that speaks about the view of biologists, "A biologist is someone who can construct a real-world counter example to any statement a theoretician makes."

Population Size

We now consider a more specific topic related to the modeling. From a large amount of existing literature, it is evident that population biology (and ecology) is perhaps the most mathematically developed area of biology (ecology) with a long history of interest by mathematician in the problems associated with the dynamics of population.

In population ecology, which (at this level) is the science that investigates the dynamics and structure of populations of given individuals, the *chief questions* of interest are the following: How is a population structured in terms of age, size, and genotype? How does this structure change both temporally and spatially, and what factors, external and internal to the population, regulate this structure (see, e.g., Hutchinson 1978)? From the point of view of a field biologist who may be interested in establishing the science and structure of a population, a standard practice is to estimate the *birth* and *death* rates for individuals in different classes, *the major difficulty being the statistical aspects*. For example, until recently, the bulk of the extensive mathematical literature on this topic had ignored the role of spatial heterogeneity as advances on the mathematical theory of spatially distributed populations had been made in recent decades (and is far from being complete).

Methods of population dynamics are also central to the understanding and application of optimal control techniques. It seems that large portions of the sciences of *forestry* rest upon the theories of population structuring. Here, the analysis of population through system of ordinary differential equations has been successful.

Let us now briefly describe a simple approach and an example when population modeling is under consideration. As a first step, answers should be found for the following questions:

1. What is the population under study?
2. What are its characteristics?
3. What are density and dispersion (e.g., spatially homogenous)?

4. Which characteristic(s) is of interest (e.g., size)?
5. What are the objectives of modeling (e.g., to predict . . .)?
6. Which approach (deterministic or stochastic)?
7. Which type (discrete or continuous, linear or nonlinear)?
8. Do populations behave nicely? etc.

When the objectives are specified, the representation of the scientific phenomenon in the language of mathematics results in a model. After the model is formulated, it must be solved either analytically or numerically. The next step is to interpret the solution.

Difference Equations

Let y_n be the number of individuals at time n (generation n), and by$_n$ and dy$_n$ denote the number of individuals who were born and who died during this period. Then

$$y_{n+1} = y_n + by_n - dy_n = (1+b-d)\, y_n = ry_n$$

where r is the net growth rate.

For realistic models of growth processes, certain restrictions are necessary, since, for instance, for constant r we have $y_{n+1} = r^n y_0$ and $y_n \to \infty$ ($r > 1$) as $n \to \infty$. The following figure describes the asymptotic behavior of this simple model.

unbounded oscillation	damped oscillation	exponential decay	exponential growth
-1	0	1	r

If $r = r(y_n)$, we have a density dependent growth rate. When the action of a previous generation determines the growth of a population, the resulting situation might be governed by second-order difference equation. A simplest model is a general second-order linear difference equation with constant coefficients—namely,

$$y_{n+2} = by_n + 1 + cy_n = 0.$$

Logistic Model

Logistic equation described in this section is considered by many investigators an appropriate model for records, especially the best times in track and field. It is an example of nonlinear first-order difference equation. It is also a model that has been used frequently for population growth and is a well-recognized name in population dynamics. In sport, population could be the participants or just fans of a given game or sport. Most people tend to regard the elongated "S shaped" logistic curve of population dynamics as somewhat exotic. Because of its appropriateness in modeling, a time series such as best annual times (in men's one-hundred-meter dash, say) software such as Minitab has included that in the list of options for time series analysis and modeling.

A difference equation of the form

$$Y_{n+1} = G(Y_n) \quad n = 0,1,2,\ldots$$

where G is a nonlinear function, is called a first-order nonlinear difference equation. For example, suppose that in equation the rate of increase is not a constant but a function of Y_n —that is, $r = r(Y_n)$ —then the equation becomes a nonlinear equation of the form

$$Y_{n+1} = (1 + r(Y_n)) \, Y_n.$$

An important case is when r is a linear function of Y_n. In that case, the equation is called a logistic equation. As mentioned earlier, logistic equation is used frequently as a mathematical model for population growth. In fact, when applied to population growth problem, a logistic equation is often presented in the form

$$Y_{n+1} - Y_n = r^* \, Y_n \, (1 - Y_n/k)$$

where r^* presents the intrinsic rate of growth, and k is the environmental carrying capacity. Note that this model corresponds to the following choice of $r(Y_n)$

$$r(Y_n) = r^* (1-Y_n/k).$$

It is a reasonable model, since in population growth, an unbounded growth is unrealistic, and, therefore, models need to take to account factors such as limited resources for reproduction. In sports, we cannot expect that records grow forever—that is, one should expect to reach an ultimate record (k).

The asymptotic behavior of a logistic equation; the interested readers will examine this by selecting different values of r^* and plotting the equation.

1.   If $O < r^* \leq 1$, then the population steadily approaches k without overshooting it.
2.   If $1 < r^* \leq 2$, then the population overshoots but undergoes damped oscillations as it approaches k.
3.   If $2 < r^* \leq 2.449$, then the population settles down to a two-point cycle.
4.   If $2.449 < r^* \leq 2.570$, then the population achieves a stable cycle with 2m points, $m > 1$, the values of m depending on r^*.
5.   If $r^* < 2.570$, then the population size varies in a wholly unpredictable manner (chaos) with different outcomes resulting from different values of initial size Y_0.

Note that if models with *chaotic* behavior are relevant to populations, then wild oscillations of population *need not* necessarily be the consequence of random environmental fluctuations but might be intrinsic to the population. Thus, an important question is "Do biological populations exhibit chaotic behavior?" This question is unresolved at present.

The continuous version of classical logistic equation—namely,

$$dx/dt = ax - bx^2$$

is a first-order Bernoulli differential equation. Here, x denotes the population size at time t. Logistic equation can be generalized to take consideration, for example, harvesting, deteriorating environment, and limitations special to population under consideration. It is worth noting an important distinction between first-order difference and differential equations with constant coefficients. The latter allows no oscillation.

Recurrence Relations

Many interesting real-life problems can be posed in a recursive manner where a present value of whatever is measured, Y_{n+1} say, can be expressed in terms of its past values, $Y_n, Y_{n-1}, Y_{n-2},\dots$ So recursion or recurrence relation is an appropriate name. Recursions are particularly suitable when dealing with a class of probability problems known as waiting times problems. We start with simple examples describing simple situations and build upon those.

Simple Interest

Suppose that a certain amount of money is deposited in a savings account. If interest is paid only on the initial deposit (and not on accumulated interest), then the interest is called simple. For example, if $40 is deposited at 6 percent simple interest, then each year, the account earns 0.06(40), or $2.40. So the bank balance accumulated as follows:

Year	Amount	Interest
0	$40	$2.40

1	$42.40	$2.40
2	$44.80	$2.40
3	$47.20	

Suppose that you are asked to develop a formula for Y_n, the amount in the account above at the end of n years. Also find the amount at the end of ten years.

Let y_n = the amount at the end of n years. $y_0 = 40$. Moreover, amount at end of $n + 1$ years = amount at end of n years + interest.

$$Y_{n+1} = Y_n + 2.40$$

Compound Interest

When interest is calculated on the current amount in the account (instead of on the amount initially deposited), the interest is called compound. Often interest is compounded more than once a year. For example, interest might be stated as 6 percent compounded semiannually. This means that interest is computed every six months, with 3 percent given for each six-month period. At the end of each such period, the interest is added to the balance, which is then used to compute the interest for the next six-month period. Similarly, 6 percent interest compounded six times a year means 0.6/6 or 0.01 interest each interest period of two months. Or, in general, 6 percent interest is compounded k times a year, and then 0.06/k interest is earned k times a year. This illustrates the following general principle:

Example. Suppose that the interest rate of 6 percent compounded monthly. Find a formula for the amount after n months.

Solution: Here, $r = 0.06$, $k = 12$ so that the monthly interest rate is i = 0.005. Let Y_n denote the amount after n months. Then, reasoning as in the preceding section,

$$Y_{n+1} = Y_n + .005 \, Y_n = 1.005 \, Y_n$$

The solution of this difference equation is obtained easily by replacing for Y_n-1.005 Y_{n-1} and repeating the process. This leads to

$$Y_n = Y_0 (1.005)_n$$

Summary

If Y_0 dollars is deposited at interest rate i per period, then the amount after n interest periods for simple and compound interests are

Simple interest: $Y_n = Y_0 + niY_0$
Compound interest: : $Y_n = Y_0 (1+i)^n$

Simple and Compound Interests in Poems

Mathematics is the foundation of science as we all know
It has applications in all other disciplines especially now
Some we all use in our daily lives, some we hardly do
Some we like to apply but we do not really know how
A good example is when we borrow money or invest
For those of us who do not know basic
mathematics the deal could end in sorrow
Let's see if we can together review the basic idea involved
We do it step by step and hopefully learn, let us see how
First, when you borrow money you may end up
paying back twice as much, you owe
A similar case is when you invest for a fixed
period but you make an early withdraw
The important terms are principle, interest,
rate, and the time period
Their technical names are interest, time, and future value
If you invest $1,000 with an annual rate of 6% for 3 years now
At the end of this period you will have 1000
+ 1000 x 0.06 x 3 = 10,180 or so

This you may consider a good investment,
but these days the interest is low
You can, of course, buy stocks but then you
could lose it even if you are a pro
Rather than simple interest you can ask for
what is known as compound interest
Where your money is invested for shorter periods and together
with its interest reinvested you will have more to show
For instance, after the first period/year you will have
1000 + 1000 x 0.06 x 1 = 1060, though not much
The next period, you get 1060 + 1060 x 0.06 x 1 = 10123.60
a little more than 10120 with simple interest now
If you increase the number of periods and
reinvest for a fraction of a year my friend
Compound interest kicks in and your money grows
faster, an upward trend, you know how
So, you see that it is better to consider an
aggressive form of compound interest
Finally, if you have a low tolerance for losing, do
not buy stock that could result in a blow
Stocks are very risky if you invest for a short period of time
You could get a green light or red, but mostly yellow
That is the end of the today's lesson my dear fellow
Stocks, if have time or are an aggressive
investor. Banks if you are mellow
Compound interest if you do understand
and have money you do not need
Simple interest if you do not care or you really need to borrow
Investment is a part of life you need to have a plan for your future
You certainly become rich if you have a feel for
it. No need to start from zero, from hello.

Euler's e (=2.718281...), an Exceptional Number

The most powerful force in the universe is compound interest
—Albert Einstein

Mathematics is filled with curious, interesting, and amazing numbers. There are now several books about the four most famous numbers in the world of mathematics—namely, π, e, i, and 0; the quantities that appear in probably the most famous and certainly a most amazing equation $e^{\pi i} + 1 = 0$ known as Euler's identity (Euler's or beautiful equation), named after the Swiss mathematician Leonhard Euler. Some call it beautiful equation or equality. Here, e is Euler's number, the base of natural logarithms; i is the imaginary unit, which satisfies $i^2 = -1$; and π is pi, the ratio of the circumference of a circle to its diameter. The equation is considered to be an example of *mathematical beauty*, perhaps a supreme example as it shows a profound connection between the most fundamental numbers in mathematics. Only one does not yet have its book. Given the trend, we can look forward to its appearance.

Indeed, since π has been so well taken care of in popular literature, we may safely turn our attention to other numbers—the number e with unique and amazing mathematical properties, for example. In retrospect, a book on/about e was a natural. With the longevity of *A History of Pi* so plainly in view, one wonders that so many years passed before such a book appeared.

Persuading readers that the number e is a natural subject for study is no easy task, but historical exposition is perfect for the job. Think about banking and investment. The leap from annual to semiannual, quarterly, monthly, and daily compounding to continuous compounding would have been astounding at the time the number e was introduced. Yet not even one case of any type of subannual compounding is documented. Furthermore, although records of compound interest date back to antiquity, the appearance of e was exactly contemporaneous with the introduction of the Napierian logarithm.

About *e*

Although most people find the pi fascinating, most mathematicians find the e even more fascinating. Pi is very easy to understand, as it is simply the ratio of circle's circumstance to its diameter; *e*, on the other hand, is not so easy to describe or explain. Both numbers appear in the complex equation representing the bell curve or normal distribution. However, e appears in significantly more places.

e and Compound Interest

An account starts with $1.00 and pays 100 percent interest per year. If the interest is credited once, at the end of the year, the value of the account at year-end doubles to $2.00. What happens if the interest is computed and credited more frequently during the year? If the interest is credited twice in the year, the interest rate for each six months will be 50 percent; so after the first six months of the year, the initial $1.00 is multiplied by 1.5 to yield $1.50. Reinvesting this by the end of the year, it becomes $1.50 x $1.5, yielding $1.00×$1.5^2 = $2.25 at the end of the year. Compounding quarterly yields $1.00×1.25^4 = $2.4414 . . . and compounding monthly yields $1.00 × (1+1/12)^{12} = $2.613035. Compounding weekly (n = 52) yields $2.692597 . . ., while compounding daily (n = 365) yields $2.714567 . . ., just two cents more. The limit is the number that came to be known as e; with continuous compounding, the account value will reach $2.7182818. Bernoulli in year 1683 was the first person to notice this.

Does the number e have any real physical meaning, or is it just a mathematical convenience? The answer is, yes, it has a physical meaning. It occurs naturally in any situation where a quantity increases at a rate proportional to its value, such as a bank account producing interest or a population increasing its size or even a piece of radioactive material going through decay. Obviously, the quantity will increase more if the increase is based on the total current quantity (including previous increases) than if it is only based on the original quantity (with previous increases not counted). How much more? The number e answers this question. To put it another way, the number e is related to how much more money you will earn under compound interest than you would under simple interest.

The more mathematics and science you encounter, the more you run into the number e. Since its discovery, it has shown up in a variety of useful applications including (but definitely not limited to) solving for voltages, charge buildups, and currents in dynamic electrical circuits, spring/damping problems, growth and decay problems, Newton's laws of cooling and heating, plane waves, compound interest, in hiring the best applicant, waiting time to the new record, etc.

Bouncing Ball

In sports, almost everything is dynamic—that is, it changes with time. In fact, most sports-related quantities change at each moment during the period when we study them—for example, if we may drop a basketball and study its behavior from the moment we let it go to the moment it finally set on the ground. Recall that in basketball, players try to control the basketball by bouncing it through choosing appropriate angle and pressure. In tennis, understanding the court condition in terms of how the ball would bounce plays a key role. In fact, many players who are the best in one type of court (e.g., grass) may not be the best in other types of courts.

Modeling Process

The balls that are used in different sports have a different amounts of bounce depending on factors such as the type of the interface they hit, age of the ball, covering, and amount of their kinetic energy that is converted to heat. Some balls retain more of their kinetic energy than others and, therefore, bounce higher. One way to measure the bounciness of a ball is through a quantity known as the coefficient of restitution (R), which is defined as square root of the rebound height to the initial height. For example, a basketball dropped from a height such that the bottom of the ball is six feet from the floor should rebound to a height such that the top of the ball is forty-nine to fifty-four inches above the floor. Since a basketball has a diameter of 9.5 inches, we can calculate that the minimum acceptable value of R is square root of (49 - 9.5)/72 = 0.74, and the maximum acceptable value is square root of (54 - 9.5)/72 = 0.79.

Let Y_n and Y_{n+1} denote respectively the heights of the ball (e.g., its top point) in n-th and (n + 1)-th bounces. Then, theoretically, we have

$$Y_{n+1} = R(Y_n) \text{ or } \qquad Y_{n+1} - R(Y_n) = 0.$$

This is a simple example of a first-order linear difference equation. Here, one value of a function, Y_{n+1}, is expressed in terms of another (past) value of the same function, Y_n. So recursion or recurrence relation is an appropriate name.

Let Y_n and Y_{n+1} denote the number of players in years (seasons or periods) n and $n + 1$, respectively. Suppose that in year $n + 1$, the number of people playing this sport has increased some percentage r (e.g., 10 percent). Then we have

$$Y_{n+1} = (1+ r)Y_n = Y_n + rY_n \text{ or } Y_{n+1} - Y_n = rY_n.$$

Here, r could be a constant or a variable—for example, a function of Yn. The simplest model is when r is constant. This case results in a linear difference equation of the first order.

Summarizing these, a difference equation describes the "change" of quantities. The change may be due to one or many different factors.

Participation

Records are often affected by the size of population in general and the size of participants in that sport in particular. Suppose that we start with a given year and call it the initial year. We may use Y_0 to denote the population size for that year. Similarly, we may use Y_1, Y_2..., Y_3, ... to denote the population size for the first year, second year, third year, and so on. Assuming that the increase in population is proportional to the size of population with a fixed rate r, then we have

$$Y_1=Y_0+rY_0=(1+r)Y_0$$
$$Y_2=Y_1+rY_1=(1+r)Y_1$$
$$Y_3=Y_2+rY_2=(1+r)Y_2$$
$$Y(n+1)=Y_n+rY_n=(1+r)Y_n$$

This equation is another example of a linear difference equation of first order. In general, a first-order linear difference equation takes the form

$$Y(n+1)=aY_n+b \qquad n=0, 1, 2,...$$

where a and b are constants and Y_0 is a given initial value.

Example. In recent years, there have been many initiatives encouraging employees to save for their retirement. A company who participates in one such initiative has a matching company pension plan for its employees. An employee elects to belong to the plan by allowing the company to deduct a fixed sum from his or her gross pay and deposit it directly into an account yielding interest at an annual rate of 6 percent compounded monthly. The company matches each deposit on a 50 percent basis—that is, for every dollar the employee contributes into the account, the company deposits 50 cents. What is the difference equation governing the employee's monthly balance? (Assume paychecks are distributed monthly.)

Solution: The initial balance is the first of the equal-sized monthly contributions. Let this amount be p, then $Y_0=1.5p$. Thereafter, each monthly balance is related to the previous balance by

Current balance	=	Previous month's balance	+	Interest earned	+	Employee's contribution	+	Company's contribution
Y_{n+1}	=	Y_n	+	$0.005Y_n$	+	p	+	$0.5p$

so that

$$Y_{n+1} = 1.005Y_n+1.5p$$

is the difference equation we seek. Note that a 6 percent annually is 0.5 percent monthly.

Higher-Order Linear Difference Equation

In the first-order linear difference equation, Y_{n+1} depends only on Y_n. There are many instances where Y_{n+1} may depend on more than one past value. For example, the number of people watching or participating in a certain sport may depend on what happened in the past two years or seasons rather than just last year or season. In fact, many sports (and organizations) experience increase and decrease in popularity because of success of an individual or a team or failure of generating excitement or producing a celebrity or a star.

To demonstrate this, let us consider a game involving two players or teams, G (George) and H (Hank), say. We assume that during any given play of the game, there is a probability p that G wins a point (or certain amount of money) and a probability q that H wins a point. These probabilities satisfy the following conditions:

$$0 < p < 1, \qquad p+q=1$$

Note that this is the famous gambler's ruin problem already. We assume that G begins the game with A points (or A dollars), and then H has B points—that is, we start analyzing the game when it is A points versus B points, and the game stops when G has either won B points or lost A points. When G has lost A points, he is eliminated.

Let $N = A + B$ be the total number of points, and let n denote the number of points that G has at any time during the play. Denote by Y_n the probability that G will be eliminated, given that he has n points. We will now derive a difference equation for Y_n.

If G has n points, then on the next play he will have $n + 1$ points if he wins and a probability of $Yn+1$ of being eliminated. Likewise, if G loses, he will have $n - 1$ points and a probability of Y_{n-1} of being eliminated from that situation. Because these are the only two possibilities, we have

$$Y_n = pY_n + 1 + qY_{n-1}$$

which is a linear second-order difference equation. Its solution gives the required probability of the player's elimination. However, to obtain a complete solution of this equation, two conditions must be specified. First, if $n = 0$, then G is already eliminated, and no further play takes place. Also, if $n = N$, then G has taken all the points, and again play stops.

Difference Equation Models

This section presents an analysis of the models that are frequently used for estimating the future records for athletic events. It considers an innovative approach to modeling based on difference equations representing models used for prediction in the literature.

Introduction

Apart from intrinsic interest, there are several medical and physiological reasons why one would like to know how fast a human being could, for example, run a medium distance such as four hundred meters. While there is a general agreement among the physiologists and physical educators about existence of an upper limit for such a speed (or a lower limit for the time needed to run this distance), the limit is not known at the present time. Because of a great interest in this question, apart from physiological research, there have been some attempts to estimate (predict) the limits through mathematical and statistical modeling. A general approach employed in these studies has been based on predictive mathematical models via examining relationships between the past and present records.

Smith(1988) has applied a so-called exponential-decay model to the records for mile and marathon races. Unfortunately, when applying the model to estimate the limiting record (best attainable time), the standard errors were so large that the estimates were meaningless. Chatterjee and Chatterjee(1982) have applied a similar model to the winning times in the men's one-hundred-, two-hundred-, four-hundred-, and eight-hundred-meter runs in the Olympic Games. They have obtained two sets of estimates for the parameters of interest and for the ultimate winning record using nonlinear model fitting algorithm and jackknife procedure.

Noubary (1994) has shown that the use of an alternative model involving logarithms of the annual records could improve the analysis of the winning times and has used the results for short-time prediction of the men's four-hundred- and eight-hundred-meter run.

Most of the successful models proposed so far are based on the following form:

$$Y(t) = Z(t,\theta) + x(t)$$

where $Y(t)$ represents records, $Z(t,\theta)$ is a deterministic function (smooth part) or trend, and $x(t)$ is the random component (rough part) representing the fluctuations.

The section suggests consideration of an innovative approach for modeling and estimating the records based on difference equations whose complementary solution yields models used frequently. Since for each model we have one difference equation and vice versa, it is possible to carry the analysis using difference equations alone or in combination with above models. Since all the above models present smooth functions, such analysis could also be performed based on differential equations measured at the integer time values.

Derivation of Difference Equations

In this section, we derive difference equations. Let $y(n)$ represent, for example, the fastest time for four-hundred-meter run at time n or at year n. The linear model (straight line) $\theta 1 - \theta 2t$ satisfies the following difference equation

$$y(n) = y(n-1) - \theta 2.$$

Note that for this case, $y(n) - y(n-1) = \theta 2$, which represents a constant annual improvement. This implies that $y(n) = y(1) - (n-1) \theta 2$, resulting in negative values when n gets large. To avoid this, we may consider the following assumptions:

(I)   The annual improvement, denoted by $y(n-1) - y(n)$, should in general decrease and eventually vanish.

(II) If $\lim y(n) = y^*$, then in general $\{y(n)\}$ decreases monotonically to y^* and, hence, $y(n) - y^*$ decreases monotonically to zero.

Based on these assumptions, we will consider a few difference equations corresponding to the models frequently used. First, we derive a difference equation for the exponential model. Suppose that for all n

$$y(n-1) - y(n) = r(y(n-2) - y(n-1))$$

where $0 < r < 1$ is a constant if we rewrite as

$$y(n) - (1 + r)\, y(n - 1) + ry(n - 2) = 0.$$

Introduction

Sports provide an inexhaustible source of fascinating and challenging problems, and many books and journals are devoted to their data analysis and modeling.

Most sports can be related to mathematics in general and several specific subjects in particular. Many can be utilized to enhance teaching of mathematical concepts. In fact, sports and the data related to them offer a unique opportunity to teach mathematics and statistics and to test the methods they offer.

This section presents a probabilistic analysis of a table tennis game with a view toward its possible use as an aid for demonstrating steps of mathematical modeling. The materials presented are suitable for further analysis and study. We begin by describing the history and the rules of the game.

History and Rules

Table tennis, also known as ping-pong, has its origins dating back to medieval tennis. Beginning as a mere social diversion, table tennis became popular in England during the latter part of the nineteenth century. It did not take long for the popularity of this game to spread. As early as 1901, table tennis tournaments were being organized, books about table tennis were written, and even local table tennis associations were formed. In 1902, an unofficial "World Championship" was held. Table tennis was widely popular in Central Europe from 1905 to 1910. A slightly altered version was introduced prior to this to Japan, which then spread to China and Korea.

Today, table tennis has become a major worldwide sport, with approximately thirty million competitive players and countless others who enjoy playing the sport at a leisurely level. Presently, the

International Table Tennis Foundation represents 140 countries. In 1988, it became an Olympic sport.

Table tennis is an object sport in which each competitor tries to control an object while the other competitor is in direct confrontation. The object of the game is to hit the ball into the opponent's table. If the ball is not returned after one bounce, a point is won. The only other way to win a point is when the opposing player commits an error by striking the net or by hitting the ball off the table. The table must be 2.74 meters (9 feet) long and 1.525 meters (5 feet) wide. The top of the net must be 15.25 centimeters (6 inches) above the playing surface, while the posts that secure the net are 15.25 centimeters beyond the sideline. The spherical ball must weigh 2.7 grams (0.09524 ounces) with a diameter of 40 millimeters (1.574 inches). The ball may be white or orange, in either case not glossy, and it can be made of celluloid or similar plastic material. While most rackets seem to be the same shape, it can actually be of any size, shape, or weight as long as the blade is flat and rigid. Natural wood must comprise 85 percent of the blade by thickness. Neither can be thicker than 7.5 percent of the total thickness or 0.35 millimeters, whichever one is smaller. A game is won by the player who either reaches eleven points by a margin of two or gains a lead of two points after both players have scored ten points. This scoring scheme was implemented in 2001. A match is simply the best of any odd number of games. A new serving system was also put into place in 2001. Each player serves for two points and then alternates with the other player until the end of the game. If the game reaches ten all, then the number of consecutive serves for each player is reduced to one. Concerning the serve, another new rule was put into effect in 2002. This rule states that the ball must be visible at all times; meaning, the server cannot block the receiver's view of the ball in any way (e.g., body, clothing, and table).

Figure 1 displays possible outcomes (states) of a game of table tennis under the new rules. Prior to 2001, each game consisted of twenty-one points instead of eleven, and five serves instead of two.

Figure 1. Table tennis scoring illustration (new rules).

0-0
1-0 0-1
2-0 1-1 0-2
3-0 2-1 1-2 0-3
4-0 3-1 2-2 1-3 0-4
5-0 4-1 3-2 2-3 1-4 0-5
6-0 5-1 4-2 3-3 2-4 1-5 0-6
7-0 6-1 5-2 4-3 3-4 2-5 1-6 0-7
8-0 7-1 6-2 5-3 4-4 3-5 2-6 1-7 0-8
9-0 8-1 7-2 6-3 5-4 4-5 3-6 2-7 1-8 0-9
10-0 9-1 8-2 7-3 6-4 5-5 4-6 3-7 2-8 1-9 0-10
11-0 10-1 9-2 8-3 7-4 6-5 5-6 4-7 3-8 2-9 1-10 0-11
11-1 10-2 9-3 8-4 7-5 6-6 5-7 4-8 3-9 2-10 1-11
11-2 10-3 9-4 8-5 7-6 6-7 5-8 4-9 3-10 2-11
11-3 10-4 9-5 8-6 7-7 6-8 5-9 4-10 3-11
11-4 10-5 9-6 8-7 7-8 6-9 5-10 4-11
11-5 10-6 9-7 8-8 7-9 6-10 5-11
11-6 10-7 9-8 8-9 7-10 6-11
11-7 10-8 9-9 8-10 7-11
11-8 10-9 9-10 8-11
11-9 10-10 9-11

A's game Advantage A Tie Advantage B B's game

Probability of Winning a Game

Table tennis can be related to many mathematical topics. We start exploring these relationships by first focusing on some probability questions.

Suppose that a game is played by two players, A and B. Let the probability of A winning the point be denoted by x, and the probability of B winning the point by y, where $x + y = 1$ and both are independent of the present score. In practice, x and y may be estimated from player's past games against each other.

Before proceeding further, we need to comment on question of possible advantage associated with serving and its effect on probability of winning or losing a game. The assumption of fixed probability of winning a point is a standard one and is usually made for simplicity. However, it is not a serious one, and its effect on probability of winning or losing a game is not as pronounced as it is for tennis, since the server has to hit her own side of the table first. In fact, as is pointed out in an article titled "How Long Is an 11 Point Game?", djmarcusetasc.com (Wed. Feb. 10 10-39-01 1993), in most cases, the assumption of fixed probability has no effect on conclusion. Also, in a more detailed study titled "Does It Matter Who Serves First?", djmarcusetasc.com (Wed. Feb. 10 10-39-02 1993), which appeared on page 31 of Jan/Feb 91 *TT Topics*, the answer to the question posed in the title is stated as no based on the following argument. If the game goes tie (deuce), then it does not matter who served first, since no matter who wins, each player will have served the same number of times. For other cases, consider the following modifications of the rules. Rather than stopping when one player reaches eleven, keep playing until twenty points have been played. If player A wins the game under modified rules, then she must win at least eleven of the twenty points and, hence, would have won the game under standard rules. Similarly, if A loses under the modified rules, she also would have lost under the standard rules. But under the modified rules, both players serve ten times, and so it doesn't matter which one served first.

In the rest of this article, we will assume that x is fixed. However, we will return to this question later and make further comments before closing. For simplicity, we will only calculate the probability of A winning a game. Following the new rules of the game, A can either win by reaching a score of eleven with a margin of two (case 1) or, if players tie at 10:10, by winning extra points to have a margin of two points (case 2). Thus, the probability of A winning a game is the sum of these two probabilities—that is,

$$P(A \text{ winning a game}) = P(\text{Case 1}) + P(\text{Case 2}).$$

Let us discuss each case in more detail.

Case 1: In this case, A can win the game by any of the following scores: 11:0, 11:1, 11:2, 11:3, 11:4, 11:5, 11:6, 11:7, 11:8, or 11:9. For these, A must win ten out of the respective ten, eleven, . . ., nineteen points played, plus the last point. Using binomial distribution, it is easy to see that the probability of A winning with the score of 11:j $\binom{10+j}{10} x^{11} y^j$, $j=0,1,...,9$. Thus, the probability of A winning the game by reaching a score of eleven with a margin of two is

$$P(\text{case 1}) = \sum_{j=0}^{9} \binom{10+j}{10} x^{11} y^j = \sum_{j=0}^{9} \binom{10+j}{10} x^{11}(1-x)^j.$$

Case 2: In this case, A and B must first reach the score of 10:10. Then A must gain a lead of two points to win the game. The probability of this event is, therefore,

$$\textbf{P(Case 2)} = P(A \text{ and B both reach 10})$$
$$\cdot P(A \text{ wins by a margin of 2}).$$

First, we calculate the first term in the right-hand side. Since to reach 10:10, a total of twenty points must be played, we have

$$P(A \text{ and B both reach 10}) = \binom{20}{10} x^{10} y^{10}.$$

Next, to win the game, A must gain a lead of two points. This can be done in infinitely many ways until A finally gains a lead of two. Let p denote the probability that A wins the game after reaching the score of 10:10. Then we have

$$p = x^2 + 2xyp.$$

This is because starting from 10:10, A can either takes the next two points with probability x^2, or each player takes a point with probability of 2xy and restart from the score of 11:11. The probability of A winning a game restarting from 11:11 is the same as starting from the score of 10:10. Solving for p, we get

$$p = \frac{x^2}{1-2xy}$$

Putting these together, the probability that A wins a game after reaching a score of 10:10 is ·

$$P(\text{Case 2}) = \binom{20}{10} x^{10}y^{10} \frac{x^2}{1-2xy}.$$

Replacing $1-x$ for y, and $x^2 + (1-x)^2$ for $1-2xy$, the probability of case 2 is

$$P(\text{Case2}) = \binom{20}{10} \frac{x^{12}(1-x)^{10}}{(x^2+(1-x)^2)} = \frac{184756(1-x)^{10}x^{12}}{1-2x+2x^2}.$$

Finally, g(x), the probability of A winning a game, is obtained by adding the probabilities for the two cases—that is,

$$g(x) = P(A \text{ wins}) = P(\text{Case 1}) + P(\text{Case 2})$$

$$= \sum_{j=0}^{9} \binom{10+j}{10} x^{11}(1-x)^j + \binom{20}{10} x^{12}(1-x)^{10}/(x^2+(1-x)^2)$$

$$= x11(1-x)j + x12(1-x)10/(x2+(1-x)2)$$

Under the old rules, the probability of A winning a game denoted by $g_0(x)$ is

$$G_0(x) = \sum_{j=0}^{19} \binom{20+j}{20} x^{21}(1-x)^j + \binom{40}{20} x^{22}(1-x)^{20}/(x^2+(1-x)^2).$$

Analysis Using Difference Equations

Let us once more consider the event that A and B reach score of 10:10. We can use difference equations to find the probability that A wins the game starting from this score or, in fact, any other possible score. Here, we use the tie situation for demonstration.

Suppose that $g(i, j)$ represents the probability of A winning the game starting from a score of $i{:}j$—that is, A has won i points and B has won j points. Since either player can win the next point, this implies that

$$g(i, j) = xg(i + 1, j) + yg(i, j + 1).$$

Now, consider the case of a tie—that is, the score of 10:10. Using this relationship, we get

$$g(10,10) = xg(11,10) + yg(10,11),$$
$$g(11,10) = xg(12,10) + yg(11,11),$$

and

$$g(10,11) = xg(11,11) + yg(10,12).$$

Substitution for $g(11,10)$ and $g(10,11)$ in the first equations yields

$$g(10,10) = x[xg(12,10) + yg(11,11)] + y[xg(11,11) + yg(10,12)]$$
$$= x2g(12,10) + 2xyg(11,11) + y2g(10,12).$$

If the game reaches a score of 12:10, then A has won the game. Therefore, $g(12,10) = 1$. Also, if the game reaches a score of 10:12, then B has won the game. This means that $g(10,12) = 0$. Finally, if the game reaches a score of 11:11, the situation is no different from the score of 10:10. Therefore, we have $g(11,11) = g(10, 10)$. Replacing for these, we obtain

$$g(10,10) = x^2 + 2xyg(10,10)$$

or

$$g(10,10) = x^2/(1 - 2xy).$$

The value of $g(10,10)$ represents the probability that A wins the game starting from a score of 10:10. Note that this is the same as what we obtained in the previous section. Following the same lines, we can obtain the probabilities for other starting scores. For example,

$g(11,10) = xg(12,10) + yg(11,11)$
$\quad =x+yx^2/(1-2xy)=x(1-xy)/(1-2xy)$
$g(10,11) = xg(11,11) + yg(10,12) = xg(10, 10) = x^3/(1-2xy)$
$g(9,9) = g(10,10) = x^2/(1-2xy)$
$g(10,9) = g(11,10), g(9,10) = g(10,11)$

The Fibonacci Sequence (Revisited)

The well-known Fibonacci sequence,

$$0, 1, 1, 2, 3, 5, 8, \ldots,$$

is described by the recurrence relation $c_{n+2} = c_n + c_{n+1}$, with the initial conditions $c_0 = 0$ and $c_1 = 1$. An interesting thing you can do is to create what is called a generating function.

Fibonacci Numbers and the Golden Section

The Fibonacci numbers are
0, 1, 1, 2, 3, 5, 8, 13, . . . (add the last two to get the next)
The golden section numbers are
0·61803 39887 . . . = phi = φ and 1·61803 39887 . . . = Phi = Φ
The golden string is 1 0 1 1 0 1 0 1 1 0 1 1 0 1 0 1 1 0 1 . . . a

sequence of 0s and 1s that is closely related to the Fibonacci numbers and the golden section.

The Golden String

The golden string is also called the *infinite Fibonacci word* or the *Fibonacci rabbit sequence*. There is another way to look at Fibonacci's rabbits problem that gives an infinitely long sequence of 1s and 0s called the golden string,

$$1\ 0\ 1\ 1\ 0\ 1\ 0\ 1\ 1\ 0\ 1\ 1\ 0\ 1\ 0\ 1\ 1\ 0\ 1\ \ldots$$

This string is a closely related to the golden section and the Fibonacci numbers.

Fibonacci Rabbit Sequence

See show how the golden string arises directly from the rabbit problem and also is used by computers when they compute the Fibonacci numbers. You can hear the golden sequence as a soundtrack too.

The Fibonacci rabbit sequence is an example of a *fractal*—a mathematical object that contains the whole of itself within itself infinitely many times over.

Fibonacci Numbers and Phi

The Fibonacci numbers in a formula for pi (π)

There are several ways to compute pi (3.1415926535 . . .) accurately. One that has been used a lot is based on a nice formula for calculating which angle has a given tangent, discovered by James Gregory. His formula together with the Fibonacci numbers can be used to compute pi. This page introduces you to all these concepts from scratch.

The Fibonomials

The basic relationship defining the Fibonacci numbers is $F(n) = F(n-1) + F(n-2)$, where we use some combination of the previous numbers (here, the previous two) to find the next. Is there such a relationship between the *squares of the Fibonacci numbers* $F(n)^2$? or the *cubes* $F(n)^3$? or other powers? Yes, there is, and it involves a triangular table of numbers with similar properties to Pascal's triangle and the binomial numbers—the Fibonomials.

Fibonacci and Phi in the Arts

Fibonacci Numbers and the Golden Section in Art, Architecture, and Music

The golden section has been used in many designs, from the ancient Parthenon in Athens (400BC) to Stradivari's violins. It was known to artists such as Leonardo da Vinci and musicians and composers, notably Bartók and Debussy.

Differential Equations Model

As most of you probably already know, the Fibonacci sequence,

$$0, 1, 1, 2, 3, 5, 8, \ldots,$$

is described by the difference equations or recurrence relation $c_{n+2} = c_n + c_{n+1}$, with the initial conditions $c_0 = 0$ and $c_1 = 1$. An interesting thing you can do is to create what is called a generating function.

This sequence comes up everywhere—for example, in biological systems describing the number of petals and the shape of broccoli. Now, the solutions to this equation is the same as the solutions to the quadratic equation:

$$x\,(x-1) = 1$$

0·61803 39887... = phi = φ and
1·61803 39887... = Phi = Φ

These, as you can probably recognize, are the golden ratio and its inverse!

Difference Equations and Change

When do things change? Many physical quantities change at each moment during the period when we study them. For example, if we drop a bowling ball from the roof of a building, we are most likely interested in the behavior of the ball from the moment we let it go to the moment it crashes on the ground. At each instant during its fall, the downward speed of the ball is different from what it is at any other instant. Its speed changes constantly as it falls. In sport, records change with time.

In this text, though, we have seen a few examples of quantities that do not change at each instant. The balance in an interest-bearing account does not change all the time but only at the times when the interest is posted to the account (or when a new deposit is made). Let us suppose that the depositor makes no further deposits after having opened the account. After the first posting of interest, we have a balance y_1. After the second, we have new balance y_2, and so on. Using this notation, it makes sense to refer to the original deposit as y_0. Thus, any account is completely described by stating the sequence of balances $y_0, y_1, \ldots, y_{11}$, and so on for that account. In other words, the value of the account is constant until it takes a sudden "leap" at the end of each interest period.

At the time interest is posted, how does the balance change? Suppose a bank account earns interest at a 12 percent annual rate (1 percent per month) compounded monthly. If the balance after three months is y_3, then the next month's balance, y_4, is y_3 (last month's balance) plus whatever interest this money earns. But the earned interest is 1 percent of that: $0.01y_3$. The equation describing this change is

$$y_4 = y_3 + 0.01y_3$$

or

$$y_4 = 1.01y_3.$$

There is nothing special about the change from y_3 to y_4 More generally, therefore, we may write

$$y_{n+1} = 10y_n$$

which applies to all values of n greater that zero. The equation that starts the process applies to the opening of the account,

$$y_0 = p$$

where p is the amount of the initial deposit. The balance after any month is completely specified by stating the equation relating successive monthly balances to each other and the amount used to open the account, y_0. In general, the difference equation for any system consists of a pair of equations—the general rule by which y_{i+1} becomes y_{II}, and the specific rule for y_0. The difference equation for this example is the pair of equations

$$y_0 = p \qquad y_{n+1} = 10y_n$$

for $n > 0$.

The Mathematics of Growth

Exponential function is hard to comprehend.

If a single bacterium is placed into a nutrient-filled bottle, it immediately begins gobbling up nutrients, and after just a short time grows so much that it divides into two bacteria. In the same manner, soon, these two each divide into two bacteria, so now there are four bacteria in the bottle. This continues until the bottle is full of bacteria. If no nutrients remain, then every one of the bacteria will die. Here, there are many interesting mathematical questions. For example, you may like to know how long it will take to have bottle half full of bacteria. To solve problems like this, you need to model the population growth.

Rabbit Population

As discussed earlier, there are many different aspects of math within art. Fibonacci sequence is an example of such connection. Recall that Fibonacci sequence starts with 0, and continues when the two previous numbers add up to the following number, so it goes: 0, 1, 1, 2, 3, 5, 8, 13, 21, 34, 55, 89, . . ., etc. When this sequence is drawn out in squares, it makes a spiral, also known as the golden ratio.

To see the connection to recursive relation, consider the Fibonacci and rabbit population. Here, there is a lag factor; each pair requires some time to mature. So we are assuming

maturation time = 1 month and gestation time = 1 month.

The pattern we see here is that each cohort or generation remains as part of the next, and in addition, each grown-up pair contributes a baby pair. The number of such baby pairs matches the total number of pairs in the previous generation. Symbolically,

$$X_n = \text{number of pairs during month n} \quad X_n = X_{n-1} + X_{n-2}$$

So we have a recursive formula where each generation is defined in terms of the previous two generations. Using this approach, we can successively calculate Xn for as many generations as we like. So this sequence of numbers 1, 1, 2, 3, 5, 8, 13, 21, . . . and the recursive way of constructing it is the solution.

Modeling Spatial Pattern of Trees

The characteristics used in forestry are commonly divided into three groups describing

1) an individual tree   2) a forest stand   3) a forest region

Trees in a stand can be characterized in terms of their size distributions and their relative locations with respect to each other. The relative locations of trees can be illustrated by the point configuration formed by the dimensionless trees on the horizontal plane, called the spatial pattern of trees. The spatial pattern of trees can also be described by using stochastic models called *spatial point processes*.

The relative spatial distribution of trees plays an important role in

1) forest research   2) forest management

Examples are inventory planning, construction of growth models of trees or stands, forest regeneration, and forest thinning. Because of difficulties with the sampling when doing research, after modeling, we can use the model to simulate forest or to estimate the number of trees in a given forest. The spatial pattern affects, through the interaction of trees, the growth of an individual tree and thereby the current timber productivity of the whole stand. The interaction of trees could be integrated into the growth models as follows.

The growth of an individual tree is estimated as a function of the soil fertility, the size and age of the tree, and the sizes and locations of the neighboring trees. The growth of the stand is obtained as the sum of the growths of individual trees. The methods of spatial analysis have been applied in forest y perhaps most frequently in *regeneration* surveys. The configuration of trees is usually compared with a configuration of independently and *uniformly distributed* trees—that is, with a configuration generated by a *Poisson process* (simplest model). If a configuration cannot be considered Poisson, the alternatives are the *regular* and *clustered* ones. These types, then, need more complicated models and mathematical tools.

The *Poisson forest* satisfies the following two conditions:

1. The number of trees in any (Borel) set A is Poisson distributed with the mean Au(A).
2. The number of trees in disjoint sets are independent.

This implies that the number of trees in two circles of the same size, one centered on an arbitrary point of the plane, and the other on an arbitrary tree, are the same.

These properties can be used to construct randomness tests based on distances. In fact, Poisson forest can be used as a basis for comparison of other models such as nonhomogeneous Poisson forest ($\lambda = \lambda(t)$), Poisson cluster processes, doubly stochastic Poisson, lattice, Markov, . . . Because of complete randomness, Poisson forests do not *interact*. In general, there are also interactions such as *repulsion* or *attraction* between trees. This can be described by *interaction processes*.

Interaction may be pairwise—that is, the interaction between two trees may depend on the relative location of these two trees only, or it may depend on the relative location of other trees too. The final interaction model consists, of course, of interaction between all the points.

Thus, in a more advanced analysis, mathematical laws for the relative locations of trees are searched. The most informative spatial data for this purpose are the mapped data (i.e., the data in which the coordinates of trees in some subarea are known). With mathematical models, one can obtain detailed knowledge about the underlying random mechanism that has generated the pattern. Further mathematical models allow the use of simulation procedures for artificial production of point patterns compatible with data.

As an illustration, when considering processes of this type in a bounded region E, the mathematical investigation proceeds by defining a neighborhood relation on E. Here, *neighbors* are a pair of points whose mutual distance is smaller than a prescribed real number r. Recall that the *Markov property* of a stochastic process in a real line implies that to predict the future of the process, it's enough to know only the present state of the process. Here, future state is replaced by the unknown configuration in the set B and the present time by the neighborhood of the set. The interaction point process are characterized by their likelihood factions f, with respect to a Poisson process. For each point configuration μ, $f(\mu)$ measures *how much more likely* the configuration μ is for that process that for a Poisson process.

If the point process generating the pattern can be assumed to be *stationary* (i.e., distributional aspects are invariant under and arbitrary translation of the origin), then it is possible to consider the second-order properties and use that for summary description. Further assumption is invariance with respect to rotations. A point process with this latter property is called *isotropic*.

We finish this section by mentioning that when analyzing the interactions (especially pairwise interactions), some ideas from statistical mechanics such as the interaction energy between two particles distances apart, which leads to Gibbs process, have proved useful.

We end by noting that the purpose of the analysis of spatial point patterns is to identify the process that has *generated* the configuration. The methods of spatial point pattern analysis can be divided into two groups depending on whether they require mapped data or not. Mapped data refers to data in which the coordinates of trees in some subarea are known. The methods for unmapped data are sometimes called field methods. These can be further divided into two groups:

1. quadrant methods (based on quadrant sampling), and
2. distance methods.

For example, in distance methods, the distance from arbitrary trees or from arbitrary points to the nearest trees are measured. The measurements can be used for hypothesis testing or the estimation of the parameters of the underlying processes.

Random Function Models

Estimation and prediction require a model of how the phenomenon behaves at locations where it has not been sampled. Without a model, one has only the sample data, and no inferences can be made about the unknown values at locations that were not sampled.

In this part, we address the issue of modeling. After a brief discussion of deterministic models, we will discuss probabilistic models.

The Necessity of Modeling

Throughout we will be using a hypothetical example. In this example, we have measurements of some variable, v, at seven regularly space locations and are interested in estimating the unknown values of v at all the locations we have not sampled. Though this example is one dimensional for graphical convenience, the remarks made here are not limited to one-dimensional estimation problems.

The sample data set consists of seven locations and seven v values. By itself, this sample data set tells us virtually nothing about the entire profile of v. All we know from our samples is the value of v at seven particular locations. Estimation of the values at unknown locations demands that we bring in additional information or make some assumptions. Clearly, there is some uncertainty about how the phenomenon behaves between the sample locations. The random function models (probability models) recognize this fundamental uncertainty and give us tools for estimating values at unknown locations once we have made some assumptions about the statistical characteristics of the phenomenon.

With any estimation procedure, whether deterministic or probabilistic, we inevitably want to know how good our estimates are. Without an exhaustive data set against that we can check our estimates, the judgment of their goodness is largely qualitative and depends to a large extent on the appropriateness of the underlying model. As conceptualizations of the phenomenon that allows us to predict what is happening at locations where we do not have samples, models are neither right nor wrong; without additional data, no proof of their validity is possible. They can, however, be judged as appropriate or inappropriate. Such a judgment, which must take into account the goals of the study and whatever qualitative information is available, will benefit considerably from a clear statement of the model.

In addition to making the nature of our assumptions clear, a clearly stated model also provides us with a constant reminder of what is real and what is modeled. With the sample data set providing a very limited view of the complete profile, there is a strong temptation to replace the frustrating reality of the estimation problem with the mathematical convenience of a model, and in so doing, to lose sight of the assumptions on which our estimation procedure is based. A typical symptom of this is the reliance on statistical hypothesis tests to test model parameters. While such tests may demonstrate that the model is self-consistent, they do not prove that the model is appropriate.

Deterministic Models

As discussed earlier, the most desirable type of estimation problem is one in which there is sufficient knowledge about the phenomenon to allow a deterministic description of it. For example, imagine that the seven sample data were measurements of the height of a bouncing ball. Knowledge of the physics of the problem and the horizontal velocity of the ball would allow us to calculate its trajectory. While this trajectory depends on certain simplifying assumptions and is therefore somewhat idealized, it still captures the overall characteristics of a bouncing ball and serves as a very good estimate of the height at unsampled locations. In this particular example, we rely very heavily on our deterministic model; in fact, we could have calculated the same estimated profile with a smaller sample data set. Our deterministic model also allows reasonable extrapolation beyond the available sampling.

Using the same seven sample data, we can imagine a scenario that would produce a very different estimated profile. We can imagine that these seven samples are interest rates at a bank, measured on the Tuesday of seven consecutive weeks. Combining this with the knowledge that the bank adjusts the interest rate only once a week, on Thursdays, we can produce its estimated profile. Like our previous example with the bouncing ball, accurate estimation is made possible by our knowledge of the context of the data set. Unlike the previous example, we depend on all of our sample data, and our knowledge of the phenomenon is not good enough to allow us to extrapolate very far beyond the available samples.

From these two example, it is clear that deterministic modeling is possible only if the context of the data values is well understood. The data values by themselves do not reveal what the appropriate model should be.

Probabilistic Models

For the seven sample data shown, we can imagine many other contexts for which a deterministic description of the complete profile would be possible. Unfortunately, few applications are understood in sufficient detail to permit a deterministic approach to estimation. There is a lot of uncertainty about what happens at unsampled locations. For this reason, the statistical approach to estimation is based on a probabilistic model that recognizes these inevitable uncertainties.

In a probabilistic model, the available sample data are viewed as the result of some random process. From the outset, it should be clear that this model conflicts with reality. The processes that actually do create an ore deposit, a petroleum reservoir, or hazardous waste site are certainly extremely complicated, and our understanding of them may be so poor that their complexity appears as random behavior to us, but this does not mean that they are random; it simply means that we are ignorant.

Unfortunately, our ignorance does not excuse us from the difficult task of making predictions about how apparently random phenomena behave where we have not sampled them. For example, though earth science data are not, in fact, the result of random processes, this conceptualization does turn out to be a useful one for the problem of estimation. Though the word "random" often connotes "unpredictable," it turns out that viewing our data as the outcome of some random process does help us with the problem of predicting unknown values. Not only does it give us estimation procedures that, in practice, have sometimes proved to be very good, but it also gives us some ability to gauge the accuracy of our estimates and to assign confidence intervals to them. To take a simple but familiar example, consider the problem of estimating the sum of two dice. With a single die able to show only the numbers from one to six, the sum of two dice must be in the range from two to twelve. While an estimate of "somewhere from two to twelve" is not very satisfying, it is at least a safe start and is certainly better than avoiding the problem by claiming total ignorance. We can go beyond this safe statement, however, since some outcomes are more likely than others. A probability model in which the numbers from one to six all have equal probability of appearing on a single die allows us to predict that seven is the most likely outcome for the sum of two dice. Were we to use seven as an estimate, the probability model could tell us that we would be exactly right about 17 percent of the time, and that we would be off by more than two only 33 percent of the time. If, for some reason, we preferred to use ten as our estimate, the probability model could tell us that we would be exactly correct less than 9 percent of the time and would be off by more than two nearly 60 percent of the time.

In this example with the dice, we benefit considerably from our knowledge of the details of the random process that is generating each outcome—namely, that we are dealing with dice and that a single die shows numbers only in the range from one to six, each with equal probability. In the actual practice of estimation, we are handicapped by not knowing the details of the random process. In fact, as we have already noted, there is no random process that is generating our sample data—no dice are being thrown behind the scenes, no coins are being tossed, and no cards are being shuffled and dealt. Having chosen to view our data as the outcome of some random process, we are responsible for defining the hypothetical random process that might have conceivably generated our data.

It is possible in practice to define a random process that might have conceivably generated any sample data set. The application of the most commonly used statistical estimation procedures, however, does not require a complete definition of the random process; as we will discuss, it is sufficient to specify only certain parameters of the random process.

Why Math Modeling After All?

Quite naturally, we look for relationships and patterns among the variable features in the world around us. We attempt to devise abstract models. We do this model building for two reasons. First, for the practicality of it—in applying an abstract model that accurately mirrors an aspect of reality, we are better able to control events; and, second, even if a model leads to no practical consequence, understanding the model leads to a sense of insight, which is felt to be valuable in its own right.

Clearly, any realistic model of a real-world phenomenon must take into account the possibility of randomness. A statistical model is one in which chance and randomness play a significant role. Although there are some recognized randomly generated processes, the majority of processes are not random. Most processes that generate random behavior are so complicated, and our understanding of them is so poor that their complexity appears as random behavior to us, but this does not mean that they are random; it simply means that we are ignorant. Why do we need models? Why are they useful? How can they help us to learn and to predict? There are the concerns of this introductory article.

A. Lincoln states, "If we could first know where we are, and whither we are tending, we could better judge what to do, and how to do it."

One of the most basic ways to "know whether we are tending" is to build a model. We use models in our life. As a child, models were used to simulate objects, such as trains, dolls, and building blocks; or situations, such as doctor's office, cops and robbers, and teaching school. As adults, we use models for the same reasons—to simplify something too big, too complex, or too dangerous. But as adults, the use of models can be extended beyond physical object or situations to incorporate more abstract ideas. For example, when a person makes a decision, he or she is actually using a model. First, the person recalls all previous information that is relevant to the decision. This information is used to formulate a mechanism regarding the interdependence of the individual pieces of information. The mechanism is then used to predict the expected outcome of each of the possible choices. Finally, a decision is made based on the expected outcome, which maximizes the desired result. Once the actual results of the decision are known, we are able to evaluate the mechanism. This enables one to determine the validity of the mechanism. One means of simplifying the formulation and evaluation of the model is to incorporate mathematics.

Mathematics can be used when the behavior of the mathematical expression is similar to the properties of the data being studied. Since the rules of mathematics are clearly defined, this enables abstract ideas to be broken down into mathematical expressions with their contributing components represented by mathematical symbols and the interdependence of the components represented by mathematical equations.

Without a model, one only has the sample data, and no inferences can be made about the values at locations that were not sampled. Estimation and prediction require a model of how the phenomenon behaves where it has not been sampled. Take, for example, the hypothetical data set in figure 1. In this example, we have the measurements of some variable, u, at seven regularly spaced locations, and we are interested in estimating the unknown values of u at all the locations not sampled. As you can see, the data values themselves do not reveal what the approximate model should be. Please note that although this example is one dimensional for graphical convenience, the remarks made here are not limited to one-dimensional estimation problems.

By itself, the sample data set of figure 1 tells us virtually nothing about the entire profile of u. All we know from our sample is the value of u at seven particular locations. Estimation of the values at unknown locations demands that we bring in additional information or make some assumptions.

The most desirable information is a description of how the phenomenon was generated. In certain situations, the process that generated the data set might be known in sufficient detail so that an accurate description of the entire profile can be made from only a few sample values. In such situations, a *deterministic* model is appropriate. A deterministic model predicts a single outcome from a given set of circumstances. For example, imagine that the seven data points were measurements of the height of a bouncing ball. Knowledge of physics and the horizontal velocity of the ball would allow us to calculate the trajectory shown in figure 2. While this trajectory depends on certain simplifying assumptions and is, therefore, somewhat idealized, it still captures the overall characteristics of a bouncing ball and serves as a very good estimate of the height at unsampled locations. Because of our deterministic model, we could calculate the same estimated profile with a smaller data set. Our deterministic model also allows us reasonable extrapolation beyond the available sampling.

Using the same sample data set, imagine a scenario that would produce a very different estimated profile; the seven sampled points are interest rates at a bank measured on the Tuesday of seven consecutive weeks. Combining this with the knowledge that the bank adjusts the interest rate only once a week, on Thursday, we can produce the estimated profile in figure 3. Like our previous example, accurate estimation is made possible by our knowledge of the context of the data set. Unlike the previous example, we depend on all of our sample data, and our knowledge of the phenomenon is not good enough to allow us to extrapolate very far beyond the available samples.

From these two examples, it is clear that deterministic modeling is possible only if the context of the data is well understood.

Unfortunately, very few processes are understood well enough to permit the application of the deterministic models. Though we do know many of the fundamental processes, the variables of interest in the data sets are typically the result of vast numbers of processes whose complex interactions are not yet understood to be described quantitatively. For the vast majority of data sets, we are forced to admit that there is some uncertainty about how the phenomenon behaves. Any realistic model of a real-world phenomenon must take into account the possibility of randomness. The probabilistic models (stochastic models) recognize the fundamental concept of uncertainty and gives us tools for estimating values at unknown locations once we have made some assumptions about the statistical characteristics of the phenomenon (characteristics that hold true an average).

In the probabilistic model, which predicts a set of possible outcomes weighed by their likelihoods or probabilities, the available sample data are viewed as the result of some random process. From the onset, it should be clear that this model conflicts with reality. Although there are some recognized randomly generated processes, the majority of processes are not random. Most processes that generate "random" behavior are so complicated and our understanding of them so poor that their complexity appears as random behavior to us, but this does not mean that they are random; it simply means that we are ignorant.

Unfortunately, our ignorance does not excuse us from the difficult task of making predictions at locations that are not sampled.

To take a simple but familiar example, consider the problem of estimating the sum of two dice. With a single die able to show only the numbers from one to six, the sum of two dice must be in the range of two to twelve. While an estimate of "somewhere form two to twelve" is not very satisfying, it is at least a safe start and is certainly better than avoiding the problem by claiming total ignorance. We can go beyond this safe statement, however, since some outcomes are more likely than others. A probabilistic model in which the numbers from one to six all have equal probability of appearing on a single die allows us to predict that seven is the most likely outcome for the sum of two dice. Were we to use seven as an estimate, the probability model could tell us that we would be exactly right about 17 percent of the time and that we would be off by more than two only 33 percent of the time. If, for some reason, we preferred to use ten as our estimate, the probability model could tell us that we would be exactly correct less than 9 percent of the time and would be off more than two nearly 60 percent of the time.

In this example with the dice, we benefit considerably from our knowledge of detail of the random process that is generating each outcome—namely, that we are dealing with dice and that a single die shows numbers only in the range from one to six, each with equal probability.

In the actual practice of estimation, we are handicapped by not knowing the details of the random process; no dice are being thrown behind the scenes, no coins are being tossed, and no cards are being shuffled and dealt. Therefore, having chosen to view our data as the outcome of some random process, we are responsible for defining the hypothetical random process that might have conceivably generated our data.

It is possible in practice to define a random process that might have conceivably generated any sample data set. The application of the most commonly used statistical estimation procedures, however, does not require a complete definition of the random process. It is sufficient to specify only certain parameters of the random process. For example, a mathematical model for the amount of traffic on a certain highway cannot possibly assume that all the relevant variables are known; this would imply knowing whether each car in the entire city is on the highway. Rather, a model would make reasonable assumptions and predict the traffic up to likelihoods or various degrees of certainty.

With any estimation procedure, whether deterministic or probabilistic, we inevitably want to know how good our estimations are. Without an exhaustive data set against which to check our estimates, the judgment of their goodness largely depends on the appropriateness of the model, because there is no such thing as the best model for a given phenomenon. Models are neither right nor wrong, but in the final analysis, a model is judged using a single quite pragmatic factor, the model's usefulness. Such a judgment must take into account the goals of the study and whatever qualitative information is available. By clearly stating the model, we have a constant reminder of what is real and what is modeled—that is, what is observed versus what is expected. With the sample data set providing a very limited view of the completed profile, there is a strong temptation to replace the frustrating reality of the estimation problem with the mathematical convenience of a model, and in doing so, lose sight of the assumptions on which our estimation procedure is based.

Let us now concentrate on a specific problem—namely, modeling distribution of large incomes. Distribution of incomes has been of great interest and concern for more than a century. In 1897, Vilfredo Pareto published his influential economics textbook. In it, he observed that the number of persons in a population whose incomes exceed y is often well approximated by cy-a for some real c and some positive a. Accumulating experience rapidly pointed out the fact that it is only in the upper tail of the income distributions that Pareto-like behavior can be expected. It was concluded that income distributions have long (fat) tails, and Pareto distributions and their close relations and generalizations do indeed provide a flexible family of long-tailed distributions that may be used to model income distributions as well as a wide variety of other economic distributions.

Until 1975, Pareto's distribution and its close relatives were accepted based on empirical evidence, usually with little more justification that "this family of distributions has properties similar to those exhibited by many observed income distributions." Because of the fact that for almost eighty years the distribution of large income was accepted on empirical evidence only, it makes one wonder about the existence of a theoretical justification for Pareto's distribution.

No justification was offered until 1975. By this date, only a related method known as the "threshold method" had been introduced and used by hydrologists in its earliest form and later developed from a more rigorous mathematical standpoint by statisticians. One important step was the suggestion that different types of upper tail behavior can be distinguished from the conditional mean exceedance function (cme function).

$$M(u) = E[Y - u \mid Y > u]$$

Bryson (1974). The conditional mean exceedance $M(u)$ is the average amount by which an income exceeds a threshold u given that it is larger than u. In 1975, Pickands made a significant contribution to this field. A basic theorem proved by him became the basis of future investigations. It follows from this theorem that there are only three possible types of upper tail behavior that the cme function can exhibit: (1) an unbounded "long-tailed" distribution has cme function that is approximately linearly increasing in the upper tail; (2) and unbounded "medium-tailed" distribution has constant cme function in the upper tail; and (3) a bounded "short-tailed" distribution has acme function that is approximately linearly decreasing in the upper tail. The three distributions can be combined into a single form known as the generalized Pareto distribution (GPD) representing respectively a long-, a medium-, and a short-tailed distribution. Note that this model and theorems related to that provide a clear justification for usage of the conventional Pareto distributions for income data above a large threshold. It also demonstrates the strength of the mathematical modeling.

The conclusion that could be drawn from this is that the mathematical theory developed provides a clear justification for all empirical evidences in favor of the Pareto distribution.

As we observe the world around us, quite naturally we look for relationships and patterns in the variable features: We attempt to devise models. But by making these models mathematical, we give strength and justification.

CHAPTER 8

Mathematics and Faith

*I prefer questions that cannot be answered to the
answers that cannot be questioned.
My friend who says he does not believe in God had called
for an ambulance for his mom. I noticed that he was quietly
saying, "God, please help them to get here soon."
Somebody asked Bernard Show what he would tell
God if he faced him after death. "I would ask him why
he did not show me more evidence," he replied.
Death is not the end, it is simply walking out of the physical form and
into the spirit realm, which is our true home. It's going back home.*

—Stephen Christopher

When it's a question of money, everybody is of the same religion.

—Voltaire

*Am I accountable for the critical choices made for me: getting born,
my birthplace, country, time, gender, parents, religion, culture, etc.?*

Science and Faith

THE HISTORY OF science includes countless hard-fought battles with traditional certainties. A classic example is the fierce conflicts of belief associated with Creationism's rejection of the theory of evolution. It seems that wherever modern science goes too far and establishes itself as a kind of alternative to religion, it produces a backlash. The root of conflicts is that in the world of religion, miracles are certainly possible; in the world of science, they are not. Additionally, brain as a belief engine searches for understanding and drives us at least to believe, even if we do not actually know about God, astrology, extraterrestrials, or string theory.

All this takes place because of our ability to link cause and effect. Once we understood the connection between cause and effect, we could not stop searching for reasons why the world is as it is. Once we realized that we only have to rub two sticks together vigorously to create fire, we wanted to know the causes of other things, such as disease and death. But it proved impossible to apply the principle of "no effect without a cause" to strokes of fate of this kind without resorting to the supernatural. When our ancestors reached the limits of their understanding, they almost inevitably came to the conclusion that an invisible God must be responsible—a solution that was so compelling and has remained so for many to this day that a lot of societies developed it quite independently. Additionally, this line of thought answers our major questions simultaneously and at the same time gives us something to hold on to.

After all, we are human—weak, fragile, and not equipped with tools or intelligence to uncover the secrets of this complex world. Faced with uncertainties, we feel lost and often panic. To comfort ourselves, some of us ignore them, some deny them, some avoid them, and some try to understand them. Some of us even find believing in fate or even conspiracy more comforting than admitting ignorance.

Should I Believe in God?

For the reasons mentioned, most people believe in some form of supernatural power. Why? Consider the Blaise Pascal's philosophical argument regarding God. Think about two events: the event that God exists and the event that a person is a believer that God exists. Suppose that p is a number between zero and one, the probability that God exists. Each person has options of becoming a believer or nonbeliever. This leads to four disjoint (mutually exclusive) possibilities: God exists and person is believer, God exists and person is nonbeliever, God does not exist and person is believer, God does not exist and person is nonbeliever. Suppose that the payoffs of these options can be expressed in dollars as follows:

	God Exists	God Does Not Exists
Believer	100,000	– 10,000
Non-Believer	–100,000.000.000	100,000

To analyze this further, let us first use an example to explain what the expected value or expected return is. Think about a game where you flip a fair coin and win a dollar if head comes up and lose a dollar if tail comes up. If you play this game for few hours, you expect to break even—that is, expect neither win nor lose. This is because you expect to win half of the times and lose half of the times. Mathematically, this is calculated as $1/2 (+1) + 1/2 (-1) = 0$.

Using the numbers in the above table, the expected payoffs are 10^5 p $-10^4 (1 - p)$ for a believer and -10^{10} p $+ 10^5 (1-p)$ for a nonbeliever. These two values are equal if $p = 1/90911 = 0.000109$. Thus, the expected payoff is higher for believers if the chance that God exists is bigger than just 0.000109. In fact, Pascal's original argument uses infinite rather than -1010 and concludes that believing in God results in better payoff for any value of p, no matter how small.

Non-Euclidean Geometries and Statistical Physics

The goal of this section is to demonstrate that believing in no god, one god, or infinitely many gods as a postulate without a proof will not lead to any contradiction.

One of the frequently cited arguments in favor of revision of mathematics in light of empirical discoveries is the general theory of relativity and its adoption of non-Euclidean geometry. Recall that Euclid developed the idea of geometry around 300 BC. In his book, he starts with five main postulates, axioms, or assumptions from which he drove theorems of geometry. The postulates were the following:

1. Given two points, there is a straight line that joins them.
2. A straight-line segment can be prolonged indefinitely.
3. A circle can be constructed when a point for its center and a distance for its radius are given.
4. All right angles are equal.
5. From a point outside a line, only one line can be a parallel to it.

The last postulate is more complicated than the other four. Over the years, many mathematicians tried to derive it from the first four. However, so far, nobody had yet come up with such a proof.

Details

Geometry is the realm of mathematics in which we talk about things like points, lines, angles, triangles, circles, squares, and other shapes, as well as the properties and relationships between all these things. For centuries, it was widely believed that the universe worked according to the principles of Euclidean geometry where parallel lines never crossed. This was the geometry we learned in school and was the only kind of geometry taught. In fact, for a long time, my favorite joke was "Parallel lines have a lot in common. It is unfortunate that they can never meet."

The beginning of the nineteenth century would finally witness decisive steps in the creation of non-Euclidean geometry. Non-Euclidean geometry arises when the parallel postulate is replaced with an alternative one. Doing so, one obtains hyperbolic geometry and elliptic geometry, the traditional non-Euclidean geometries. The essential difference between the geometries is the nature of parallel lines. Euclid's fifth postulate, the parallel postulate, which states that within a two-dimensional plane, for any given line ℓ and a point A, which is not on ℓ, there is exactly one line through A that does not intersect ℓ. In hyperbolic geometry, by contrast, there are infinitely many lines through A not intersecting ℓ; while in elliptic geometry, any line through A intersects ℓ.

Another way to describe the differences between these geometries is to consider two straight lines indefinitely extended in a two-dimensional plane that are both perpendicular to a third line:

- In Euclidean geometry, the lines remain at a constant distance from each other and are known as parallels. This means that a line drawn perpendicular to one line at any point will intersect the other line, and the length of the line segment joining the points of intersection remains constant.
- In hyperbolic geometry, they "curve away" from each other, increasing in distance as one moves further from the points of intersection with the common perpendicular.
- In elliptic geometry, the lines "curve toward" each other and intersect.

In my view, the story of geometries somehow relates to the story of three statistics regarding particles in physics: Fermi–Dirac, Bose–Einstein, and Maxwell–Boltzmann. If we have indistinguishable particles, we apply Fermi-Dirac statistics. To identical and indistinguishable particles, we apply Bose-Einstein statistics. And to distinguishable classical particles, we apply Maxwell-Boltzmann statistics. Statistical physics probability often use examples involving balls and boxes to explain these. In the context of statistical physics, the balls are replaced by particles to present a more realistic situation. Let me clarify this.

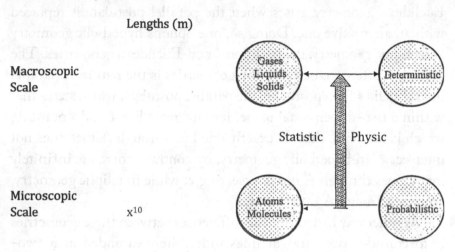

In the nineteenth century, Newton's mathematical laws enabled astronomers to predict the existence of another planet, Neptune, where it was supposed to be. This led, for the first time, to understanding the universe and the way it works. However, constant increase in amount of errors led to the breakdown of the so-called clockwork universe, and gradually it was replaced by a more reasonable view—namely, the probability/statistical model of reality.

It is now generally recognized that the ideas of randomness are central to much of modern physics and have overthrown the "clockwork universe" conceptions of earlier centuries. The laws of probability and statistics were developed by great mathematicians such as Fermat, Pascal, and Gauss, and received their first major application in physics in the kinetic theory of gases developed by Maxwell and Boltzmann.

Here, the use of statistics is necessary because the number of particles involved is too great for any realistic deterministic calculations. With the advent of quantum theory, physics seemed to be based on an essential randomness whose reality was debated by great physicists Bohr and Einstein till the end of their lives. Only recently, in the experiments of Alain Aspect, a convincing demonstration has been given that the inescapable randomness of quantum theory is a fact of nature.

In developing the kinetic theory of gases, Maxwell showed that physical theory could fruitfully exploit probabilities that represent our ignorance. Boltzmann thought that this approach could be generalized and was convinced that such generalization would open up fields of application far beyond the mere perfect gas.

This, in 1872, was the birth of statistical physics, destined for further elaboration by Gibbs, though in a more abstract fashion; that is why we confine ourselves here to Boltzmann's method, which has the advantage of reasoning directly in terms of the actual physical system.

Given n particles and $m > n$ boxes (different levels of energy), we place at random each particle in one of the boxes. We wish to find the probability p that in n preselected boxes, one and only one particle will be found.

Since we are interested only in the underlying assumptions, we shall only state the results. We also verify the solution for $n = 2$ and $m = 6$. For this special case, the problem can be stated in terms of a pair of dice: the $m = 6$ faces correspond to the m boxes, and the $n = 2$ dice to the n particles. We assume that the preselected faces (boxes) are three and four.

The solution to this problem depends on the choice of possible and favorable outcomes.

We shall consider the following three celebrated cases:

Maxwell-Boltzmann statistics. If we accept as outcome all possible ways of placing n particles in m boxes distinguishing the identity of each particle, then the required probability is

$$p = (n/m)((n-1)/m)...(1/m).$$

For $n = 2$ and $m = 6$, the above yields $p = 2/36$. This is the probability for getting (3, 4) in the game of tossing two dice, where there were thirty-six possible outcomes.

Bose-Einstein statistics. If we assume that the particles are not distinguishable—that is, if all their permutations count as one—then

$$P = (1/n)(2/(n\text{-}1))....((m\text{-}1)/(n+m\text{-}1)).$$

For $n = 2$ and $m = 6$, this yields $p = 1/21$. Indeed, if we do not distinguish between the two dice, then the possible outcomes are (1, 1), (1, 2), (1, 3), (1, 4), (1, 5), (1, 6), (2, 2), (2, 3), (2, 4), (2, 5), (2, 6), (3, 3), (3, 4), (3, 5), (3, 6), (4, 4), (4, 5), (4, 6), (5, 5), (5, 6), (6, 6), and there is only one favorable outcome because the outcomes (3, 4) and (4, 3) are counted as one.

(1,1)	*(1, 2)*	*(1, 3)*	*(1,4),*	*(1, 5)*	*(1, 6)*
(2, 1)	**(2,2)**	*(2, 3)*	*(2,4)*	*(2, 5)*	*(2, 6)*
(3, 1)	(3,2)	**(3,3)**	*(3,4)*	*(3, 5)*	*(3, 6)*
(4, 1)	(4,2)	(4,3)	**(4,4)**	*(4, 5)*	*(4, 6)*
(5, 1)	(5, 2)	(5,3)	(5,4)	**(5,5)**	*(5, 6)*
(6, 1)	(6, 2)	(6, 3)	(6,4)	(6,5)	**(6,6)**

Fermi-Dirac statistics. If we do not distinguish between the particles and also we assume that in _each box we are allowed to place at most one particle, then

$$p = n!/(m - n)!m! = (1/(n\text{-}m+1))(2/(n\text{-}m+2))...((n\text{-}m)/(n)).$$

For $n = 2$ and $m = 6$, this yields $p = 1/15$. This is the probability for (3, 4) if we do not distinguish between the dice and also we ignore the outcomes in which the two numbers shown are equal. That is the possible outcomes are (1, 2), (1, 3), (1, 4), (1, 5), (1, 6), (2, 3), (2, 4), (2, 5), (2, 6), (3, 4), (3, 5), (3, 6), (4, 5), (4, 6), (5, 6).

One might argue, as indeed it was in the early years of statistical mechanics, that only the first of these solutions is logical. The fact is that in the absence of direct or indirect experimental evidence, this argument cannot be supported. The three models proposed are actually only hypotheses, and the physicist accepts the one whose consequences agree with experience.

Example (sum of three dice). Sometime before the year 1642, Galileo was asked to find the ratio of the probabilities of three dice having a sum of nine or a sum of ten. We will go further and examine the whole probability distribution for the sum of three faces of three dice. Since the probability is not obvious, we will use elementary methods.

We begin, as usual, with the sample of the equally likely outcomes; the roll of a single die at random means that we believe each of the six faces has the same probability of 1/6. For two dice, all thirty-six possible outcomes are equally likely each with probability 1/36. The third die leads to the product space of these thirty-six by its six giving $6^3 = 216$ outcomes in the final product space, each with probability 1/216. Notice that the product space is the same whether we imagine the dice rolled one at a time or all at one time.

We do not want to write out all these 216 cases; rather, we would like to get sample space by a suitable grouping of events. If we label the sum (total value) of the three faces by S having values running from three to eighteen, then we want to find the probability distribution of S. The probability that S will have the value k is written as

$$P(S = k) \qquad (k = 3, 4, \ldots, 18).$$

In our approach for fixed k, we partition the value k into a sum of three integers, each in the range of one to six. We will first consider the canonical partitions where the partition values are monotonely increasing (or else monotonely decreasing). Once we have these, we will then ask, "In how many places in the sample space will there be equivalent partitions?" We make the entries for the canonical partitions in the table below.

Table of Canonical Partitions of k on Three Dice

k	Canonical Partitions						
3	(1,1,1)						1
4	(1,1,2)						3
5	(1,1,3)	(1,2,2)					6
6	(1,1,4)	(1,2,3)	(2,2,2)				10
7	(1,1,5)	(1,2,4)	(1,3,3)	(2,2,3)			15
8	(1,1,6)	(1,2,5)	(1,3,4)	(2,2,4)	(2,3,3)		21
9	(1,2,6)	(1,3,5)	(1,4,4)	(2,2,5)	(2,3,4)	(3,3,3)	25
10	(1,3,6)	(1,4,5)	(2,2,6)	(2,3,5)	(2,4,4)	(3,3,4)	27
11	(1,4,6)	(1,5,5)	(2,3,6)	(2,4,5)	(3,3,5) . (3,4,4)		27
12	(1,5,6)	(2,4,6)	(2,5,5)	(3,3,6)	(3,4,5)	(4,4,4)	25
13	(1,6,6)	(2,5,6)	3,4,6)	(3,5,5)	(4,4,5)		21
14	(2,6,6)	(3,5,6)	(4,4,5)	(4,5,5)			15
15	(3,6,6)	(4,5,6)	(5,5,5)				10
16	(4,6,6)	(5,5,6)					6
17	(5,6,6)						3
18	(6,6,6)						1
							Total= 216

These are the increasing canonical partitions; in how many equivalent ways can each be written? If the three indices are distinct, then there are evidently exactly $3! = 6$ equivalent sequences in the entire sample space. Finally, if all three indices are the same, then there is only one such sequence in the sample space. Thus, we have to multiply each partition on the left by its multiplication factor (6, 3, or 1) and then sum across the line to get the total number of partitions that are in the original sample space and that also have the value k. These totals are given on the right. Dividing these sums by the total 216, we get the corresponding probabilities. When we notice the structure of the table, the symmetry of the totals above and below the middle, and check by adding all the numbers to see that we have not missed any, then we are reasonably confident that we have not made any mistakes.

The answer to the question asked of Galileo, the ratio of the probabilities of a sum of nine or ten is clearly $25/27 = 0.926$. Although there are the same number of canonical partitions in these two cases, the partitions do not have the same total number of representatives in the original sample space.

At the time of Galileo, there were claims that the canonical partitions are the equally likely elements of the same space. Hence, you should review the argument we gave for the product sample space of probabilities to see if it convinces you.

The students need to be careful here. The distribution we have used in the *MaxwellBoltzmann* distribution. If we take the canonical partitions as the equally likely elements, then the distribution is the *Bose–Einstein* distribution, which assumes that it is the entries on the left-hand side of the table that are the equally likely events and they have no corresponding multiplicative factors. Thus, the right-hand column would be the sequence (1, 1, 2, 3, 4, 5, 6, 6, 6, 5, 4, 3, 2, 1, 1). The total number of equally likely cases is fifty-six.

Finally, if we consider what is known as the Pauli exclusion principle of quantum mechanics, then only the canonical partitions for which the three entries are distinct can occur, and these are the equally likely events. We then have the *Fermi–Dirac* distribution. Thus, in the table, we must eliminate all the entries for which two of the numbers are the same. When we do this, we find, beginning with the sum of six and going to the sum fifteen, the sequence (1, 1, 2, 3, 3, 3, 3, 2, 1, 1). The total number of cases is twenty.

Only the last two distributions, the Bose-Einstein and the Fermi-Dirac, are obeyed by the particles of physics. Thus, you cannot argue solely from abstract mathematical principles as to which items are to be taken as the equally likely events; we must adopt the scientific approach and look at which reality indicates. If we wish to escape the medieval scholastic way of thinking, then we must make a model, compute what to expect, and then verify that our model is (or is not) closely realized in practice (except when using "loaded dice").

2 Dice	3 Dice
36	216
21	56
15	20

It is interesting to note that in his article about the history of probability published in *Biometrika* (vol. 43, 1956), M. G. Kendall draws our attention to an early example of counting the possibilities in dice games. Bishop Wibold, about AD 960, enumerated fifty-six virtues, letting each virtue correspond to the outcome of the throw of three dice. The dice were thrown, and the thrower concentrated on the corresponding virtue (honesty, loyalty, etc.) for some time.

Counting the Options

I clearly remember the day my mother expressed her admiration for the Lord because he has given us different faces to be identified and differentiated. Although as a child I immediately thought about twins as an exception to the rule, it made me think and surprised me for years until I learned about counting techniques, which was another type of surprise. Let me explain.

Suppose that you have two books, A and B, and a small shelf that takes two books. To place the books on the shelf, you have two choices, AB and BA. With three books, your choices are ABC, ACB, CAB, ACB, BCA, and BAC, $3 \times 2 \times 1 = 6 = 3!$ (3 factorial), three choices for placing the first book, two choices for the second book, and one for the third. With four books, it is $4 \times 3 \times 2 \times 1 = 24 = 4!$ Moreover, with only fourteen books, your choices are $14! = 87,178,291,200$, more than twelve times the population of world. Amazing. Think about a big shelf, a library. The number of choices is unbelievably large and mind-boggling. Alternatively, think about a room with only fourteen seats. The first person that enters the room has fourteen choices to pick a seat; the next, thirteen; the third, twelve; and so on. Therefore, there are total of $14!$ permutations. Think about a large class, a movie theater, or a football stadium. What is even more surprising is the fact that here we are talking about integers (discrete variables). Imagine what happens with continuous variables or cases such as the human face and all the possibilities. Simply countless. I am not sure if I could explain this to my mom and, more importantly, if she would believe it. However, I am still fascinated by the fact that there are infinitely many possibilities even in discrete case. Let me be a little formal and state the main principles of counting.

Addition principle: If an event A_1 can occur in a total of n_1 ways and if a different event A_2 can occur in a total of n_2 ways, then the event A_1 or A_2 ($A_1 \cup A_2$) can occur in $n_1 + n_2$ ways provided that A1 and A2 are disjointed—that is, they cannot occur simultaneously.

Multiplication principle: If an event (operation) A_1 can occur in a total of n_1 ways and if after that a different event (operation) A_2 can occur in total of n_2 ways, then the event A_1 and A_2 ($A_1 \cap A_2$), the whole operation, can occur in $n_1 \times n_2$ ways.

Sir Arthur Eddington, an astronomer, once wrote in a satirical essay that if a monkey were left alone long enough with a computer and typed randomly, any great novel could be replicated. What he did not probably know was the fact that Earth's life is not enough to replicate even one book. To see that, consider the following example;

What is the probability that this passage could have been written by a monkey? Ignore capitals and punctuation and consider only letters and space.

"I cannot think about life without belief."

With twenty-six letters plus space, there are twenty-seven choices for each of the forty positions. Thus, the number of possibilities is 27^{40}. What we see is just one option. So the probability we are seeking is $5.5647984 \times 10^{-58}$.

Note that it is the second principle that we used earlier to enumerate the number of arrangements of a set of objects. Consider the number of arrangements of the letters *a*, *b*, *c*. We can pick any one of the three to place in the first position; either of the remaining two may be put in the second position, and the third position must be filled by the unused letter. The filling of the first position is an event that can occur in three ways, the filling of the second position is an event that can occur in two ways, and the third event can occur in one way. The three events can occur together in a $3 + 2 + 1 = 6$ ways. The six arrangements or permutations, as they are called, are abc, acb, bac, bca, cab, and cba. In this simple example, the elaborate method of counting was hardly worthwhile because it is easy enough to write down all six permutations. But if we had asked for the number of permutations of six letters, we should have had $(6)(5)(4)(3)(2)(1) = 720$ permutations to write down. It is obvious now that, in general, the number of permutations of n different objects is

$$n(n-1)(n-2)(n-3)\ldots(2)(1) = n!.$$

For example, in table tennis (using the old rules when I played), to win a standard game, a player must either be the first to reach twenty-one points with a margin of two or win two consecutive points following a tie at twenty all before the opponent does. Considering the first case, show that for the game to finish 21–0, 21–1, . . ., 21–19, the number of possibilities are those given in the following table. Note that for the game to end, for example, 21–4, the winner should take the last point plus twenty points from the first twenty-four points played. If you like challenges, try to calculate the number of possibilities for the second case (going through 20–20).

21–0	1
21–1	21
21–2	231
21–3	1771
21–4	10626
21–5	53130
21–6	230230
21–7	888030

21–8	3108105
21–9	10015005
21–10	30045015
21–11	84672315
21–12	225792840
21–13	573166440
21–14	1391975640
21–15	3247943160
21–16	7307872110
21–17	15905368710
21–18	33578000610
21–19	68923264410

We now turn to a very different topic—namely, belief comparison of the world's two major religions.

God and Mathematics

According to Galileo, "Mathematics is the language with which God has created the universe." The only question here is then God.

Throughout history, most mathematicians have been Platonist, at least in practice. We tend to think that mathematical ideas are discovered rather than invented. In more recent times, some have questioned this, claiming that mathematics is simply the brain's way of understanding how the universe is structured, and mathematics could be very different for an extraterrestrial species (see, for example, Lakoff and Núñez 2000.) Others disagree, pointing out how mathematics inexplicably predicts new discoveries. Of course, all agree that certain things such as notation, conventions, and choice of axioms are man's invention. But where do the beautiful results we admire come from? The Greeks cannot be said to have "invented" the Pythagorean theorem. Most would agree they (and other cultures) "discovered" it.

Most Christian theologians, from Augustine (AD 354–430) onward, as well as Christian mathematicians, have agreed with a Platonist perspective, believing that mathematics is in the mind of God, and we discover these eternal truths. Mathematics cannot be part of Creation because it is not a physical part of nature—it is a collection of abstract ideas. One does not physically create abstract ideas; one conceives them. And God must have always known these ideas, so they have always been part of his thoughts. Mathematics preceded Creation and is untouched by the Fall. It is perfect and beautiful and contains awe-inspiring ideas, such as Cantorian infinity, which is part of God's nature but not part of our physical universe. However, we ourselves are fallen, so our understanding and use of mathematics is imperfect.

Some modern Christian thinkers have proposed other possibilities similar to those of Lakoff and Núñez, making mathematics a human activity or only one of many possible systems of mathematics in the mind of God. Nevertheless, all Christians affirm that mathematics is not independent of God. Even if there are other possible systems of mathematics, the one we know is the one God chose for us as good, and it has always been known by God. It is not some arbitrary invention. I like to think that when I am studying mathematics, I am studying the very thoughts of God, that mathematics is part of God's attributes. God did not "create" love; God is love (1 John 4:8). Likewise, God did not "create" one and three; God is one Being in three Persons. God did not "create" infinity; God is infinite. And so on.

But whatever position you take, whatever the ontology of mathematics, it should not surprise us that mathematics is beautiful, because God is beautiful. Mathematics is indeed "an aesthetic subject almost entirely." Mathematical beauty and usefulness is a mystery only if we do not believe it comes from God (see, for example, the classic article by Wigner 1960.)

Euler's identity, relating five fundamental constants and three basic operations, is often called the most beautiful result in mathematics (Wells 1990).

Education

Though aesthetics is part of the very foundation of mathematics, it is largely neglected in math classrooms. As mathematician Seymour Papert pointed out, "If mathematical aesthetics gets any attention in the schools, it is as an epiphenomenon, an icing on the mathematical cake, rather than as the driving force which makes mathematical thinking function" (1980, p. 192). However, an increasing number of researchers (including myself) have been noting important consequences of mathematical aesthetics for how we teach mathematics at all ages.

The interested reader can turn to researchers such as Nathalie Sinclair to see how modern research has been discovering the importance of aesthetics in mathematics education. Aesthetics is a "way of knowing" mathematics prior to verbal reasoning and should be an important part of our mathematics classrooms. Indeed, Sinclair (2008) has found that good math teachers tend to use aesthetic cues in their teaching implicitly, though they may not realize it. For example, teachers who reveal a "secret weapon" or present a surprising fact or note simpler ways to express certain solutions are modeling a useful aesthetic to their students. In my own research (Eberle 2014), I have found that even elementary school children come with their own aesthetic ideas and use them in valid mathematical ways when given the opportunity to do open-ended math problems. And this is true of all children, not just those that are gifted in mathematics. Children's initial aesthetic ideas are far from those of mathematicians, but through experience they are refined. Educators from John Dewey to the present day have argued that aesthetics is important for all of education, and now we are discovering how this is true for mathematics.

Nathalie Sinclair (2006) has proposed that mathematical aesthetics has three roles in education:

1. Aesthetics gives intrinsic motivation to do mathematics. This is in contrast to the extrinsic coaxing we often use

with students. Instead of "sugarcoating" math problems by placing them in artificial contexts, we should allow students to explore the natural symmetry and patterns found in every branch of mathematics. I sometimes challenge teachers to see how many patterns they can find in the "boring" multiplication table. They are usually very surprised. Students can also engage in mathematics in a natural way by pursuing projects they themselves have suggested. Such genuine contexts are highly motivational. (See these posts by Josh Wilkerson for a Christian perspective on this idea.)

2. Just as with mathematicians, aesthetics guides students to generative paths of inquiry. When allowed to explore freely, children use their own aesthetics to find valid mathematical insights, though this may take time. Students need opportunities to pursue their own ideas and conjectures.

3. Aesthetics helps students to evaluate their results. Often math is presented as black and white with only right and wrong answers. But if students are allowed to do more open-ended inquiry or project-based mathematics, they can use their growing sense of aesthetics to evaluate the solutions found.

My Life Philosophy as a Mathematician

The world is amazingly, mind-bogglingly, and unbelievably complex. Regardless of how much progress we may make in future, it is abundantly clear that it is impossible for us to uncover its infinitely many secrets. This is because we are not equipped with the tools to enable us to reach that point. However, this is not enough to ignore our curiosity and desire to search for the truth. Apart from our collective efforts, many look for what seems reasonable and plausible to them. Most of us follow our parents and local communities using our cultural logic as reason or support. Some come up with their own views and theories. Either way, most of us accept certain postulates without proof or convincing evidence. From a very large number of ideas, thoughts, and theories presented so far, few have made sense to a larger group of people or have proved to be more useful and practical. They have passed to the next generations as a truth and have found special places in history. They have also gradually become the way of life and the law of land, and members of society have been forced to accept and act upon them. With the progress of science, rather than presenting new ideas or theories, in many parts of the world, people have started examining, questioning, and criticizing most established theories. Soon, this has reached the level that many societies demanded separation of them from government's rules and regulations. However, some believers and groups that had benefit or found comfort on these ideologies kept them alive and, in occasion, gave them a facelift to attract people. These two groups still fight using variety of strategies. In the meantime, science has, through inventions, made people's life more comfortable and pleasing. This has given people more time to do things other than cooking, washing, and cleaning, including thinking about secrets of life.

I also believe that there is really no significant difference between humans and other creatures in terms of feeding and needs. We sometimes behave differently just to make the life a little controlled and predictable. For example, our ancestors realized that to protect babies, forming a family is a good idea. And for that to work, they had to create an environment and a culture that a boy would not look at his sister or mother sexually. Interestingly, unlike Eastern countries, Western countries have extended the idea to cousins.

She Is Nice

We often hear people say, "A is nice," or "B is a terrible person." Every time I hear such a thing, I wonder what it really means or indicates. To clarify, we all know that most of us prefer a pleasant lie to a harsh truth. I recently read somewhere that an average American hears two hundred lies a day, mostly, of course, white lies.

We like people who make us feel good about ourselves by whatever including lies and politically correct statements. In short, if a person's personality or behavior falls in line with what makes the society or some of us feel good about ourselves, we classify that person as good, nice, or pleasant. If a person is quiet or a loner, we often classify them unfavorably. A more significant factor is attractiveness. We classify good-looking people as nicer.

Write about the fact that if your personality or talent matches with what society considers desirable, you are called a good person.

Good Guy, Bad Guy

We are individuals with our own likings and dislikings, desires, dreams, and fantasies. However, some of us are luckier than many. Those who like things that are readily available are luckier than those who like hard-to-get things. More importantly, those whose likings fall in line with what society has accepted or has considered a good attribute are even luckier. For example, a beautiful woman who enjoys looking good and being admired or complimented is lucky in the Western countries but unlucky in Middle East. A person who likes diversity is unlucky in Japan but lucky in many other countries who welcome or promote it. A gay person who is born in certain countries are not only unlucky but also even guilty.

Beauty Bias

Although appearance can be a source of pleasure, it can be a source of pain and suffering too as it could lead to experience stigma, discrimination, and health problems such as eating disorders, depression, etc. Women often bear a vastly disproportionate share of these costs and pay greater penalties for falling short. To understand the dimension of the problem, very recently, Iran banned "ugly" teachers: women with facial hair and men with acne, scars, and fewer than twenty teeth will not be allowed in the classroom.

Being beautiful is convenient, although "It hurts to be beautiful" has been a cliché for centuries. Studies show that good-looking people have a higher chance of being hired, offered better positions with higher pay, and even given faster promotions. Their teachers, their students, their waiters, and even their jury usually treat them better. Men get a bump for height, women are favored if they have hourglass figures, and racial minorities get points for light skin color.

Personal Account

My personal view of life is based on the premise that we humans are selfish and self-servant creatures. This is evident from the fact that we believe that the billions of billions of stars are created because of us. Every day we kill billions of animals, claiming that they are created for us. We judge everything in relation to ourselves. Everything we do in life is for personal, financial, emotional, or any other type of gain. We love others because they give us what we want or need. If that stops, our love for them stops too. To counter this, some people refer to individuals who help others in faraway places or do risky things. My answer to them is so simple; they do that because they get satisfaction from them. Some of us need to be nice. Some cannot handle conflict. Some are just followers. Whatever we are or do is for our own comfort. We join a group to feel better. We may go to church, support a football team, follow a singer as a fan, which are all the same in principle. We seek membership of a group for feeling better about ourselves. We live with lies because it is more comforting than dealing with truth. We tell our children that somebody called Santa will bring you toys and will grant your wishes. We accept ideologies with no solid supporting evidence that trivialize this amazing complexity just because they give us comfort and escape from reality.

Finally, I believe that if there is a Creator, the reason for Creation should have been love and attraction, including the love for the Creator. Many chose the latter because it is in one's head and in a perfect form. I completely understand this and realize that it is a perfect example of everlasting love. What I often wonder about is if the Creator had other options or alternative for this form of life.

To see and admit our limitations, here is my argument. Think about the numbers between zero and one on an interval of lengths one. Suppose that from this line, we can drop any number we wish. Let us a first drop the infinite sequence of numbers 1/2, 1/3, 1/4, . . . Next, drop the infinite sequence 2/3, 3/4, 4/5, . . ., and all other fractions such as midpoints between 1/2 and 1/3 (5/12), 1/3 and 1/4 (7/24), and all other fractions between zero and one. Now, what do you think will happen to that line? In other words, what will be left if any? Well, it turns out that the line will stay intact—that is, nothing will happen to that line, absolutely nothing. Why? Well, because fractions known as rational numbers, though infinite, are countable—that is, they are nothing compared to irrational numbers that are uncountable and cannot be represented as fractions. Also, their decimal representation is unlimited digits with no pattern. But what is surprising is that even though irrational numbers are, let us say, far, far more than rational numbers, very few people could write even two or three irrational numbers between zero and one. Why? Well, we simply don't understand irrational numbers. In our life, we constantly deal with what we understand and apply—that is, rational numbers. Here is another way of explaining the situation. Think about discrete and continuous variables. There is a gap between the values of the discrete variable; unlike continuous variables, where there is none, that they are compact.

As a result of all these differences, most of us have an incorrect perception about numbers, especially irrational numbers. To discuss the consequences, we note we always simplify the numbers to a level we wish. In other words, we discretize all continuous variables measured. When we say I am thirty years old, we are not really thirty. We ignore the months, weeks, days, hours, minutes, seconds, one-tenth of seconds, . . . We may know two or few friends with the same height. But is such a thing even possible? The answer is no. In fact, the chance of finding two people with the same height is zero. Why? Because between any two numbers, never mind how close, there are infinitely many other numbers. Moreover, this infinite is uncountable. The difference between the two numbers could be in 1,000, 10,000, 1,000,000, . . . decimal place. Recall that irrational numbers written in decimal format not only have no limit but have no pattern too. Pi is a good example of this.

My Views, a Summary

I believe that human beings, though wish to think otherwise, are not equipped with tools to enable them to figure out the world and reveal its infinitely many secrets. We may, however, be able to make life easier and more practical by improving the man-made tools such as computers. We are certainly able to deal with problems we face locally and in small scale, but not globally and in large scale. We may be able to comfort ourselves by simplifying this amazing complexity to ridiculously low level, but deep down know that we are really deceiving ourselves. Therefore, when it comes to analyzing, what we often do is take this rough world and smooth it to the level that we can understand, and then model the smooth part it and use it as our guide.

How Did We Get Here?

The day we became a creature known as human being who could think and reason was the beginning of an everlasting wondering about the glorious and complex world around us. Where did I come from? What am I doing here? What is this all about? Why are things the way they are? What is that bright thing up there? And so many similar questions. Once we got hungry or thirsty, needed a shelter, had to learn to find food, etc., we stopped asking and focused on taking care of our needs. Having satisfied our needs, we returned to the questions and thought about them again knowing few things about life.

Soon after, we found ourselves confronting even deeper questions and the need for explanation to survive. In the process, we learned about an important concept—namely, cause and effect. This gave us hope and direction to look at things and seek answers. Those with better insights were then named leaders, philosophers, social scientists, clergies, etc., by us. Attention and respect given to them resulted in increase in number of such individuals, even though only very few of them had convincing answers for our fundamental questions.

We then did split and followed individuals or groups whose thoughts made more sense to us. This was the beginning of classical ideologies and ideological divisions. By adding a simplified and heroic stories to our children's education, we somehow managed to create individuals and groups known as believers who named their belief the truth and followed them wholeheartedly. In other words, we became followers and believers of our thoughts and imaginations, not using a universal logic but through applying local or cultural logic.

Some of us even tried to force them on others for which we paid dearly. Some even went further and classified people outside our groups as nonbelievers and subject to certain punishments. Rather than updating our thoughts through observations and experimentation, we decided to accept only the changes that fitted our ideology and ignore others. Of course, and understandably, we had no other choice. We needed to believe in something to comfort ourselves and give meaning and purpose to this life. After all, we are so weak and fragile that we need comfort even knowing that they are not the truth. We needed to hold on to something seemingly meaningful and plausible, something to make living possible and practical.

It should be noted that some of the ideas utilized to regulate the society were smart. For example, since childhood, I was impressed by the idea of people policing themselves by believing that they are watched and are accountable for their actions, although, as we know, the idea did not work the way it was intended. Some created ideas such as Santa Claus for its reward knowing that it was only a human creation.

Although these lines of thoughts and the ideologies based on them are still followed by a significant part of population of the world. A few centuries ago, some tried to introduce new ideas and push people to take time out from the old ideas and focus mostly on the things that make our present life easier and more pleasant. This led to different sets of thoughts and lifestyles each with their own advantages and disadvantages. The main problem here was coming up with reasonable and acceptable replacements for what people were used to and how to go through the transformation. Some partially successful replacements were sports, music, more open relationships between sexes and as such. Note that although mostly sallow and temporary, they all include community membership. Just think about going to church and compare it with going to a concert or a football game. They conceptually serve the same purpose—namely, of becoming a part of something more than ourselves but by using different tools or formats enjoyable or exciting.

These thoughts led to a great deal of change, mostly good. However, they also led to a set of new social problems. Today, as is evident from recent political movements, some communities are going back to old ideas and values to counter the social problems we face.

Now, more than ever, focus on the material world has become a big part of human life and, in the same time, has brought with it many comforts as well as problems. A large part of the young population now live a speedy life with wrong or no directions. The technology has helped popular media to change people's focus to what they like to do. Their attempts has led to what is known as brainwashing to the extent that people hardly want to think. We just prefer to follow. They have split us to few political groups to which we have dedicated our amazing brain. They have made us to have presumptions about everything and think that we know better than others. In the United States, we are somehow blocked from watching media from other countries or even learning from them. Emptiness and loneliness have led to a significant increase in drug and alcohol use and abuse. So many suicides, depressions, crimes, addictions, teenage pregnancy, and as such.

My Appreciated Argument

In my life, I have received and given what I may call advice from and to friends, family, and specially my students. Some I still remember, and some have been forgotten. The ones I remember are the ones I have received significant appreciation. I have enjoyed them since. As a child, I myself did not have a mentor or advice giver and had to spend so much energy and pay so much penalty to do the thing correctly and with desirable outcome.

A student of mine who was also my relative had to decide between two options and was very worried about making a "wrong" decision and regretting that forever. The fear in her mind was magnified, partly by her mother, who was very unhappy with some the decisions she had to make in her own life, and partly by the people she received advice from. When she explained to me her problem, my response was as follows:

No decision is right or wrong when you make it. What makes it right or wrong is your future actions and efforts. For example, suppose that you have to choose between two universities to continue your education. If after selecting one you work hard, concentrate on your work, and keep your focus on that, you will most likely succeed, and that makes your decision right. If, on the other hand, you waste time, lose your focus, and get distracted, you fail, and that makes your decision wrong. In short, no decision is right or wrong at the time you make it. Your future action makes it wrong or right.

A Smart Girl

This story with a clever comeback from a young girl was emailed to me by a friend.

An atheist was seated next to a little girl on an airplane, and he turned to her and said, "Do you want to talk? Flights go quicker if you strike up a conversation with your fellow passenger."

The little girl, who had just started to read her book, replied to the total stranger, "What would you want to talk about?"

"Oh, I don't know," said the atheist. "How about why there is no God, or no heaven or hell, or no life after death?" as he smiled smugly.

"OK," she said. "Those could be interesting topics, but let me ask you a question first. A horse, a cow, and a deer all eat the same stuff—grass. Yet a deer excretes little pellets, while a cow turns out a flat patty, but a horse produces clumps. Why do you suppose that is?"

The atheist, visibly surprised by the little girl's intelligence, thinks about it and says, "Hmmm, I have no idea."

To which the little girl replies, "Do you really feel qualified to discuss God, heaven and hell, or life after death, when you don't know crap?" And then she went back to reading her book.

Hawkins, Einstein, and Shaw

After the passing of Steven Hawkins, to everybody's surprise on Pi Day, I thought he probably knows now whether he was right about God or not, assuming, of course, there are things to know. "What I have done is to show that it is possible for the way the universe began to be determined by the laws of science. In that case, it would not be necessary to appeal to God to decide how the universe began. This doesn't prove that there is no God, only that God is not necessary." I also thought about Albert Einstein's view about the same subject. But what kept me thinking was Bernard Shaw's answer about the question, "What would you ask the Lord if you actually meet him after you die?" He said, "I would ask him why he did not provided me with more evidence."

Happiness

Studies show that a perceived quest, being part of something bigger, is a reliable indicator of happiness. For some, this could be achieved by joining a fan club. Some studies have shown how religion relates to happiness. Causal relationships remain unclear, but religion is more prevalent in happier people. This correlation may be the result of community membership and not necessarily belief in religion itself. Believers are happier people because belief means hope, and people who have hope think positively. Surveys show that those who are confident in their beliefs are more resilient to life's adversities. In addition, many ideologies are linked to intensely experienced, intoxication-like states of happiness.

How can one increase his or her likelihood of happiness? Well, all one needs is to find out what to do and what to avoid. So there is, indeed, a link between happiness and knowledge—the knowledge of what we can do to be happy. Consequently, answers to the question of happiness may be found precisely where you would least expect them—in science. That may surprise many, because to most people, science and happiness are not compatible. Science is objective, cold, cerebral, and calculating; happiness, on the other hand, is subjective, warm, and a gut feeling. For example, science helps us to understand better why desiring something is not the same as owning or consuming it. Positive psychology helps to scientifically study the strengths and virtues that enable individuals and communities to thrive.

Just Smile

A pastor died and went to heaven. The arrangement was made for him to live in an apartment. Soon after, he found out that a New

York taxi driver was living in a big mansion nearby. Unhappy about it, he sent a letter to God asking for explanation. "My son, when you were giving a sermon, everybody was falling sleep. When he was driving, everybody was praying."

Are We Truthful in Our Beliefs?

Is truth still a virtue? More importantly, is there such a thing as truth, or do we just agree upon certain things as truth?

We live at a time when, more than ever, spurious realities are manufactured by the media, politicians, and political groups. Major news agencies present selected news and analyses based on their version of "truth" and leave readers and viewers wondering what to believe or whom to trust. Of course, what is happening is not entirely new. Deceptions have been used by many including politicians and the media since the dawn of Western civilization. The history of humankind is full of well-known lies and crafty seasoned liars, although, as is believed by many, notably Napoleon Bonaparte, "History itself is nothing but a set of lies agreed upon." Even academic science—a world largely inhabited by people devoted to the pursuit of truth—has been shown to contain a rogue's gallery of deceivers.

Role of the Lies

According to experts, deceit and falsehoods lie at the very heart of our culture. Learning to lie is a natural stage in childhood development. Some children become sophisticated liars as they age. Some grow up to believe that lies make the world a better place, a thought described by Katherine Dunn as "the truth is always an insult or a joke, lies are generally tastier. The nature of lies is to please whereas truth has no concern for anyone's comfort." Finally, many grow to find the self-deception more comforting than self-knowledge.

Why Lies?

Why we lie elucidates the essential role that deception and self-deception have played in evolution. It shows that the very structure of our minds has been shaped from our earliest beginnings by the need to deceive. By examining the stories we tell, the falsehoods we weave, and the unconscious signals we send out, we learn more about our minds and ourselves. These tales of deception are so enthralling because they speak to something fundamental in the human condition. The ever-present possibility of deceit is a crucial dimension of all human relationships, even the most central one, our relationships with our own selves. It is believed that humanity has utilized self-deception as a survival mechanism and discovered that our capacity for dishonesty is as fundamental to us as our need to trust others. According to Y. Bhattacharjee, "Being deceitful is woven into our very fabric, so much so that it would be truthful to say that to lie is human."

White Lies

Minor lies are known as white lies. A white lie is an unimportant lie, especially one uttered in the interests of tact or politeness, an untruth told to spare feelings or from politeness. A lie with good intentions. A lie told with the intent of sparing someone's feelings. A lie about something trivial, one for which there will be few consequences if caught. A lie to prevent an argument or bad feelings about something generally meaningless. Everyone tells a white lie on occasion; it is just a question of why. Some white lies save relationships, some ease a hectic situation, and others buy us time. The list could go on forever. Stretching the truth is a natural component of human instinct because it is the easy way out. We all do it, so there is no reason to deny it.

What Is the Limit?

At a personal level, as long as we are not hurting others or breaking the law, "innocent" white lies may make life a little more pleasant. They can absorb potential friction among our varying personalities and vacillating moods as we nudge into one another in our daily routines. Sometimes white lies cushion us from ourselves. We just need to be careful not to harm or destroy the trust. As is pointed out by Al David, "Most lies have the power to tarnish a thousand truths." Finally, lying has shown signs of being detrimental to health as it takes so much energy out of the liars. After all, as pointed out by Abraham Lincoln, "No man has a good enough memory to be a successful liar."

At the community level, lies can have severe consequences. Societies work under the premise that those who represent them speak with truth and integrity and consider these as prerequisites of having such responsibilities and privileges. Leaders and the media reporting them need the trust and loyalty of their followers, readers, or listeners. Consistent and intentional utterance of untruths removes all integrity from what should be the bastion of truth and justice. Lies tear the very fabric that holds people and communities together and threaten the health of democracies.

CHAPTER 9

Mathematics and Medicine

I N RECENT YEARS, mathematics have become a major player in medical sciences. Most indicators of health are quantified, and decision processes are developed based on them. Doctors now have to make decisions based on data about their patients, and this requires knowledge to do inference. This is, of course, what we expect and hope for. However, in the real world, things are different and this is what we intend to discuss here. We plan to emphasize medical tests and the problems we face using them.

Am I Fit?

Two of the major questions asked by individuals and exercise physiologists are (1) how to define fitness and, more importantly, (2) how to measure it. It is well-known that factors such as maximum heart rate and the time it takes for it to return to normal upon cessation of exercise are related to the person's physical condition. For example, the fitter the person is, the faster his or her pulse rate will return to normal after exercise. Since it is very easy to measure the pulse, many fitness tests have been developed based on recovery time after exercise.

The pulse ratio tests, as they are called, usually involves stepping up and down at a specified rate off a box of a specified height. One such test is the Harvard step test, originally designed for use with male university students. It is a straightforward test to administer and has been adapted for groups other than male students. The values used for the height of the box and rate are dependent upon the age and sex of the person.

The test for adult males is as follows: the subject steps up and down off a box twenty inches high at a rate of thirty times per minute. When the subject steps up onto the box, he must attain a position in which the body is erect; crouching is not permitted. The stepping procedure involves four stages: left foot is placed on the box, the right foot is placed on the box, and then the left foot is placed on the floor, and the right foot is placed on the floor. The person is permitted to change the order of the feet provided that order of the four stages and the rate of stepping are maintained.

The stepping continues for five minutes unless exhaustion is reached previously. Either case, the duration of stepping is recorded. Immediately after completion, the person sits down in a chair and three pulse counts are taken (at the wrist) at following times: 1 to 1½, 2 to 2½, and 3 to 3½ minutes after stepping ceased. The person's Harvard Index is then obtained from the formula

Harvard Index = 100 x (Duration of exercise in seconds) / (2 x Sum of three pulse counts during recovery).

The physical condition, or fitness, of the subject is then determined according to the following scale. If the Harvard index is greater than 90, level of fitness is excellent; 80–89 is good; 55–79 is average; and less than 55 is poor. As an example, suppose that a Harvard step test using a highly trained marathon runner as a subject gave the following three pulse counts: 51, 48, and 43. The Harvard Index is, therefore,

$$(300 \times 100) / 2(51 + 48 + 43) = 105.63.$$

So we conclude that the subject was extremely fit. Marathon runners can be expected to score high values of the Harvard index since the index is a measure of recovery from prolonged exercise, and the marathon is certainly a prolonged exercise.

The manager of sport teams often record their players indices prior to and on completion of a period of training to see if training has improved the players' fitness.

Are Diagnostic Tests Useful?

Diagnostic tests are frequently used to detect the presence or absence of a disease. Ordinarily, a *positive* test result indicates presence, and a *negative* test result indicates absence of the disease. But how often do test results match with reality—that is, what percentage of people who test positive actually have the disease, and what percentage of people who test negative actually do not have the disease? It turns out that, in some cases, the answer to these questions surprises everybody including medical professionals. In fact, when only a small percentage of population have the disease, the probability for a patient with a positive test result to actually have that disease can be surprisingly low, even for a test with acceptable accuracy. Hard to believe? Here are few demonstrating examples.

Mammogram

Mammogram is used to detect breast cancer in women. It has been estimated that of women who get a mammogram at any given time, only 1 percent truly have breast cancer. Note that 1 percent presents the probability of breast cancer in a onetime test and should not be confused with the probability of breast cancer in women's lifetime. Mammograms typically produce positive test results for 86 percent of women with breast cancer and negative test results for 88 percent women without it. With these specifications, the probability of having breast cancer for a woman that has tested positive is surprisingly less than 8 percent.

Why is this probability so low? Consider a random sample of one hundred women having mammograms at a given time. Here, one woman is expected to have breast cancer, 1 percent of the sample. For the woman with breast cancer, the chance of detection is 86 percent. So we would expect a positive test result for her. For each of the ninety-nine women without breast cancer, the likelihood of a negative test result is 88 percent. So, for this group, we would expect about 0.88 x 99 = 87 negative and 12 positive (false positive) test results. This means that out of the thirteen women with positive test results, only one actually has breast cancer (1/13 < 8%).

ELISA Test

Next, consider HIV testing using the standard Wellcome ELISA test. According to the Food and Drug Administration, the ELISA test picks up approximately 99.3 percent of HIV positives and 99.99 percent of HIV negatives (an accurate test). The incidence of HIV positive in the general population without known risk factors is estimated to be twenty-five in one million—that is, in a group of ten million people without known risk factors, about 250 are expected to actually be HIV positive. Applying the ELISA test to this group of 250, about 250 x 0.993 = 248 of them are expected to have positive tests. Also from the remaining 9,999,750 true HIV negatives, about 9,999,750 x (1 - 0.9999) = 9,999,750 x 0.0001 = 1,000 are expected to have a positive tests (false positive). This means that about 1,000/1,248 (80 percent) of the positive test results would actually come from people who are not HIV positive!

Now, before judging the ELISA test as a waste of time, we should note that the test has dramatically increased the likelihood of detection; while only 0.0025 percent of people with no known risk factors are HIV positive, the test will identify almost 20 percent of them correctly (an increase by a factor of almost 8,000!). Still, the number of false positives is high. This, as explained earlier, is the problem with testing for a rare disease. A study in the United Kingdom in the late 1980s confirms the numbers given here. Out of 3,122,556 blood samples taken from people without known risk factors for HIV, 373 tested positive for HIV based on the ELISA test. These samples were then retested using the much more specific (and expensive) western blot test, and 64 cases were confirmed. This means that 83 percent of the positive test results were, in fact, false positive.

PSA

Finally, consider the prostate specific antigen (PSA) blood test for prostate cancer in men. Autopsy studies suggest that about half of all men over fifty have cancerous cells in their prostate, but only about 2.4 percent of them die of prostate cancer. The PSA test has high rates of false positive and false negative (about 50 percent of the men with high PSA readings do not have prostate cancer). When a person tests positive, other tests are usually considered, which are often inconclusive. Given that cancerous prostate cells might never pose a health threat, a decision as to whether PSA tests should be given routinely or not is not an easy one. The current recommendation by the American Cancer Society is that men over the age of fifty should have an annual PSA test, along with a digital rectal exam, although a statistical analysis published in the 1994 issue of the *Journal of the American Medical Association* suggests that the benefits of such screening are marginal at best.

Overuse of the Diagnostic Tests

In a recent communication, the Choosing Wisely campaign, which represents some 375,000 doctors and includes *Consumer Reports*, has acknowledged that many popular medical tests, though critical, are overused. One good example are the tests used to detect heart diseases. According to Elliott Fisher, MD, a director of the Dartmouth Institute, which studies health care issues, EKGs and stress tests are among the most overused tests for people over the age fifty with no symptoms. A 2010 Consumer Reports survey found that 44 percent of people with no symptoms of heart disease had a screening test—an EKG, an ultrasound, or an exercise stress test. That means millions of healthy people each year get tests that they really don't need. According to Steven Nissen, a prominent cardiologist at the Cleveland Clinic, for those at low risk for heart disease, these tests are ten times more likely to show a false positive than a real problem. Moreover, like EKGs, stress echocardiograms, which create pictures of the heart during exercise frequently, yield misleading results. In fact, although for several years cardiology guidelines have discouraged use of these tests for people who have no symptoms, their use has been very common, says James Fasules, MD, of the American College of Cardiology.

The Lesson

It seems that both patients and doctors rely too much on tests and their outcomes. Doctors often order tests to satisfy patients and to protect themselves from malpractice suits. Patients often come to expect tests and not to be told that they don't need any. According to Elliott Fisher, MD, a director of the Dartmouth Institute, which studies health-care issues, an estimated thirty thousand Medicare patients die each year from unnecessary care. He also estimates that up to one-third of spiraling health-care costs is wasted on unnecessary tests and treatments. Based on such information, the

Choosing Wisely campaign concluded that by not choosing wisely, doctors could actually hurt more people than help.

Medical Errors: The Third Leading Cause of Death in the United States

Did you know that the medical errors is the third leading cause of death in the United States? Hard to believe, but here are some supporting findings:

- According to a recent publication in the *Journal of Patient Safety*, as many as 440,000 people die each year from medical errors.
- Only heart disease and cancer kill more Americans than preventable medical errors in hospitals, a Senate panel was told on July 17, 2015.
- The United States has the most expensive health care in the world. We spend more on health care than the next ten biggest spenders combined: Japan, Germany, France, China, United Kingdom, Italy, Canada, Brazil, Spain, and Australia. If our health-care system were a country, it would be the sixth-largest economy on the entire planet. Still, compared to the rest of the world, our health care is ranked about average—that is, our high spending is not buying us particularly safe care, said Dr. Ashish Jha of the Harvard School of Public Health.
- According to some experts, the two significant causes of such a mediocre performance are overreliance on technology and a poorly developed primary care infrastructure. We are second only to Japan in the availability of technological procedures such as MRIs and CAT scans, but unlike our expectation, this has not translated into a higher standard of care. We have one of the best systems for treating acute surgical emergencies, but our system is an unmitigated failure at treating chronic

illnesses. It seems that the conventional medicine, with its focus on diagnostic tests, drugs, and surgical interventions for most ills, harms an unexpectedly large number of patients. The lethality of such system is in part due to side effects, whether "expected" or not, but preventable errors also account for an absolutely staggering number of deaths. According to a study published in the *Journal of Patient Safety*, the problem may also be linked to the "cascade effect," where diagnostic procedures lead to more treatment, more symptoms, and, hence, more complications and deaths.

- Some managers think that it is our tort law that adds to the staggering cost of medical care. But, according to Tom Baker, a professor of law and health sciences at the University of Pennsylvania School of Law and author of *The Medical Malpractice Myth,* making the legal system less receptive to medical malpractice lawsuits will not significantly affect the costs of medical care. Others think that the problem is partly due to a large number of unnecessary medical and surgical procedures (around 7.5 million annually) and hospitalization (around 8.9 million annually).

Remarks

1. Most studies do not blame the physicians for all that is happening. In fact, some argue that, in many ways, physicians are just as victimized by the deficiencies of the health-care system as patients and consumers are. With increased patient loads and mandated time limits for patient visits set by HMOs, as well as the required paperwork, most doctors are doing the best they can do to survive.

2. Some find the outcome not at all surprising. For example, homoeopathic practitioners believe that the medical science has gone too far in the wrong direction, and the more they go down that path, the worse things are sure to happen.

3.   Some do not see a need for change, as they see no cause/effect relationship in published reports, only conjecture based on statistics. They argue, for example, that if ten thousand people die while undergoing medical treatment, it cannot be taken to imply that these people died because of undergoing medical treatment. It is just as likely or even more likely that these same people would have died if there had not been any medical treatment.

Is There an Alternative?

Most experts agree that there isn't any easy fix. Some argue that any profession is, and always will be, within the confines of some for-profit system. To suggest that there is a solution outside of this reality is naïve. Within the medical field, someone should always profit. The question is how to put this in line with patients' interests or benefits.

Some suggest minimizing interactions with the conventional system, which, at least in the case of chronic disease, has little to offer. Some blame the conventional strategies, as they often target the symptoms and not the underlying cause of the disease. They suggest a gradual transition to so-called integrated medicine based on mind-body connection. Integrated medicine combines the most scientifically validated and least harmful therapies from both high-tech and holistic medical practices. It seems that some doctors and patients alike are bonding with this philosophy and its whole-person approach—designed to treat the person, not just the disease. They think that therapies that take advantage of the subtle interactions between a person's state of mind and basic physiological functions in their body are a reasonable alternative. This approach includes the mind-body medicine that uses relaxation techniques and the power of thoughts and emotions.

Drug Overdose, a New Epidemic

Drug overdose has become a serious problem in the United States and around the world. According to the Centers for Disease Control and Prevention (CDC), on average, ninety-one Americans die every day from overdose of opioids such as heroin, the synthetic opioid fentanyl, prescription painkillers such as OxyContin, and now the elephant tranquilizer Carfentanil. Drug use is now the leading cause of accidental death in America. In 2016 alone, more Americans (around sixty thousand) died from drug overdoses than ever died of AIDS, guns, or car crashes in a single year. In 2015 statistics, the state of my residence, Pennsylvania, with 3,264 deaths, was ranked only second to California, with 4,659 deaths. This is significant noting that the population of Pennsylvania is less than a third of population of California.

According to the estimates, in the United States alone, more than twenty million people twelve and older struggle with a substance abuse disorder, of which two million struggle with addiction to prescription painkillers, and six hundred thousand suffer from heroin addiction. Since 1999, the amount of prescription opioids sold on the market has quadrupled.

Overdose

An overdose is defined as the intentional or accidental ingestion of a drug over the normal or recommended amount. The body responds with severe symptoms because it is overwhelmed and is unable to metabolize the drug quickly enough. An overdose can cause a person to fall into unconsciousness, enter a state of psychosis, or experience painful symptoms. Each type of overdose poses significant health risks, including contributing to a person's death.

History

The opioid epidemic is not only about illegal drugs. It began, in fact, with legal drugs. Back in the 1990s, doctors were persuaded to treat pain as a serious medical issue. Pharmaceutical companies took advantage of this concern. Through a big marketing campaign, they got physicians to prescribe products like OxyContin and Percocet in droves—even though the evidence for opioids treating long-term chronic pain was very weak and the evidence that opioids cause harm in the long term was very strong.

So painkillers proliferated, landing in the hands of not just patients but also teens rummaging through their parents' medicine cabinets, other family members and friends of patients, and readily available in black market. As a result, opioid overdose deaths trended up—sometimes involving opioids alone, other times involving drugs like alcohol and benzodiazepines typically prescribed to relieve anxiety.

Noticing the rise in opioid misuse and deaths, officials have cracked down on prescription painkillers. Law enforcement, for instance, threatened doctors with incarceration and the loss of their medical licenses if they prescribed the drugs unscrupulously. Though helpful, many people who lost access to painkillers were still addicted. So some who could no longer obtain prescribed painkillers turned to much cheaper, more potent opioids—heroin and fentanyl, a synthetic opioid that is often manufactured illegally for nonmedical uses. This pushed the number of victims to a new high. In addition, according to a 2016 report by the surgeon general, only 10 percent of Americans with a drug use disorder obtain specialty treatment partly because of the shortage of treatment options.

Worldwide Statistics

According to the international bodies, in 2014, there were about 207,000 drug-related deaths, with overdose accounting for up to a half of them and with opioids involved in most cases. China accounted for 49,000, and Oceania, including Australia and New Zealand, had around 2,000. In European Union countries, more than 70,000 lives were lost to drug overdoses in the first decade of the twenty-first century.

Final Words

It seems that overdosing has taken the place of moderation in our culture. Every day, people are dying from overdose, and the numbers are growing at alarming rates. According to the reports, for opioid epidemic, the years 2014, 2015, and 2016 have been the deadliest. These are some explanations:

1. The cost of heroin is roughly five times less than prescription opioids on the streets.
2. Painkiller addiction (people who abuse or who are dependent on prescription painkillers are forty times more likely to abuse heroin).
3. Shortage of treatment options.

Considering these,

- we need to look at the problem deeper, especially the social and behavioral side of it;
- we need to investigate why about one hundred million Americans suffer from chronic pain despite the fact that we spend more money for health care than the next ten big spenders combined; and
- we need to seriously pursue policies that would curb this growing problem.

We now turn to a very different subject—namely, the role of mathematics in what may be called the foundation of medical science, biology.

Mathematics and Biology

Mathematical reasoning has contributed to biology in many ways. It enabled the biologists to obtain quantitative estimates in situations in which information has previously been only qualitative, for example, in estimates of the number of genes involved in the inheritance of a character. Other potential topics are the application of mathematics to genomics, phylogenetic, and the topology/geometry of proteins and macromolecules.

When the data are quantitative, a mathematical model can be used for detailed analysis. Here are some examples applied to a few familiar situations.

Tomography, which is used heavily nowadays as a diagnostic tool, relies on rather deep mathematics to work properly. There is in particular the Radon transform and its inverse, which are useful for reconstructing a three-dimensional visualization of body parts from "slices" taken by a CAT scanner. Medical imaging is an another example. The genetic code is an interesting piece of combinatorics in itself, and I cannot help but mention genetic algorithms, which are a beautiful example of biology inspiring mathematics, rather than the converse.

The field of mathematical biology is vast. Here are a few ideas, in no particular order.

The Hodgkin-Huxley equations in neurobiology provide an incredibly accurate quantitative description of action potentials in neurons/myocytes/excitable cells.

The logistic equation is a simple model of population growth, and the Lotka-Volterra equation describes population growth in a predator-prey situation.

The basic local alignment search tool (BLAST) is one of the most widely used bioinformatics programs, developed by mathematicians (Altschul, Gish, and Lipman) in the most cited paper of the 1990s (and the most cited biology paper of all time). BLAST is used in medical resequencing, and genome sequencing is having an increasing impact on medicine. The mathematical content is stochastic analysis.

One area in which combinatorics interacts with the biological sciences is in the Ewens's sampling formula of population genetics. Population genetics generally relies heavily on mathematics.

Related to J. M.'s answer, one neat area is source localization of EEG. The idea is similar to other brain scanning methods such fMRI or PET, but instead of measuring blood flow in the brain or nuclear physics, EEG data are collected from a subject, and electrical sources of these surface voltages are reconstructed using inverse problem approaches. Inverse problems are a venerable area of applied math with a long history spanning many disciplines. While this is more interesting from a research standpoint instead of clinical, this is a cool area of work for understanding brain activity.

On top of what has already been said in other great answers, Bernoulli's equation and the Navier-Stokes equations come up often when studying blood-flow.

3D surface curves are used to model tumors to correctly apply heat treatments. This particular application is the motivation to Larson's *Calculus*'s thirteenth chapter, "Multiple Integrals."

There exists an entire volume devoted to "Fuzzy Logic and Medicine." The link there can tell you more, as well as a Google search.

For application of chaos theory or nonlinear dynamics to heart rate, you can check the paper by A. L. Goldberger titled "Non-linear dynamics for clinicians: chaos theory, fractals, and complexity at the bedside" (link).

If you're interested in applications of differential equations in the biological/medical area, I suggest looking at Clifford Taubes's *Modeling Differential Equations in Biology*.

I also read a while ago that fractals are used in cancer detection, since the fractal dimension of tumors is different from the fractal dimension of normal tissue.

Obesity Epidemic in Numbers

Obesity is now approaching epidemic proportions globally. Today, more than 1.8 billion adults are overweight and 320 million are obese. A recent comprehensive study including 19.2 million adults in 186 countries revealed that in the last forty years, the average global body mass index has risen by the equivalent of 3.31 pounds per person, per decade. High-income English-speaking countries account for some of the biggest rises. They also account for more than a quarter of severely obese people in the world. As expected, the United States together with China have the most obese people in the world, with the United States having the most severely obese people of any country. In fact, based on the data collected last September, in the United States, the obesity rate exceeds 35 percent in three states, 30 percent in twenty-two states, 25 percent in forty-five states, and 20 percent in every state. The annual health costs of obesity are now around 21 percent of our total medical costs. This is mainly because there is a link between obesity and more than sixty chronic diseases, including diabetes and cardiovascular diseases. The only bit of good news is that the trend of increasing obesity in the United States has slowed since the year 2000.

As for what is driving America's chronic weight problem, there are no definite answers. Many theories are out there including contribution of genetics, age, and lifestyle. So far, the preponderance of evidence points to the two causes most of us already suspect: too much food and too little exercise. The role of food and diet is not only obviously major but also complex. We receive mixed messages when it comes to what to eat and how much. As a result, we have developed a culture of looking for fast food and, at the same time, fast weight-loss options. We spend more time at work and less time in our homes and kitchens than our parents. We consume 31 percent more calories now than forty years ago. We walk less than people in any other industrialized country. This is even true for our youngsters. Presently, less than 15 percent of our school-aged children walk or bike to school, compared to 48 percent that did in 1969. So, as expected, 13.9 percent of our high school students are obese, and an additional 16.0 percent are overweight.

So, considering several negative effects of obesity, we all should be concerned not only for our own heath but also for the sake of our children, grandchildren, as well as our country and the world. Research shows that many problems related to obesity can be avoided by simple changes in lifestyle. This includes staying active and eating healthier food and smaller portions of it.

Surprising Fact about Body Fat

You may not like your fat, but your body certainly does.

For most of us, body fat has a bad reputation. It is something we agonize over, scorn, and try to exercise away. For scientists, on the other hand, fat is one of the most fascinating and least understood component of the human body.

Benefits of Fat

Fat serves a variety of important purposes. As such, our body is endowed with many self-defense mechanisms to hold on to it. Other than storing excess calories and releasing hormones to control metabolism, fat can use stem cells to regenerate, increase appetite if threatened, and use bacteria, genetics, and viruses to expand itself. It triggers puberty, enabling the reproductive and immune systems. It even affects brain size. We need fat for warmth and insulation, for cushioning our bones and internal organs, for energy, and even to think. Fat also bolsters immunity. It helps produce a slew of hormones crucial to our health, including estrogen, leptin, angiotensin, and tumor necrosis factor-alpha to help keep cells healthy.

Little Fat Does the Body Good

Research shows that being slightly overweight actually increases longevity. Overweight people with certain chronic diseases often live longer and do better than normal-weight people with the same conditions. They actually have a 6 percent lower risk of dying from any cause. Some experts believe that fat helps guard the body from damage, particularly as we age. Anytime body fights illness or deals with a chronic disease, it requires more energy stored as fat. The positive effect of fat seems particularly true if it is located around the butt and thighs, according to one Oxford University study. The thought is this fat traps potentially harmful fatty acids that can travel through the bloodstream to the heart.

Different Types of Fat

While the amount of body fat does matter, the type and where it is stored matter much more.

White Fat

White fat is much more plentiful than other types of fat. It usually settles around hips and thighs, but it can be found anywhere. White fat sits right underneath the skin, providing padding and creating curves. Its second role is to burn energy and produce a hormone called adiponectin. Research suggests adiponectin has anti-inflammatory and "insulin-sensitizing" properties that may help reduce risk for developing type 2 diabetes. When people become fat, the production of adiponectin slows down or shuts down, setting them up for disease. Fortunately, regular exercise increases adiponectin.

Brown Fat

This so-called good fat acts more like muscle, which means it burns energy even when we are inactive. When activated, brown fat burns white fat. Lean people tend to have more brown fat than overweight or obese people do. Brown fat can be grown with regular aerobic exercise.

Visceral ("Deep") Fat

Visceral fat is the toxic kind. People with a large waist or belly have visceral fat. It drives up the risk of diabetes, heart disease, stroke, and even dementia. Visceral fat accumulates around the organs deep inside the abdominal cavity. Thin people with no apparent "belly fat" can have dangerous amounts of visceral fat.

Visceral fat causes problems because it churns out stress hormones like cortisol and inflammatory substances called cytokines that affect the body's production of insulin. The result is increased risk of both type 2 diabetes and heart disease.

Subcutaneous Fat

Subcutaneous fat is found directly under the skin. In terms of overall health, subcutaneous fat in the thighs and buttocks, for instance, may not be as bad and may have some potential benefits.

Belly Fat

Belly fat is both visceral and subcutaneous. Abdominal fat is viewed as a bigger health risk than hip or thigh fat.

Concluding Remarks

The full story of body fat reveals that the shape or composition of the body is not definitive of how healthy the person is. An overweight or slightly obese person can certainly be fit and metabolically healthy. However, the picture changes once the person becomes obese. In other words, the level of health has everything to do with the way we treat our body and not its size.

Experts believe that we should not stress too much over fat; rather, stay committed to our activities and nutrition. We should not also be dissuaded by the commercials and images that show perfectly proportioned clones. All we need is to understand how our body works and set achievable goals for ourselves.

Make Aging an Investment, Not an Expense

As we age, living well demands taking care of several dimensions of wellness: physical, emotional, mental, social, intellectual, occupational, spiritual, environmental, and financial. It requires understanding the age-related issues and trends and a plan to stay active, healthy, happy, and useful.

What Is Aging?

Aging is the progressive accumulation of changes associated with increasing susceptibility to health issues. In addition to wrinkles, turkey necks, and gray hairs, aging could lead to many other changes. Although we cannot stop the aging process, we can minimize its effects through making appropriate choices. It is well established that including physical and social activities, avoiding loneliness, securing family love and support, having positive attitude, staying curious, finding ways to motivate oneself, welcoming opportunities to learn new things, laughing as much as possible, and having a four-legged friend help to slow down the process. Other things we can do include taking care of:

Cardiovascular System

As we age, our blood vessels and arteries become stiffer, causing our heart to work harder to pump blood through them. This often leads to hypertension and other cardiovascular problems. To avoid, eat healthy, do not smoke, manage stress, and get enough sleep.

Bones, Joints and Muscles

With age, bones tend to shrink in size and density, which weakens them and makes them more susceptible to fracture. To avoid, get adequate amounts of calcium and vitamin D, avoid substance abuse, and cut back on pain pills.

Digestive System

Many factors such as a low-fiber diet, not drinking enough fluids, and lack of exercise can contribute to digestive problems. To avoid, eat healthy, do not ignore the urge to have a bowel movement, drink whole milk, drink lots of water, eat veggies, eat like the Greeks, live like the Amish, and eat nuts.

Bladder and Urinary Tract

Loss of bladder control is common with aging. To avoid, go to the toilet regularly, maintain a healthy weight, do not smoke, do Kegel exercises, avoid bladder irritants, and avoid constipation.

Senior Moments

The brain and nerve cells are the only cells in the body that cannot regenerate. Once brain cells are damaged, they are not replaced. As a result, memory might become less efficient. It might take longer to learn new things or remember familiar words or names. To avoid, eat healthy, stay mentally active, be social, lower your blood pressure, and quit smoking. Find your purpose, embrace your faith, take vacation, consider mountain life, watch your grandkids, read, make peace with family, and drive less.

Eyes and Ears

With age, we become more sensitive to glare and have trouble adapting to different levels of light. Our hearing might diminish. To avoid, schedule regular checkups, wear sunglasses and hat when you are outdoors, and use earplugs when you are around loud machinery or other loud noises.

Teeth

Our gums might pull back (recede) from our teeth, and as a result, our teeth and gums might become slightly more vulnerable to decay and infection. To avoid, brush and floss and schedule regular checkups.

Skin

With age, our skin thins and becomes less elastic and more fragile with a simultaneous decrease of fatty tissue just below the skin. To avoid, bathe in warm, not hot water, use mild soap and moisturizer, use sunscreen and wear protective clothing, and do not smoke.

Weight

Maintaining a healthy weight is more difficult as you get older. To avoid, eat healthy and watch the portion sizes.

Sexuality

With age, sexual needs and performance might change. To avoid, share your needs and concerns with your partner, get (or stay) hitched, and talk to your doctor.

Final Words

For centuries, humans have dreamed of defying old age with tales of the mythical fountain of youth. Today, scientists are a bit closer to unlocking the secrets of longevity. According to the United Nations, by 2050, 20 percent of the world's population will be over sixty; in regions like North America and Europe, the proportion will be even higher, almost a third. As such, aging could pose a great challenge to the society unless seniors take on responsibility for their own well-being and healthy aging.

Good Things Come to Those Who Sweat

The human body is equipped with an amazingly effective cooling system. Our skin is covered with approximately two million to five million sweat glands that run like ductwork in an attic. Sweating indicates that our cooling system is doing its job. How much a person sweats is determined by physiological characteristics such as age, gender, and weight, and factors such as temperature, the level of exertion, and how anxious we feel. On average, a healthy human can produce up to one to three liters (two to six pints) of sweat per hour. As we age, our skin changes and the sweat glands produce less sweat. This sometime makes it harder to cool off and may increase the risk of heat stroke.

According to some studies, women are at a disadvantage when it comes to cooling off during heavy bouts of exercise or during hot conditions. This is because, compared to men, women carry less body fluid and may sweat less to prevent dehydration. For this reason, women are advised to take more precautions in extreme heat conditions or during long bouts of exercise.

Our Cooling System

The human body hosts two different types of sweat glands: eccrine and apocrine. Eccrine sweat glands work like ducts on the skin's surface and produce a watery substance. These glands are mostly concentrated on the brow, hands, and feet, and they function primarily as the body's air-conditioning unit. Apocrine sweat glands are found in the hair follicles located in and around the scalp, armpits, anus, and genitals. Apocrine sweat glands produce a thicker, plasma-like substance that also contains fatty acids and protein byproducts, including urea and ammonia. Emotional stress triggers apocrine glands to expel fatty sweat into the skin, where bacteria break it down. This turns into an odorous fatty acid substance, which makes certain types of sweat smell and causes unsightly underarm stains. These glands remain inactive until puberty, which explains why preteen children suddenly smell after recess.

Sweat is primarily made up of water, but it also contains salt and, depending on a person's diet, other chemicals. One of the highest mineral concentrations in sweat is sodium, which explains why sweat tastes salty. In addition, sweat contains moderate amounts of potassium, chloride, calcium, and magnesium as well as small amounts of trace minerals including copper, zinc, and iron.

Hyperhidrosis

For the most part, sweating is necessary and healthy. However, sweating more than "normal" may be caused by a condition called hyperhidrosis. Hyperhidrosis occurs when the body sweats unnecessarily, dripping perspiration from the head, feet, palms, or underarms. Though it is not harmful, excessive sweating can put a damper on daily activities and cause added stress, which can lead to even more sweating.

Benefits of Sweating

Sweating has a number of incredible health benefits:

- **Healing powers:** Sweat glands play a role in the wound-healing process, including recovery from scrapes, burns, and ulcers.

Happiness defenders: The act of sweating alone does not ward off bad moods, but a good sweat in the gym or outdoors increases endorphin levels—those feel-good hormones that contribute to a runner's high. Endorphins are related to positive mood and an enhanced sense of overall well-being.

Kidney protectors: Sweating limits the accumulation of salt and calcium in the kidneys and urine, which can reduce the risk of kidney stones. In addition, it increases thirst and water consumption, making it less likely for kidney stones to form.

So, knowing that our body is equipped with an amazing cooling system with healing power, it is wise to take advantage of it and avoid some critical health problems with a little perspiration.

CHAPTER 10

Coincidences

An unusual day is a day where nothing unusual happens.

MOST PEOPLE FIND coincidences fascinating. Coincidences are often so unlikely that they are not expected to occur. For example, many stories about the United States presidents are so unlikely or surprising that are classified as coincidences. A most well-known and well-studied one is the Abraham Lincoln and John F. Kennedy connections that have occurred exactly a century apart. Here are a few examples out of many listed later in this chapter.

- Abraham Lincoln was elected to Congress in 1846; John F. Kennedy in 1946.
- Lincoln was elected president in 1860; Kennedy in 1960.
- Andrew Johnson, who succeeded Lincoln, was born in 1808. Lyndon Johnson, who succeeded Kennedy, was born in 1908.

- John Wilkes Booth, who assassinated Lincoln, was born in 1839. Lee Harvey Oswald, who assassinated Kennedy, was born in 1939.

Presidency and July 4

One strange historical "coincidence" concerns the early presidents of the United States. Among the first five presidents, Washington, Adams, Jefferson, Madison, and Monroe, three of them, Adams, Jefferson, and Monroe, died on the same day of the year. And the date was none other than the *Fourth of July*. Of all the dates to die on, that must surely be the most significant to any American. The probability that three people out of five die in the same day of the year is about one in five million. The fact that the day was July 4 and the five people were presidents make it even less likely and more surprising. To the early presidents, the anniversary of independence meant so much, and, as such, they were really keen to hang on until they had reached it. This is what happened to Thomas Jefferson, the third president. John Adams, the second president, actually died a few hours after Thomas Jefferson.

Does this historical event support the theory that we could postpone our death day by looking forward to something important such as a milestone? I usually ask students to express their views about this and the related issues. What I have noticed during the years is that this example surprises more students than the Abraham Lincoln and John F. Kennedy connections.

Coincidences

A coincidence is a surprising concurrence of events, perceived as meaningfully related, with no apparent causal connection. Most people are fascinated and puzzled by coincidences and, as such, sometimes overreact to them. Perhaps it is a part of our biology that conspires to make phenomena such as coincidences more meaningful than they really are. Some look for reason, explanation, or pattern. Some consider them insignificant, arguing that the world is so large that anything could happen and so small that "weird" things happen all the time. Those who look for pattern do so based on the fact that human beings are "pattern-seeking animals." We quickly forget "uninteresting" patterns but pay attention and remember the interesting ones. For instance, basketball fans notice patterns such as a sequences of successful shots (streak or hot hand) and ignore many other equally unlikely and uninteresting occurrences. People who win the lottery often attribute their winning to some special number pattern they chose and ignore the fact that all numbers are equally unlikely and it is merely luck that a particular person holds the winning entry. In short, many events or incidences classified as coincidences are bound to happen, even though the particular form of, let us say, a coincidence is unpredictable. Additionally, often "coincidences" are not as unlikely as we think. A classic example is the birthday match, which surprises some people so much that they refer to it as a coincidence. We need 366 people in a room to guarantee a birthday match. With only twenty-three people in one place, the chance of two or more people having the same birthday is more than 50 percent. With forty, it is almost 90 percent. Also, the number of people needed to have a 50 percent chance of 3, 4, 5, and 6 people with the same birthday are just 88, 187, 313, and 460, respectively.

If we want a 50 percent chance of finding two people born within one day of each other, we only need fourteen people. If we are looking for birthdays a week apart, the magic number is seven. If we are looking for an even chance of finding someone having a specific person's exact birthday, we need 253 people.

So from the above introduction, a coincidence is a surprising concurrence of events, perceived as meaningfully related, with no apparent causal connection. Though puzzling, one thing is clear. Coincidences are rare events with a very small probability of occurrence plus some other conditions. And this is where mathematics joins the fun.

Let me start by few clarifying statements from Lisa Belkin.

a. Religious faith is based on the idea that almost nothing is coincidence.
b. Science is an exercise in eliminating the taint of coincidence.
c. Police work is often a feint and parry between those trying to prove coincidence and those trying to prove complicity.
d. Without coincidence, there would be few movies worth watching, and literary plots would come grinding to a disappointing halt.
e. Coincidences feel like a loss of control. Believing in fate, or even conspiracy, can sometimes be more comforting than facing the fact that sometimes things just happen.

Given that there are 330 million people in the United States, at least 330 times a day, a one-in-a-million shot is going to occur—that is, something will occur to one person out of a million. Also, because people do many different things in a given day, some of the occurrences are in fact even less likely.

Here are some other unlikely occurrences that makes one wonder. Think about September 11 (9/11). The number 911 is used for emergency cases. The Twin Towers looked like the number eleven. So perhaps all the things that happened had something to do with the number eleven: $9 + 1 + 1 = 11$, the first flight to hit the Twin Towers was flight 11 with ninety-two people on board, $9 + 2 = 11$. September 11 is the 254th day of the year, $2 + 5 + 4 = 11$ (also $365 - 254 = 111$). Eleven letters in "New York City," "Afghanistan," "the Pentagon," and "George W. Bush." New York is the eleventh state admitted to the union. The area code 119 is for Iraq and Iran. The flight that crashed in Pennsylvania had sixty-five people on board, $6 + 5 = 11$. There are exactly 911 days between this and the September 11 (2001) attack, and the March 11 (2004) attack in Spain. These are all interesting, but do they mean anything? More importantly, are coincidences really as unlikely as we think? Here are a few examples helping to answer this question.

Lottery: People who win the lottery often attribute their winning to some special system they used in choosing a lottery number. Yet someone was bound to win, and it was merely luck that a particular person held the winning entry. So we could conclude that many events classified as coincidences are bound to happen, even though the particular form of, let us say, a coincidence is unpredictable.

Poker: The chance of getting a royal flush is very low (1/650,000). Thus, if we were to get a royal flush, we would be very surprised. But the chance of any hand in poker is very low. Here is the problem: we just do not notice when we get all the others, but we notice when we get the royal flush. The same is true about hot hands in sports.

Bridge Hands

In the card game bridge, there are 635,013,559,600 different thirteen-card hands possible. This number of possible hands could be realized if all the people in the world played bridge for a whole day. Any given hand held by a player is equally probable, or rather, equally improbable, as its probability is 1/635,013,559,600. In other words, any hand is just as improbable as thirteen spades. This is an example of the daily occurrence of very improbable events.

Lincoln and Kennedy Connections

Here is the complete list of Lincoln-Kennedy connections. Would you consider these as coincidences?

- Abraham Lincoln was elected to Congress in 1846. John F. Kennedy was elected to Congress in 1946.
- Abraham Lincoln was elected president in 1860. John F. Kennedy was elected president in 1960.
- Both last names contain seven letters.

- Both were particularly concerned with civil rights.
- Both wives lost their children while living in the White House.
- Both presidents were shot on a Friday.
- Both were shot in the head.
- Lincoln's secretary was named Kennedy. Kennedy's secretary was named Lincoln.
- Both were assassinated by Southerners.
- Both were succeeded by Southerners.
- Both successors were named Johnson.
- Both were Democrats and former senators.
- Andrew Johnson, who succeeded Lincoln, was born in 1808. Lyndon Johnson, who succeeded Kennedy, was born in 1908.
- Both names have thirteen letters.
- John Wilkes Booth (JWB), who assassinated Lincoln, was born in 1839. Lee Harvey Oswald (LHO), who assassinated Kennedy, was born in 1939.
- Both assassins were known by their three names.
- Both names comprise fifteen letters.
- Replacing 1 for A, 2 for B, etc., JWB = LHO = 35.
- Booth ran from the theater and was caught in a warehouse. Oswald ran from a warehouse and was caught in a theater.
- Lincoln was assassinated in Ford's Theatre. Kennedy was assassinated in an automobile made by the Ford Motor Company.
- To cap it all off, Booth and Oswald were assassinated before their trials.
- A week before the assassination, Lincoln was in Monroe (Maryland). A week before the assassination, Kennedy was with Monroe (Marilyn).
- Both presidents wore glasses.
- The digits of 11/22 (November 22) add to six, and Friday has six letters.
- Take the letters FBI, shift each forward six letters in the alphabet and you get LHO.

- Oswald has six letters. He shot from the sixth floor of the building where he worked.
- The triple shift of FBI to LHO is expressed by the number 666, the infamous number of the beast.

More on United States Presidents

- 1840: William Henry Harrison (died in office)
- 1860: Abraham Lincoln(assassinated)
- 1880: James A. Garfield(assassinated)
- 1900: William McKinley(assassinated)
- 1920: Warren G. Harding (died in office)
- 1940: Franklin D. Roosevelt (died in office)
- 1960: John F. Kennedy (assassinated)
- 1980: Ronald Reagan (survived assassination attempt)

Teaching Values

After talking about above examples in the classroom, we remind students that there have always been rumors based on skewed historical facts that have gradually become part of the story simply because we enjoyed talking about them. Consider the list of coincidences mentioned above that supposedly linked the deaths of Presidents Lincoln and Kennedy. In popular literature, it goes, in part, like this: The two men were elected one hundred years apart; their assassins were born one hundred years apart (in fact, 101 years apart); they were both succeeded by men named Johnson; and the two Johnsons were born one hundred years apart. Their names each contained seven letters; their successors' names each contain thirteen letters; and their assassins' names each contain fifteen letters. Lincoln was shot in a theater and his assassin ran to a warehouse, while Kennedy was shot from a warehouse and his assassin ran to a theater. Lincoln, or so the story goes, had a secretary named Kennedy, who warned him not to go to the theater the night he was killed (for the record, Lincoln's White House secretaries were named John Nicolay and John Hay, and Lincoln regularly rejected warnings not to attend public events out of fear for his safety, including his own inauguration); Kennedy, in turn, had a secretary named Lincoln (true, Evelyn Lincoln), who warned him not to go to Dallas (he, too, was regularly warned not to go places, including San Antonio, the day before his trip to Dallas). I leave it to you to decide how to look at these all.

Note that it is not always possible to calculate the likelihood of occurrences of the links mentioned above. We can only do this subjectively.

Coincidences or Not

Would you consider the following as number coincidences, or do you see a reason behind them?

One Pattern

- 1 times 1 = 1
- 11 times 11 = 121
- 111 times 111 = 12321
- 1111 times 1111 = 1234321
- 11111 times 11111 = 123454321
- 111111 times 111111 = 12345654321
- 1111111 times 1111111 = 1234567654321
- 11111111 times 11111111 = 123456787654321
- 111111111 times 111111111 = 12345678987654321

Permutation Pattern

- 1 times 142857 = 142857
- 3 times 142857 = 428571

- 2 times 142857 = 285714
- 6 times 142857 = 857142
- 4 times 142857 = 571428
- 5 times 142857 = 714285
- 7 times 142857 = 999999

What about This?

- I stop in a restaurant while traveling far away from home, and another family from my hometown is there.
- I happen to be sitting right next to a person whom I met at a professional football game.
- I happen to bump into a friend I have not seen for months at the grocery store.
- I happen to hold the ticket that wins the Powerball jackpot.
- I meet my future spouse at a get-together at a friend's house.

The History of Coincidences

History, in general, and history of mathematics, including probability theory, usually invoke students' interest. Clearly, people had a different view and attitude toward coincidences throughout history. A recent book (June 2003) titled *The History of Coincidences* lists all coincidences from 2500 BC to the present. It is written by Dylan Wynn and is about the discovery of coincidences in history. There are many coincidences in history and in important events each year. There are also coincidences similar to Lincoln/Kennedy in many major events in history. For a writing assignment, we sometimes ask students to research coincidences and find out what people in other countries think about coincidences in their history and the history of the world. For example, people in China relate the occurrence of an earthquake to important political events. The following section discusses a cultural aspect of coincidences and its relation to the teaching of probability and statistics.

Teaching Probability Using Coincidences

This section describes an idea for motivating students in an introductory probability and statistics course. The motivating theme is coincidences, and the implementation begins with a questionnaire that samples attitudes of students toward coincidences and involves them in an analysis of such events. The questionnaire serves as a context for the instructor to develop the technical concepts of probability and statistics and provides an opportunity to increase student motivation and involvement. A representative sample of student views regarding coincidences, together with several famous and popular coincidences, is included for possible use. These examples are useful for class discussion, as they are related to the concepts covered in introductory probability and statistics courses. The goal of the article is not to decide how these materials should be used by instructors but to merely provide resources for possible application. We think that the sample questionnaire presented here should be modified by the instructor to offer more insight based on the level and type of the class.

Introduction

Teaching the basic concept of probability presents a challenge to instructors of introductory courses on probability and statistics. Most students approach these courses with reluctance and are suspicious of results obtained by the methods discussed in the class. Additionally, many are easily befuddled by complex formulas. Helping them to overcome these obstacles—and to like these subjects—is, thus, a difficult task.

Many textbooks try to motivate students by presenting applications. This addresses students' apparent desire for relevance of their studies to the outside world and their skepticism about the practical value of the methods presented to them. For students who have not yet made a commitment to a career field, this may not always be very helpful. In fact, examples that lie in a future field of application may fail to motivate because they are not of *immediate* concern to the majority of students who often do not see far beyond their present experiences.

This article describes a strategy for drawing students into probability and statistics—the use of a questionnaire that asks students to respond to the questions regarding coincidences. The questionnaire itself serves as an introduction to the coincidences and its ties to the probability concepts. The student responses serve as a base to motivate and build lesson plans to teach the concepts of the probability and statistics. This connection to life experiences makes difficult concepts more relevant and comprehensible.

To introduce students to the probability and statistics, coincidences are a very good course-theme. Most students are fascinated by coincidences and their analysis. This is true whether or not they have any prior experience with probability or statistics. As their teachers, we can relate the concepts we teach to these experiences and thereby help them to become more comfortable with the material. When students see a connection between what they study and a familiar subject that fascinates them, they usually show more interest and pay more attention. Additionally, students like others are intrinsically interested in analysis of information about their own views and those that excite others.

Based on our experience, in addition to laying the ground work and easing the difficulties of basic concepts, the questionnaire helps students to develop a better attitude toward probability and statistics. Discussions stemming from the questionnaire increase student class participation. We have noticed that they feel pride in seeing how their own experiences contribute to the development of the course. Many mentioned that their participation helped them to see firsthand that probability and statistics are disciplines with relevant applications and deep meanings. We think that this approach is adaptable at all levels at which probability and statistics are taught.

Coincidences, a Unifying Theme

The coincidence-centered teaching begins with a questionnaire, a version of which is shown below. On the first day of class, we give each student the assignment of completing the questionnaire for the next class period. To assure that students will carefully participate in this activity, we promise to count it as a quiz. When handing out the questionnaire, we ask students to read it over and to make sure that they understand each question. We speak with students individually and offer clarifications. Some questions may seem a little vague and confusing. However, every time we used them, they led to good class discussion—including discussion of the need for precise, agreed-upon definitions.

Over several semesters, responses to the questionnaire have given us a clear picture of the high degree of student naiveté about concepts related to probability and statistics. More importantly, it has provided us with information to which we can help them attach the concepts that we seek to teach.

In what follows, we first present a sample questionnaire. The same questionnaire may be given to the students in the end of semester to see whether their responses remained the same or changed.

Opening-Day Questionnaire

1. What is a coincidence? Explain.
2. Can you name a popular coincidence?
3. Why do people consider the event in question 2 a coincidence?
4. Could you name a coincidence that happened to you?
5. Why do you consider the event in question 4 a coincidence?
6. What makes a coincidence interesting?
7. Do you think coincidences are events with a very low probability of occurrence? If your answer is yes, how low?
8. What does the word "random" remind you of? Do you think we all perceive the randomness the same way?

9. Do you think our intuition about coincidences is closely related to our understanding of randomness?

10. People react to coincidences according to whether they are willing to accept most of them as insignificant or insist on always finding a meaning behind them. Do you agree?

11. Did you know that
    a. from the first five presidents of the United States, three died on July 4? Would you classify these as coincidences?
    b. a woman named Evelyn Marie Adams won the New Jersey Lottery twice, in 1985 and in 1986?

12. Did you know that from the first five presidents of the United States, three died on July 4? Would you classify this as a coincidence?

13. Suppose that you are sitting in a class with thirty students. If you find that one or more students have the same birthday as you, will you be surprised? Will you consider this event a coincidence?

14. Could you come up with an estimate for the probability of the event in question 12? Explain how you arrived at your estimate. Also, if you used words like "low" or "very low," could you convert that to a numerical value?

15. Do you think that other students would come up with the same estimate for the probability in question 12?

16. What kind of background information does a person need to make reasonable numerical estimates of the probability in question 12?

17. What branches of science are involved in the process of estimating or calculating the probability in question 12?

18. Two important elements of coincidences are (1) their probability of occurrence and (2) the degree of surprise if they occur. Do you agree?

19. If you flip a coin five times and observe heads every time, would you think that a tail is due now?

20. If you answer to question 18 is no, would you conclude that the coin is not fair?

What do you think about the following definition?[1]
A coincidence is a surprising concurrence of events, perceived as meaningfully related, with no apparent casual connection.

21. According to the *Webster's New Collegiate Dictionary*, coincidences are the occurrences of events that happen at the same time by accident but seem to have some connections. What do you think about this definition?

22. Which of the following do you consider a reasonable statement?[2]

   a. Religious faith is based on the idea that almost nothing is coincidence.

   b. Science is an exercise in eliminating the taint of coincidence.

   c. Police work is often a feint and parry between those trying to prove coincidence and those trying to prove complicity.

   d. Without coincidence, there would be few movies worth watching, and literary plots would come grinding to a disappointing halt.

   e. Coincidences feel like a loss of control. Believing in fate, or even conspiracy, can sometimes be more comforting than facing the fact that sometimes things just happen.

   f. The really unusual day would be one where nothing unusual happens.

---

[1] Definition suggested by Diaconis and Mosteller, *J. of Amer. Stat. Assoc.* 1989, v. 84.

[2] Discussions presented in the article by Lisa Belkin, *New York Times,* August 11, 2002.

Examples of Coincidences and Their Analysis

This section includes the discussion we present after collecting students' responses to the questionnaire. Referring to question 1 and students' general view toward coincidences, we usually start the discussion by mentioning that people often relate coincidences to the unexpected or unusual occurrence of an event or series of events. We ask students if they think this is the case. We then let them express their opinions and discuss opposing views. We continue by pointing out that people also think that coincidences are events with a very low probability of occurrence. We ask if they consider this line of thought reasonable. Usually, a large number of students consider this reasonable. We ask what they consider a very low probability. The answer varies from one in one hundred to one in one million. We then ask them to express their opinion about the following argument. Suppose that a coincidence is defined as an event with a one-in-a-million chance of happening today. Then given that there are 290 million people in the United States, at least 290 times a day (over 100,000 in a year), a one-in-a-million shot is going to occur. Given that there are 290 million people in the United States, at least 290 times a day, a one-in-a-million shot is going to occur—that is, something will occur to one person out of a million. Also, because people do many different things in a given day, some of the occurrences are even less likely. We usually let students express and discuss their views. We encourage them to think about examples of this type. In response to this, one student mentioned that while driving to the university, his car started shaking on the highway. He stopped to check on the problem. While doing this, he found a hundred-dollar bill in the middle of nowhere.

Next, we push toward more deep concepts by arguing that if we examine any significant event carefully, we would find many unusual occurrences and patterns that may seem like a coincidence. Again, we ask students to think of examples for this. The student who found a hundred-dollar bill mentioned that the shaking of the car was a signal for him to stop because he needed the money badly.

As a more universal case, we usually present examples such as the following discussed earlier. Think about September 11 (9/11), for instance. The number 911 is used for emergency cases. The Twin Towers looked like the number eleven. So perhaps all the things that happened had something to do with the number eleven. Let us look at a few of these. First, $9 + 1 + 1 = 11$, the first flight to hit the Twin Towers was flight 11 with ninety-two people on board, $9 + 2 = 11$. September 11 is the 254th day of the year, $2 + 5 + 4 = 11$ (also $365 - 254 = 111$). There are eleven letters each in "New York City," "Afghanistan," "the Pentagon," and "George W. Bush". Also, New York was the eleventh state admitted to the union, 119 ($1 + 1 + 9 = 11$) is the area code to both Iraq and Iran, and flight 77 that crashed in Pennsylvania had sixty-five people on board, $6 + 5 = 11$. Finally, recall the March 11 (2004) attack in Spain. There are exactly 911 days between this and the September 11 (2001) attack. These are all interesting, but do they mean anything? At this point, we let students express their views about patterns of this type and their relations to the coincidences.

After the preliminary discussions, we continue by posing the question, "Are coincidences as unlikely or as surprising as we think?" We ask students to think about this example. People who win the lottery often attribute their winning either to their amazing luck or to some special system they used in choosing a lottery number. Yet someone was bound to win, and it was merely luck that a particular person held the winning entry. So we could conclude that many events classified as coincidences are bound to happen, even though the particular form of, let us say, a coincidence is unpredictable. We then ask students to tell us what they think about the question and the example. This topic usually creates a lot of discussion among students. We end this part by referring to a question regarding Evelyn Marie Adams, who won the New Jersey Lottery twice, in 1985 and 1986. Newspapers widely reported that this event had a probability of one in seventeen trillion. We ask students to think about the following statement: given enough tries, the most outrageous things are virtually certain to happen (the law of large numbers). After discussing this, we return to the supposed one-in-seventeen-trillion double lottery winning and argue that this figure is the right answer to the wrong question: What is the probability that a *preselected* person who buys just *two* tickets for separate lotteries will win on both? The more relevant question is, what is the probability that *some* person, among the many millions who buy lottery tickets (most buying multiple tickets), will win twice in a lifetime? It has been calculated that such a double winning is likely to occur once in seven years, with the likelihood approaching certainty for longer periods.

Next, we turn to an example used in the questionnaire. One strange historical coincidence concerns presidents of the United States. Three of the first five presidents of the United States died on the same day of the year. And the date? It was none other than the Fourth of July. Of all the dates to die on, that must surely be the most significant to any American. This might, of course, be part of the explanation as to why the coincidences happened. You can imagine the early presidents being really keen to hang on until the anniversary of independence, a date that meant so much to them, and giving up the ghost as soon as they knew they had reached it. This is apparently what happened to Thomas Jefferson, the third president. John Adams, the second president, actually died a few hours after Thomas Jefferson. We ask students to express their views about this example and if they think we have a strong reason to believe that this was a coincidence. It is interesting to note that this event surprises many students even more than the September 11 example.

Sometimes we continue with an example from the world of sports such as the professional American football hall of famers. What is the probability of having two hall of famers from the same high school, we ask? Do you think that this is a rare event? Here are some examples: Al Davis and Sid Luckman, George Halas and Leo Nomellin, Mel Renfiro and Arnie Weinmeister, Elroy Hirsch and Jim Otto, High McElhenny and Bill Walsh, and Bobby Layne and Doak Walker.

We then proceed by asking what they think about the examples discussed so far and whether they think these are really examples of coincidences. We also ask if they still think that coincidences are events with a very low probability of occurrence (rare events). The majority of responses are yes. So we continue with the next question. We ask if they know anything about poker. If not, we explain the basics and continue with an example of a hand of poker. We point out that the chance of getting a royal flush is very low (1/650,000). Thus, if we were to get a royal flush, we would be very surprised. But the chance of any hand in poker is very low, we point out. Here is the problem: we just do not notice when we get all the others, but we notice when we get the royal flush. We ask if they can show that this is the case. We ask them to try this as homework. We usually go over this and the birthday problem described below at some stage during the course.

We usually point out that some of the arguments we are using may seem counterintuitive. We explain that they stem from the facts that, on one hand, the world is so large that any event is likely to happen, and on the other hand, it is so small that it makes occurrences of unlikely events so surprising that they may classify as coincidences. We state that one relatively simple example of this is the birthday problem. Students have already responded to the question posed regarding this problem. We usually refer to some of the responses and also the following facts. There are 365 days in a year, and so we need to assemble 366 people in a room to guarantee at least one birthday match. But how many people do we need in a room to guarantee a 50 percent chance of at least one birthday match? We ask students to guess a number. Usually, the numbers vary a great deal. We let them discuss or defend their guess. We then point out that, to anybody's surprise, the answer is twenty-three. Next, we talk about further facts about birthday matches that often surprise them. Here are some examples: With only forty-one people, the chance of at least one common birthday is more than 90 percent. With eighty-eight people, the chance of at least three common birthdays is more than 50 percent. After covering the laws of probability, we show how these are calculated.

If students show interest, we then continue as follows. If we want a 50 percent chance of finding two people born within one day of each other, we only need fourteen people; and if we are looking for birthdays a week apart, the magic number is seven. Incidentally, if we are looking for an even chance of finding someone having a specific person's exact birthday, we need 253 people. We then ask, why, despite numbers like these, are people constantly surprised when they meet a stranger with whom they share a birth date or a hometown or a middle name? Again, we let students discuss this. Next, we change the question and put it in the following format. Why do you think most people are amazed by the overlap of this type and yet conveniently ignore the countless things they do not have in common with others? At this point, we ask students to discuss this further and try to find other examples of this type, and if they like, present it in the class.

To emphasize the point, we next present the following simple example involving probability. We ask students to calculate the probability that two randomly selected individuals have birthdays, for example, on May 25 and October 13. We usually choose two students in the class and use their birthdays. The answer (after providing hints) they should find or suggest is $\left(\frac{1}{365}\right)^2$. We then change the dates to, say, June 9 and June 9 (again, we choose someone's birthday in the class). Not very sure, they eventually produce the same answer. Then we remind them that in both cases, the probability is less than 1/133,000. We ask them if they consider this a small probability. Most students answer yes. Next, we ask a question regarding the degree of surprise. We ask which of these two events they find more surprising. All of them, of course, refer to the latter. This reveals that, although these two events are equally likely, the latter creates a great deal of surprise and, therefore, may be classified as a coincidence. We then ask students if they could think of other examples of this kind and, more importantly if they are convinced that the two main components of a coincidence are probability and the degree of surprise.

Next, we turn to a discussion regarding our perception about coincidences. Why do we classify some events as coincidences? According to John Allen Paulos human beings are "pattern-seeking animals." It might just be part of our biology that conspires to make coincidences more meaningful than they really are. We point out that a good example of this is the hot hand in sports such as basketball, where we notice certain pattern and ignore many equally probable others. In fact, we sometimes spend a class period on discussion of hot hands in basketball, starting with the following questions: Do you think there is such a thing as hot hand in basketball? What do you need to see to decide that a player has a hot hand? If I flip a coin and get five heads in a row, could I claim that I have a hot hand for heads? Students, especially male students who play or watch basketball, show a lot of interest in this topic and present many useful discussions. We ask interested students to write about this and count that as an extra credit.

Next, we turn to present some classical and popular examples of coincidences. Our goal is to get the students' attention and increase their level of interest. We start with the well-known set of similarities between the life and death of two United States presidents.

Where Should I Sit?

Consider a classroom with only thirty seats. In the first day of class, students entering the class can choose where they like to sit. The first student has thirty choices, the second student has twenty-nine choices, the last (thirtieth) student has only one choice, etc. The total number of possible selections for thirty students, $(30)(29) \ldots (2)(1) = 30!$, is astronomical. Thus, the probability of any particular set of selections is practically zero. This is another example of the daily occurrence of highly improbable events. Next, we continue with occurrences or coincidences that are interesting but are not well known.

Personal Coincidences

As pointed out earlier, different people have different perceptions about coincidences. The author of the book on history of coincidences, Dylan Wynn, considers the following events in history that happen on or near his birth date of May 31, 1970, as coincidences. Joan of Arc was burned at the stake on May 30, 1415. William Shakespeare's first daughter was baptized on May 26, 1583. Austrian composer Wolfgang Mozart's father died on May 25, 1761. Pres. George Washington's wife died on May 29, 1787. The first steamship to cross the Atlantic Ocean from the United States to England was on May 27, 1819. The first message transmitted on a telegraph line was on May 24, 1844. Adolph Hitler's brother, Gustav, was born on May 17, 1885. Mount Everest, the tallest mountain in the world, was first climbed on May 29, 1953. The first woman to drive in the Indy 500 car race was on May 30, 1977. The first black woman to become a U.S. ambassador and to hold a cabinet post, Patricia Harris, was born on May 31, 1924. Serial killer Jeffrey Dahmer was born on May 21, 1960. Nicole Simpson, ex-wife of O. J. Simpson who was murdered in 1994, was born on May 19, 1959. Ted Kaczynski, the famous Unabomber, was born on May 22, 1942. Serial killer Robert Lee Yates, arrested in 2000, was born on May 27, 1952. Eric Robert Rudolph, guilty of the 1996 bombing in Atlanta, was arrested on May 31, 2003.

We end this part by looking at a couple of number patterns that we sometimes use in the class. Several of our students consider these as coincidences. Some find these more surprising than some of the coincidences presented earlier. We usually remind students that noticing coincidences has played an important role in scientific discoveries. The ability to see obscure connections (those that most people miss) is believed by some to be among the more important abilities a scientist can have. A good example of this is stories about Isaac Newton and his discoveries.

Coincidence in International Settings

Many of the motivating themes described thus far may only appeal to American students as they invoke incidents in U.S. history and sports. It is important to realize that cultures vary in their perception of coincidences, chance, and randomness. The most effective examples are those that touch on these perceptions and use themes that are relevant to the students' lives. We have learned this through responses to the questionnaire by our international students.

In West Africa, for example, the use of coincidences involving the weather is a good way to engage students. In this region of mostly subsistent farmers, the lack of rain can be devastating. Perhaps because of this, precipitation has come to have mystical characteristics. For instance, in the country of Burkina Faso, many people believe that if it rains on the day of a person's death, it is the doing of the spirit of the deceased person. By using the number of rainy days in the month of the death in the previous year or years (most farmers know this information off the top of their heads), one can calculate the probability of rain on the day of the death. The point is not to take away cultural significance of the event but rather to show how statistics can be used to predict the likelihood of it. Other region-specific examples used by the authors are perceived coincidental connection between political events and natural disasters in the earthquake-prone region of Central Asia.

Our experience has been that using coincidence examples as a motivating theme in probability courses are only effective if the example used are easy for students to relate to and understand. As such, examples used should be adapted to the cultural context of the students. This requires the instructor to have a certain level of understanding of his or her students' cultural and socioeconomic environment. The questionnaire that appears in this article can be used as a tool to help in this effect and could be modified by the instructor to offer more insight.

Analysis of the July 4 Coincidence

A few years ago, I wrote something about presidents and July 4 in our local paper and, later, more explanations about it. Here is a summary of these explanations. It is known that concepts involving uncertainty and probability are often confusing.

Here, a source of confusion is the use of probability as additional *evidence,* not as a proof (see the last paragraph of my letter). In my letter, I used the example that Presidents Jefferson, Adams, and Monroe all died on July 4, the anniversary of the signing of the Declaration of Independence, and examined two hypotheses:

H1: Just a coincidence (null hypothesis) and **H2**: More than just a coincidence (alternative hypothesis).

Clearly, all we can do is to provide *evidence* (a key word) for or against each hypothesis and draw a conclusion using a prespecified level of significance (type one error), which, in this case, is the probability of rejecting H1 while it is actually true. It is important to remember that a hypothesis may not be rejected at all levels of significance. In fact, a hypothesis rejected at one level of significance may not be rejected in another level of significance.

I showed that the probability of what happened assuming that H1 is true is low. This probability is known as the P-value and plays a key role in decision-making. Here, low means less than the chosen level of significance. Low probability provides evidence that what occurred was not a commonplace event.

To clarify, consider the following. Suppose you are sitting in a class with thirty-five students. Finding out two or more matching birthdays surprises most students because they think that this is not a commonplace event. But once they learn that its probability is more than 80 percent (a commonplace event), they accept that nothing special is happening in that class. Note that unlike the presidents, here, students have no control over the outcome. The same applies to example of three attorneys in Aaron's letter.

As for the monkey example, from the time I was a student, I considered this example artificial and misleading. One line of this letter in my computer has at least one hundred characters. There are more than thirty choices for each position. So there are thirty times itself one hundred times different possible permutations. This is more than ten times itself eighty times, the number of atoms in the universe. More importantly, the distribution of characters used in a book like *Hamlet* is not uniform—that is, like most books written in English, some characters are used more frequently than others. But let us, for the sake of example, say that a monkey could manage to produce the first few lines of the *Hamlet*. Then which conclusion seems more logical—: pure luck or an alternative that maybe the monkey knew a few things?

CHAPTER 11

Mathematics and Computer

NUMBERS AND NUMERATION systems have played a key role in human development. Throughout history, our understanding of numbers has gone through a dramatic change and has reached a high level of sophistication. A number is an abstract concept used to describe a quantity. A numeration system is a set of basic symbols called numerals, such as 0, 1, and 2, and some rules, such as addition and multiplication, for making other symbols from them. The invention of a precise, workable numeration system is one of the greatest accomplishments of humanity. The numerical digits that we use today are based on the Hindu-Arabic numeral system developed over a thousand years ago. Other than decimal system/base 10 system, which uses the ten symbols (0, 1, . . ., 9), a most popular system is the binary or base 2 system, which uses only the two symbols, 0 and 1. Because computers use a sequence of switches that can be only on or off (bit), base 2 works well there.

This chapter is an attempt to explore certain aspects of the binary system that relates to mathematical sciences. It focuses on the role of such systems and tries to analyze and visualize the world where decimal system is replaced by binary. It, whenever appropriate, uses what is known as Boolean logic.

Binary and I

When I work with binary numbers, I feel
at home, somewhere I have been.
When I do mathematical manipulations, I feel
comfortable, something I have seen.

When I think about binary numbers, I feel I
am watching a black-and-white movie.
It brings back both bad and good memories, the
times I either felt depressed or groovy.

I feel that I am living in a world with my dear
friends known as And, Or, and Not
Where everything has two faces, just present
or absent, things between are cut.

Binary System

The binary system furnishes a powerful tool for both calculation and communication. Today, most computer and smartphone users are familiar with terms such as "megabytes" (MB) and "gigabytes" (GB). A medium-sized novel contains about 1 MB of information. One MB is $2^{10} = 1,024$ kilobytes, or 1,048,576 (1024 x 1024) bytes, not one million bytes. Similarly, one GB is 1,024 MB, or 1,073,741,824 (1024 x 1024 x 1024) bytes, all powers of 2.

It was not long ago that gender was considered binary, male or female. A "nonbinary" person only existed outside the "masculine" and "feminine" gender norms. It meant that, walking down the block, the person will get called both "sir" and "ma'am," and neither was right." This chapter is an introduction to the binary system based on 0 and 1. Of course, this is not how the world works in its entirety. For example, the binary gender framework currently used by the IOC to categorize athletes as male or female—with the goal of ensuring fairness, replicability, and harmonization—does not consider the complexity of gender and, as such, faces certain problems. In fact, there are various ways gender is understood, including cultural, anatomical, biological, hormonal, chromosomal, and personal approaches. Clearly in the domain of uncertainty, directional thinking that requires to take care of many moving parts is more suitable. However, by making our picture of the world more and more complex, we will need to consider models with so many factors that often lead to inability to deal with it.

Decimal and Binary Systems

For most of us, the commonly used decimal number system feels so natural and useful that it is hard to think of any other numbering system including the binary system, even though it is a positional system similar to decimal system with a difference that it uses 2 as a base rather than 10—that is, unlike the decimal system with ten symbols, 0, 1, 2, 3, 4, 5, 6, 7, 8, 9, for representation, the binary system uses only two symbol/digits, 0 and 1, for representation. As such, it is a more "efficient" system. The system works with powers of 2 just as the decimal system works with powers of 10. For example, the binary representation of the numbers 0 to 10 are respectively 0, 1, 10, 11, 100, 101, 110, 111, 1000, 1001, and 1010. Operations in the binary system are no different from operation in the decimal system.

Recall the well-known principle of positional numeration. From the point of view of algebra, the principle rests on the simple idea of presenting any whole number as a *polynomial* developed along the powers of the *base*. Thus, for example,

465 at the base *ten* means	$=4 \cdot 10^2 + 6 \cdot 10 + 5;$
1233 at the base *seven* means	$=1 \cdot 7^3 + 2 \cdot 7^3 + 3 \cdot 7 + 3;$
1110100001 at the base *two* means	$=1 \cdot 2^8 + 1 \cdot 2^7 + 1 \cdot 2^6 + 1 \cdot 2^5$ $+ 1 \cdot 2^4 + 0 \cdot 2^3 + 0 \cdot 2^1 + 1.$

The *digits* of the number are the coefficients of the polynomial, and the only feature that distinguishes the expression from a general polynomial is the restriction that those coefficients must be positive integers less than the base. The conversion of a number expressed in any system to another numeration system is quite a simple matter, and so is the inverse operation. In what follows, we present more detail about the binary system such as how information is stored in binary form in computers. Computers don't understand words or numbers the way humans do. Modern software allows the end user to ignore this, but at the lowest levels of your computer, everything is represented by a binary electrical signal that registers in one of two states: on or off. To make sense of complicated data, your computer has to encode it in binary.

Binary and Your Hands

Recall that decimal system was developed in reference to our ten fingers. Suppose that you want to count in the binary system using fingers on one of your hands. It is easy to see you can go up to 31, 00000 = 0 up to 11111 = 31.

Computers and Binary

There is nothing more fundamental to computer technology than the binary system. All the elements of computer operations, from software to portable devices to servers, are built from ones and zeros. In a world with only a binary number system, very little would change operationally with computers, as computing technology is rooted in that system. Nineteenth-century mathematician Charles Boole developed a system of logic that would later serve as the foundation for both computer software and hardware. The basic operations, AND, OR, and NOT, can form simple statements with binary properties and can be combined to create more complex logical constructs.

A great example of binary operations in computer hardware is the use of AND, OR, and NOT to form gates. For instance, an inverter, a NOT gate would take an input of 1 and produce an output of 0. An OR gate outputs a 1 if either or both of the two inputs are 1, and it outputs a 0 if neither input is a 1. Systemically combining multiple gates, you have a computer. Boolean gates were utilized in early computers with electromechanical relays that were either on or off. In more modern machines, a similar implementation can be seen with transistors, which also have binary on/off, open/closed properties.

Functionally, even if the only numbering system we had were the binary number system, it would be entirely likely that computing technology would have continued along the current evolutionary track. The complexity of the binary number system would come in two areas: network operations and the actual functional process, the user interface of the computer.

The functional interface we have now such as the standard keyboard and, in many cases, an accompanying ten-key keypad would be somewhat problematic to design under the binary number system. Unchanged would be the alphabetic keys and symbols. Since there would only be the numbers "0" and "1," there would be little need for the keypad function beyond the "1" and "0," as well as the top row of the standard keyboard.

Comment: Transistors have been called the most important invention of the twentieth century. Vacuum tubes were unreliable and generated lots of heat, so scientists searched for some kind of device that could serve as a switch that can change state millions of times per second and has no moving parts. It sounds like magic. The American team that developed this switch, the transistor, was awarded the Nobel Prize in physics for their efforts.

Why Binary?

One could perhaps artificially construct *advantages* of binary, such as it being much easier to multiply by two, since you just add a zero (like multiplying by ten in decimal). But perhaps we should first start by understanding *advantages for whom?*

All number systems are merely a model for representing and describing quantities. Both binary and decimal are positional number systems in that the same symbol or glyph represents different order of magnitude based on its position. Move a digit to the left one position means either ten times (in decimal) or two times (in binary) more than it did before.

There are many different bases that can be used for positional number systems. The Babylonians used base 60 (although their symbols were composite entities with each "digit" being represented by quantities of two glyphs in a base 10 system within the base 60 system). Today, there are only a handful of bases in common use in Western culture. Those non-decimal bases extant are related to computing science (primarily bases 2, 8, and 16). Which brings us back to "advantages for whom?"

The key advantage of any positional number system is that it enables calculation or computation in a way that a non-positional system, such as Roman numbers, does not. So how do you choose a base for a number system? What are the trade-offs?

One of the most immediately apparent is the *number of symbols* versus the *complexity of the arithmetic*. Base 10 is a natural choice for people with ten fingers. The number of symbols is not overwhelming to memorize, and it keeps numbers compact enough to be easily recognizable and ascribe quantitative meaning while still making basic arithmetic something that most people can do in their head.

If you deconstruct the ways you were taught to do arithmetic for larger numbers, you will find they are procedures that can be described and generalized to provide a quantifiably correct answer by using rules for manipulating symbols.

Consider addition. We can add very large numbers together by breaking the task up into a series of additions of individual digits, and "carrying" any rollover (numbers bigger than nine that can't be represented in a single digit) to the next order of magnitude column. There are fifty-five (10 + 9 + 8 + 7 + 6 + 5 + 4 + 3 + 2 + 1) different addition combinations for two base 10 digits valued 0–9. When we first learn basic addition of single digits, we might count up one finger at a time from the first digit until we reach the sum. Eventually, we learn those fifty-five combinations and can add up "in our head" so we don't have to.

Compare that now to binary addition. The *complexity* of adding digits is reduced to just three combinations (2 + 1). That means every digit addition operation requires fewer resources, but the trade-off is how quickly we roll over into a new order of magnitude by carrying a digit, creating much longer strings of digits very quickly.

For most, it's much easier to recognize and assign quantitative significance to the decimal number 349 than its binary representation, 101011101.

As others have answered, the main *reason* for using binary is that it is the underpinning of modern computing because electronic circuitry and storage are binary—that is, it is entirely founded on distinguishing between two states, whether they are voltage, charge, or magnetic polarization. With just two states, the implementation of logic is reduced considerably, as in the example of addition above. The *resources* required to store and perform computation operations are fewer with fewer symbols or states. This is the advantage *for computers* and, I guess by extension, the engineers who design them. :-)

Just to round out the discussion, the other non-decimal bases we mentioned at the start, octal (base 8) and hexadecimal (base 16), are advantageous because they form a bridge for people between binary and compactness. Base 10 does not translate very well to binary because ten is not a power of two. By using base 8 or base 16, there is a 1:3 or 1:4 mapping of digits, respectively. Using decimal 349 as an example again to demonstrate this mapping:

| Base 2: | 101 011 101 | Base 8: | 5 3 5 |
| Base 2: | 1 0101 1101 | Base 16: | 1 5 D |

This means people who work better with compactness of representation, enabling quick assignment of meaning, and can handle more combinations in basic arithmetic operations can work faster with less chance of error yet still translate back and forth with relative ease to binary.

Aside from the obvious trade-off between manual computational convenience on the one hand and, on the other, compactness of expression, I cannot think of much to say. In the end, these are just two different alphabets with the same expressive power.

We do, however, have a thought regarding the choice to use binary instead of ternary logic in computers. We are aware of the fact that Soviet scientists did build computers in the 1950s or '60s using ternary logic. And as we discussed, this has implications for fault tolerance: a ternary logic gate must be sufficiently sensitive to distinguish three levels of electric current; whereas, a binary logic gate only has to distinguish current from its absence.

Here's another thought. Given that humans are naturally disposed to base 10, thanks to their ten fingers, there is a certain sense in which binary is more intuitive than ternary: any fraction that has a finite representation in base 2 also has one in base 10; but there is no number that has a finite representation in both base 3 and base 10 (since three and ten are co-prime).

Binary Instead of Decimal?

The binary system is used in almost all computers today as the most basic way to represent data. Computers are, at their most basic level, a vast network of transistor that can only have two states: on or off. Most digital files like mp3, jpeg, and mpeg also use binary to store data.

Binary calculations take more time and space but are easier as we only use digits 0 and 1. In a world without advanced technology, of course, it is faster to perform calculations in the decimal system. After all, most of us learn counting using the ten figures of our hands, although we can learn binary using only our thumbs. Also, for most of us, it is easier to recognize the significance of the decimal number 2,435 than it is binary representation, 1111 0110 0111 1101. Overall, operations in binary are easier and can be done almost instantly using a computer. It is the best because it is so easy/cheap to build circuits that only take on two values. But the decimal system is much more practical. By the way, if you use your fingers as binary digits—extended means 1, folded means 0—then you can count to 1,023 on your fingers ("four," or "00100," looks like a rude gesture, however).

Someday, the binary system might become obsolete because of the introduction of quantum technology. Presently, the binary powers computer systems and enables us to deal with complex tasks.

Finally, like me, you may be curious about the binary representations of well-known irrational numbers such as π = pi=3.14... As expected, like other irrational numbers, they are mathematically an infinite sequence of 0s and 1s with no discernible pattern. The first twenty-two bits of π are

$$11.0010\ 0100\ 0011\ 1001\ 0101\ 1000$$

One interesting observation: unlike the decimal system where, using calendar dates, March fourteen (3.14) is designated as Pi Day, the same cannot be done in binary world.

And, last, to convert texts into binary, codes are assigned to letters and characters used in writing. For example, a = 01100001, b = 01100010, c = 01100011, and, finally, z = 01111010.

Summary of the Advantages

Binary is extremely simple to implement. Any system that has an "on" and "off," or "high" and "low" state can be used to encode and/or manipulate data. It is easily implemented in an electronic system by using on/off signals. Binary devices are simple and easy to build. Binary signals are unambiguous. The implementation in electromechanical and electronic hardware is simple. Immunity to noise is another advantage.

1. Binary is the lowest base possible that is practical (base 2), and, hence, any higher counting system can be easily encoded (e.g., decimal, octal, hexadecimal, etc.).

2. Binary logic is easy to understand and can be used to build any type of logic gates (AND, OR, NAND, XOR), which can be used to build higher-order components (counters, multiplexers, adders, etc.) and ultimately the computers and other tech devices our world now relies on.

3. Binary data is extremely robust in transmission because any noise tends to be neither fully "on" nor "off" and is easy to reject. On/off signals provide very reliable operation. They have extremely low possibility of errors when transferring data because it is easy to decide the difference between the on and off state of the circuit, and some noise or disruption during the transfer process will not disrupt that decision.

4. Binary would be the most effective way to attempt to communicate with any type of alien civilization. Just as "math" is a type of universal language (any alien civilization would understand a sequence of prime numbers, for example), binary is a universal alphabet.

In sum, computers are binary. Physically, computers work entirely in terms of switches that can be either on or off. As such, binary numbers are more appropriate than non-binary numbers for representing computer logic. The bit, the smallest unit of information, is also a fundamental unit of information theory. Combined with the physical nature of computers, this will lead to binary numbers coming up naturally when dealing with things like information storage and compression.

When debugging dumps of memory, disks, or data streams on modern microcontrollers and computers, hexadecimal provides a compact format that is easy to break into bytes and convert into bits.

Binary can help search for an element very efficiently as we pass through it. This is because for a tree having m branches for each node and having N nodes in total, the time taken to search for an element is (assuming the elements are kept in an order suitable for searching) $m (\log (N)/\log (m))$. This value is minimized when $m = 2$. In short,

- Binary codes are suitable for computer applications.
- Binary codes are suitable for digital communications.
- Binary codes make the analysis and designing of digital circuits easier.
- Since only 0 and 1 are being used, implementation is easy.

How Are Letters Stored in Binary?

To store letters as binary data, computers use a system called UTF-8, which stands for "Unicode Transformation Format 8 bit" and is an extension of the earlier ASCII (American Standard Code for Information Interchange). The UTF-8 system designates a one-to-four-bytes-long number to each character. For example, the numbers 0–9 are actually represented by decimal numbers 48–57. Uppercase letters are represented in alphabetical order as 65–90, and lowercase letters are 97–122. In a computer, these numbers are stored as digits in binary. The first byte of data can represent all of characters needed to write in English, and three bytes can hold most of the characters used in any language. All four bytes can hold everything previously mentioned as well as symbols used in mathematics, emojis, and some more uncommon Asian characters.

Characters are usually represented as a hexadecimal number for ease of viewing, but when they are stored by a machine, they are stored as a special extension of binary called binary UTF-8. Binary UTF-8 has two key differences from regular binary. First, the largest byte of each character in binary has a marker that takes up a number of the leftmost bits and represents how many bytes the character uses. A one-byte character is preceded by a "0," a two-byte character is preceded by "110," a three-byte character is preceded by "1110," and a four-byte character is preceded by "11110." Second, each character that uses more than one byte has a "10" at the beginning of each byte to the right of the largest to signify that it is a continuation byte.

For example, a character like {, the left curly bracket, using only one byte, has the marker of just "0" and a binary UTF-8 code of 01111110. An example of a two-byte character is the Japanese symbol for yen, ¥, which is character 10100101 in normal binary. In binary UTF-8, this character is 11000010 10100101. The byte farthest to the right has the smallest six bits of the normal binary "100101" code preceded by the continuation marker "10." The remaining two bits "10" are in the larger byte preceded by the byte marker "110" and padded with three extra zeros. Three- and four-byte-long characters follow the same conventions as two-byte long codes with the only differences being the larger byte marker and another continuation byte marker of each byte to the right of the largest.

Computing in Binary World

Humans have naturally adopted the decimal number system for counting. If we only had two fingers, many aspects of our life would be completely different. One major difference is how we would interact with computers differently.

Computers already use binary. This is because if computer hardware was in decimal, it would be much slower and much harder to implement with transistors. Base 3 was proven to be the most efficient, but we still use binary for computers for simplicity. So, what would change if humans naturally used the binary number system instead? The first thing that comes to mind is programming would be completely different. Currently, if you need to do any simple counting, you would use an integer. But if we used base 2, would we still need to use integers? Surely, we would still know about the decimal system, but how often would we use it in programming? I assume if we counted in binary, then octal or hexadecimal would be more natural to us than integers.

One seeming benefit is that we wouldn't have to worry about floating point conversion or rounding errors because we wouldn't naturally think in decimal. I also like to believe that humans would understand how computers work more if we were taught the binary system from birth. It would be more relatable and easier to grasp, since it would follow along with what we already know. Another thing I wonder about is what our computer keyboards would look like. Right now, we have ten digits, 0–9. In a binary world, would our keyboards have only 0 and 1? We would also need a way to type large numbers fast, so the way we type would look much different. I would guess the keys would be binary shortcuts to the powers of two: 1, 2, 4, 8, 16, 32, 64, 128, 256, or every other power of two. The way the computer represents time would be different. In fact, all the ways the computer represented numbers would line up with the way we naturally represent numbers. We wouldn't have to constantly convert numbers to binary in programming or when using a computer.

Overall, using the binary system would completely change almost every aspect of our lives, and computers being a large part of people's lives would be affected as well, from the way we program to the way we type and how we represent numbers and probably letters. Using a computer would present a completely different experience—in some ways for the better and some ways for the worse. But for now, I don't see us switching from decimal any time soon.

Mathematics and Neurobiology

The number of patterns nerve cells could
form is 1 followed by 800 zeroes.

This chapter presents where artificial intelligence has gone recently. For example, in recent years, face recognition has become more feasible. Companies such as Apple have already started using face recognition as a substitute for a password on the newest iPhones. In a study of artificial intelligence, the hot topic is now neural networks. The neural networks, which were originally inspired by biological models, are composed of relatively simple computational cells arranged in a network and trained to perform whatever task is being attempted. There is a significant amount of mathematical theory behind all these studies.

Brain versus Supercomputer

Most of us are impressed by the abilities of the modern computers. A supercomputer can add, multiply, or compare a million large numbers in less than one second. Currently, the fastest supercomputer in the world is the Sunway TaihuLight, a Chinese supercomputer. It is capable of ninety-three petaflops, where one petaflop equals four quadrillion calculations/second. The second fastest is the Tianhe-2, which was the world record holder for three years, from 2012 to 2015. It was a computer capable of 33.86 petaflops. Several million transistors can be accommodated in a square centimeter, and each can switch in less than one-billionth of a second. Steady advances are being made in both hardware and software.

Compared with the capabilities of the human brain, even the very latest supercomputers are still stupid. The brain's high ability is primarily attributable to the fact that all its nerve cells operate in mutual association as a "neuronal network." Physicists have now succeeded in quantifying the intercooperation of these neurons on a mathematical basis. Given appropriate examples, a network of this nature can learn to store and retrieve information independently. Both neurobiology and computer science can profit from this knowledge.

So it is all the more surprising that a child possesses abilities that a modern supercomputer cannot even approach. It is no problem for us to recognize the passing face of someone we know as we rush through a busy street. In a fraction of a second, we have recognized the person, quite independently of the crowd of people, or any movements, size, and orientation, light or shade, or his new hat. At the same time, memories and many other items of information on this person awaken within us. Our brain achieves all this without any effort; even a child can do it. A computer, on the other hand, although it can record images electronically, is only able to understand their content in very simple cases and after a long calculation. Why is this?

- How do primates, including humans, tell faces apart? Scientists have long attributed this ability to so-called face-detector (FD) neurons, thought to be responsible for distinguishing faces, among other objects. But no direct evidence has supported this claim. Now, using optogenetics, a technique that controls neural activity with light, MIT researchers have provided the first evidence that directly links FD neurons to face discrimination in primates—specifically, differentiating between males and females.

- Researchers have demonstrated how to decode what the human brain is seeing by using artificial intelligence to interpret fMRI scans from people watching videos, representing a sort of mind-reading technology. The advance could aid efforts to improve artificial intelligence and lead to new insights into brain function. Critical to the research is a type of algorithm called a convolutional neural network, which has been instrumental in enabling computers and smartphones to recognize faces and objects.

The material basis of man's ability to recognize what he sees are the nerve cells. Our brain contains some hundred million such neurons; the grey matter comprises a dense, extremely homogeneous, and apparently random network of strongly ramified cells. These neurons, just like the transistors in computers, can be regarded as tiny electronic switches; each cell emits electrical pluses and, thus, controls other cells.

One essential difference between a computer and the brain is the mode of operation of the switch elements. In a computer, orders previously fed in by an operator are processed step by step. This involves the participation of only a few of the available transistors. The neurons, in contrast, function as a network; a vast number of nerve cells are simultaneously activated to process the information.

The current theory is that the basic difference lies in this network structure, and, using simple models, we are now beginning to understand why the properties of a network differ so radically from those of a conventional computer.

A neural network comprises more than a hundred billion neurons. Each of these neurons is directly linked to some tens of thousands others. The point of contact between the nerve cells—the so-called synapses—is responsible for learning and memory.

In recent decades, neurobiologists have made important discoveries regarding the functioning of the neurons so that we now have a very sound understanding of the individual cell, and research is being concentrated on the complicated biochemical processes that occur in the points of contact of the nerve cells, the synapses. However, the higher functions of the brain are still mostly a closed book to us, even today. How do neurons cooperate to recognize a person? Is there a single cell that sends electrical pulses only if you see your grandmother? How is your grandmother stored in your brain—in one synapsis or distributed over the whole brain? How does the brain call up all the associations that manifest themselves when you think of someone you know? All these points are still essentially unknown; there are even some scientists who doubt that the brain can even understand itself.

Nevertheless, some new results on the cooperation of neurons have been obtained in recent years—and not through experiments on rats or monkeys but by using *mathematical models*. And not biologists but physicists solved these models. The brain, of course, is far too complicated to be described in purely mathematical terms. But it is possible to concentrate on just a few essential mechanisms and to try to describe how the neurons intercooperate in these. The initial steps in this direction have been successful. We are now beginning to understand how intercooperation between a multitude of simple individual elements can produce complex new abilities. Many scientists and engineers are even convinced that we could already apply this new knowledge to the design of new computers we have.

The model is confined to the following mechanisms: each neuron has only two states—either it "fires" (emits pulses) or it is at rest. As the pulses move via the synapses to the adjacent neurons, they either gain or lose strength depending on the nature of the contact point (exciting or inhibiting synapses). Each synapse is described by a figure denoting the strength of the electric potential created in the receptor cell by the pulse entering it from the emitter cell. A neuron collects all the incoming signals and adds them together according to their strengths. If the sum of the signal exceeds a given threshold value, the neuron fires; otherwise, it remains at rest. The instantaneous information is the activity pattern of all the nerve cells; thus, as in a computer, it is written in binary form as a sequence of 1 (= neuron fires) and 0 (= neuron rests).

In this way, the network of nerves adopts a mathematical form. The exchange of signals between the nerve cells via the synapses thus becomes a series of nonlinear mathematical equations with a large number of unknowns and feedback. Such equations are normally very difficult or impossible to solve, but in this case, physicists have succeeded by applying the methods used in theoretical solid-state physics. We can now convert the properties of the network into a mathematical form that will permit many things to be quantified and understood.

The mathematical solution reveals that a neuronal network can operate as a so-called associative memory, which means that a mass of data can be stored by selecting the appropriate synaptic strength. The system is now able to complete incomplete information and, thus, to associate data. A faded picture can now be fully restored. Alternatively, the system is independently capable of finding all the personal data relating to a name even if it is incorrectly written. They function simultaneously with a vast number of images. Each image provides a pattern for the activation of the neurons into which the complete network can be firmly "locked." Any deviations from this return automatically into the stored pattern.

The remarkable thing is that a network operates quite differently from a computer. A computer stores data like in a storage depot—each shelf has a number, and anyone searching for data must know the number or he will have to search through all the shelves individually. If the data are incomplete, all the shelves will have to be compared with the available information. The entire operation will be controlled step by step by a program and has no numbered drawers. All data are stored over the entire system, with each synapse containing a small portion. In other words, the store extends over the whole network. Retrieval of incomplete or incorrect data is effected according to content; the network completes and corrects information itself. Doing this, all the neurons work simultaneously in accordance with the simple rules of the model.

But the model can do more than this, and once again, it is the mathematical solution that gives us the properties. It is able to learn certain tasks by itself without requiring any programmer to tell it precisely what it is expected to do. This learning is achieved—as in the biological nervous system—by changing the synaptic strengths; a synapse becomes more permeable the more it is used. Hence, the system can learn tasks by means of adaptable synapses. One possible task would be to store a mass of data associatively. Once again, the competition between inhibiting and exciting contact points enables the special properties of the network to develop.

The laws of learning for such networks were postulated back in the early '60s under the aegis of cybernetics. But only now have we succeeded in calculating general laws of this learning process—using the methods of theoretical solid-state physics.

We can now try to find applications for these abilities, or we can use the model to improve current descriptions of biological processes. During the first years of life, a large proportion of the synapses disappears despite the fact that at this time we have important elementary things to learn. So our hereditary information does not hold the detailed circuitry of our brain, but initially a surplus is produced, which is later trimmed down by the learning process. The question to be addressed to the model now is whether the death of synapses is the only process a network requires to learn. The mathematical equations show how this works. The applications have been demonstrated so far only in the form of network stimulation using a conventional computer. However, electronic chips are in the process of production that will soon give us the "neurocomputer" too as a piece of hardware. In this manner, a network has learned to read aloud a written English text. Of course, we already have conventional programs that can do this, but the network has learned to do it by itself, without a program, using simple mechanisms, quickly and without any expert knowledge, simply by learning from examples.

Other networks, again, have learned to make medical diagnoses in emergencies, to play backgammon, to assess bank clients, or to drive a lorry and trailer backward onto a ramp. These are all first attempts and demonstrations. Research is currently being conducted in this field all over the world, and fresh discoveries and applications are certain to emerge.

This example illustrates how surprising interdisciplinary linkages can arise in science. Who would have thought ten years ago that models and theoretical methods used in solid-state physics would have produced important foundations for neurobiology and computer science?

We end by mentioning that a twist on neural networks intended to make machines better able to understand the world through images or video. In one of the papers posted recently, Hinton's capsule networks matched the accuracy of the best previous techniques on a standard test of how well software can learn to recognize handwritten digits.

CHAPTER 12

Mathematics of Games and Competition

I N THIS CHAPTER, we wish to get a feel for how mathematics might be employed to help situation that involves conflict or competition. It is mostly about situations where one person's decision depends on other people's decisions. Think about games like chess or tennis, where your move is determined, at least in part, by what you think the other players could or would do. In a larger scale, the cold war was the stage for one of the original and most important applications of what is known in mathematics as the game theory. The basic idea of game theory is quite simple and mostly familiar.

What Is Game Theory?

Competitions and conflict have been a central theme throughout human history and literature. It arises whenever two or more individuals, with different values and interests, compete to control

619

the course of actions or events. Game theory being relatively new uses mathematical tools to study situations that involve both conflict and cooperation. The theory is useful for solving many problems of modern life. In some respects, it is the science of strategy, or at least the optimal decision-making of independent and competing actors in a strategic setting. The key pioneers of game theory were mathematicians John von Neumann and John Nash, as well as economist Oskar Morgenstern. Game theory has a wide range of applications, including psychology, evolutionary biology, war, politics, economics, business and economics, and life.

In game theory, every decision maker must anticipate the reaction of those affected by the decision. In business, this means economic agents must anticipate the reactions of rivals, employees, customers, and investors. What economists call game theory, psychologists call the theory of social situations, which is an accurate description of what game theory is about.

Despite its many advances, game theory is still a young and developing science. There are two main branches of game theory: cooperative and non-cooperative game theory. In addition to game theory, economic theory has other main branches that are closely connected to game theory. Decision theory can be viewed as a theory of one-person games or a game of a single player against nature. General equilibrium theory can be viewed as a specialized branch of game theory that deals with trade and production, and typically with a relatively large number of individual consumers and producers. Let us start with a life example.

Relationship Game

"Why do people stay together in monogamous relationships? Love? Fear? Habit? Ethics? Integrity? Desperation?" In an interesting essay, Mark Colyvan has considered a rather surprising answer that comes from mathematics. It turns out that cooperative behavior, such as mutually faithful marriages, can be given a firm basis in game theory. He has suggested that faithfulness in relationships is fully accounted for by narrow self-interest despite the fact that faithful behavior is usually thought to involve love, ethics, and caring about the well-being of your partner. He has also considered the philosophical upshot of the game theoretic answer to see if it really does deliver what is required. While in this topic, here is an another example of mathematics in matters of the heart.

Stable Matching

Suppose that there are n women and n men. How can we match them with each other into partnerships? Imagine that each man and woman can rank all members of the opposite sex by how attractive a partner they are. Can a matching be made so that no two people would prefer to be with each other than their partners? It turns that an algorithm exists that always produces such a "stable" matching. But there are multiple such stable matching, and it turns that the people making proposals to potential partners are the ones that get the best outcome in this algorithm, teaching us all an important lesson about not being afraid to approach that one guy/girl.

- Wikipedia's description of the stable marriage algorithm, which also describes the optimality of the solution for the men and women.
- Wikipedia on the stable roommate algorithm, wherein you don't have to match people by gender. This problem is much harder—for example, it need not even have a stable solution. There is also a much more complicated algorithm to find a stable matching if it exists.
- A research project at the University of Glasgow, which investigated various algorithms for stable marriage and which stable matchings they choose.

Analysis of Games

Knowledge of probability and statistics can aid an individual decision maker when there are no competitors whose decisions could alter the decision maker's analysis of a situation. As mentioned earlier, competitive decision-making situations are the subject of the game theory. Here are more competitive situations: (1) coach's decision of stating team their positions; (2) union's decision when negotiating a new contract with management; (3) army's strategy when participating in a war.

The players in a game, who may be individuals, teams, organizations, or even countries, choose from a list of options available to them—that is, courses of action they might take—which are called strategies. The strategies chosen by the players lead to outcomes, which describe the consequences of their choices. We assume that the players have preferences for the outcomes; they like some more than others. In this setting, game theory analyzes the rational choice of strategies—that is, how players select strategies to obtain preferred outcomes. For example, how can a team decide which set play (predetermined play) to perform when the play might have varying degrees of success that depends upon what the opposing team does? Example for such a situation occurs in soccer when a direct or an indirect kick close to the penalty area (box) is awarded to a team. Soccer fans know that in recent years, many goals were scored in this way, and there are many famous players who specialize in this area. Among other areas to which game theory has been applied are bargaining tactics in labor-management disputes, resource-allocation decisions in political campaign, military choices in international crises, and the use of threats by animals in habitat acquisition and protection. As far as learning is concerned, the theory provides an excellent review of many of the topics including studies in probability as expected.

Unlike the subject of individual decision-making, which researchers in psychology, statistics, and other disciplines study, game theory analyzes situations in which there are at least two players who may find themselves in conflict because of different goals or objectives. The outcome depends on the choices of all the players. In this sense, decision-making is collective, but this is not to say that the players necessarily cooperate when they choose strategies. Indeed, many strategy choices are non-cooperative, such as those between combatants in warfare or competitors in sports. In these encounters, the adversaries' objectives maybe at cross-purposes—a gain for one means a loss for the other. But in many activities, especially in economics and politics, there may be joint gains that can be realized from cooperation. In any case, the game involves a payoff of some amount after each play such that one player's win is usually the other player's loss and no fee is paid to a third party.

There are also many cases where interactions may involve a delicate mix of cooperative and noncooperative behavior. In business, for example, firms in an industry cooperate in gain tax breaks or reduce advertising costs even as they compete for shares in the marketplace.

In what follows, we present several simple examples of two-person games of total conflict in which what one player wins, the other player loses, so cooperation never benefits the players. We distinguish two different kinds of solutions to such games.

Finally, as we are all aware, chance usually plays an important role in our lives. In fact, many man-made games, backgammon being an excellent example, also depend upon results of a randomizing device. Such games can be viewed as having an additional player, called chance, with its moves and their consequences specified just as for the other players. In previous chapters, we discussed many problems and situations that involved uncertainty and chance. In this chapter, we concentrate on new ideas and, therefore, do not discuss games with chance moves.

Strategies

1. Consider a very simple example regarding a contract based on negotiation or renegotiation of a professional player. There

are various outcomes of this game situation. For example, it is possible to organize this into a payoff table where the player or his agent have two strategies (e.g., different types of contracts and the threat of a holdout and/or of becoming a free agent) and the manager has three (e.g., length of contract, residual payment, no-cut/no-trade clauses and off-season promotion work). The following table is an example of such payoff table.

Strategies	*Manager*		
	A	B	C
Player 1	$250,000	$205,000	$150,000
Player 2	$300,000	$200,000	$100,000

Here, the payoff table is organized so that the player who is trying to maximize the outcome of the game is on the left (row) and the player who is trying to minimize the outcome is on the top (column). Note that, here, we refer to participants as players and assume that the payoff table is known to both players. If the player selects strategy 2 and the general manager selects strategy B, then the outcome is a $200,000 gain for the player and $200,000 loss for the manager. Summarizing these, a strategy is a play of action to be followed by a player. Each player in a game has at least two strategies of which only one is selected for each play.

2. Consider two stores that sell ski jackets, store 1 and store 2. These are the only two stores carrying ski jackets, and each is trying to decide how to price its most competitive brand. A market analyst supplies the following information:

Store 1		*Store 2*	
		$150	$175
	$150	60%	75%
	$175	45%	60%

The entries in the matrix indicate the percentage of the business store 1 will receive—that is, if both stores price their jacket at $150, store 1 will receive 60 percent of all the business (store 2 will lose 60 percent of the business but will get 40 percent); if store 1 chooses a price of $150 and store 2 chooses $175, store 1 will receive 75 percent of the business (store 2 will lose 75 percent of the business but will get 25 percent); and so on. Each store can choose its own price but cannot control the price of the other. The object is for each store to determine a price that will ensure the maximum possible business in this competitive situation.

This marketing competition may be viewed as a game between store 1 and store 2. A single play of the game requires store 1 to choose (play) row 1 or row 2 in matrix (i.e., price its ski jacket at either $150 or $175) and simultaneously requires store 2 to choose (play) column 1 or column 2 (i.e., price its jacket at either $150 or $175). It is common to designate the person(s) choosing the rows by 1 or R, for row player, and the person(s) choosing the column by 2 or C, for column player. Each entry in matrix (5.1) is called the payoff for a particular pair of moves by 1 and 2. Matrix (5.1) itself is called a matrix game or a payoff matrix. This game is a two-person zero-sum game, since each store may be considered a person, and store 1 wins the same amount that store 2 loses, and vice versa.

3. Penalty kick, a non-strictly determined game. Consider a game that involves two soccer players (a goalie R and a player C). The player takes a penalty kick either to the right or left of the goal, hoping to score. Independently, the goalie jumps to the right or left with the aim of saving the goal. Let us label right and left as 1 and 2. Here, the payoff (in dollars) is the sum of the numbers to R if the sum is even (correct guess) and the sum of the numbers to C if the sum is odd.

Player C

right left

Right $\begin{bmatrix} 2 & -3 \\ -3 & 4 \end{bmatrix}$
Left

Would you rather be a row player or a column player, or does it matter? (Think about this for a moment before proceeding.

To answer this question, we start by checking to see whether the game is strictly determined. Using the circle and square method described in the preceding section, we have

$$\begin{bmatrix} \boxed{2} & \boxed{-3} \\ \boxed{-3} & \boxed{4} \end{bmatrix}$$

How should the players play? If player R continued to play row 2 (because of the large payoff of $4), it would not take C long to detect R's strategy, and C would obviously play column 1. Then R would play row 1 and C would have to shift to column 2, and so on. It appears that neither player should continue playing the same row or same column but should play each row and each column in some mixed pattern unknown to the other player.

Clearly, for this game, pure strategies are not the best choice (pure strategies are always used in strictly determined games, and the game above is not strictly determined). This indicates that mixed strategies should be used. But what mixed strategies? Are there optimal mixed strategies for both players? The answer happens to be yes.

We conclude by stating without proof the deep and beautiful theorem that every two-person zero-sum game in normal form has a solution in either pure or mixed strategies. Methods are developed for how such solutions can be obtained through saddle points or (when each person has only two strategies) mixed strategy probability calculations. Computation of mixed strategy solutions when players have more than two strategies requires more sophisticated mathematical techniques from a branch of mathematics called linear programming.

A game for which the payoffs to the various players do not always sum to zero is called a nonzero-sum game. Such games have a rich mathematical theory but are always subject to considerable philosophical discussion and varied interpretations.

4. Let us introduce the theory further through an example. Two soccer clubs in England compete for fans in the same large metropolitan area (e.g., Manchester United and Manchester City). To increase attendance, each club decides each year to designate a percentage of profits to exactly one of the following: advertising, player salaries, stadium improvements, or special promotions at the ballpark. Club U(United) has finished high in the standings in recent years and has several players with expensive long-term contracts. A stadium renovation project that included the construction of luxury boxes was recently completed. The owner is reluctant to spend more money on player salaries, will not spend any more on stadium improvements, but suspects the club could do a better job of promoting its successful team through advertising and special promotions at the ballpark.

Club C (City) has been less successful on the field and plays its games in an old stadium badly in need of modernization. Its owner debates between designating profits for player salaries, hoping to sign a couple of free agents, and using the money for stadium improvements. He doubts, however, the effectiveness of additional advertising or special promotions.

For this problem, we may like to answer these questions: What are the optimum strategies for each club? Is this a fair game? How do the optimum strategies correspond with the owners' inclinations? What is the expected value of the game if U plays its optimum strategy, but C always chooses stadium improvements? always chooses player salaries? What is the expected value of the game if C plays its optimum strategy, but U chooses advertising, player salaries, and special promotion, each with probability 1/3? These and many other questions may be answered using results of game theory.

5. Consider a simplified analysis of the game of tennis. In professional tennis (especially men's tennis), the game is dominated by the serve. Here, we take this a bit to the extreme. The server has three basic options for the location of the serve. He can serve it outside (W), he can hit it directly at his opponent in an attempt to jam him (J), or he can serve it "straight down the t" (T) in the middle of the court. Similarly, his opponent can guess in advance which one of these three actions to focus on in his attempt to return the serve. In our simplification of the game, we make the assumption that the winner of the point is determined by the serve. This two-person zero-sum game can thus be represented by a three-by-three matrix in which the entries (payoffs) give the probabilities of the server winning the point given his choice of service location and the opponent's choice of primary service defense.

		Serve		
		W	J	T
	W	0.4	.07	0.8
Defense	J	0.7	0.4	0.6
	T	0.8	0.6	0.5

Here, again, we may like to determine optimal strategies for both the server and his opponent. It is clear that T is a reasonable choice for the server (why?).

6. Consider the game played between the opposing goalie and a soccer player who, after a penalty, is allowed a free kick. The penalty taker can elect to kick toward one of the two goalposts of the net or else aim for the center of the goal.

The goalie can decide to commit in advance (before the kick) to either one of the sides or else remain in the center until he sees the direction of the kick. This two-person zero-sum game can be represented as follows, where the payoffs are the probability of scoring a goal.

		Goalie		
		Breaks left	Remains Center	Breaks right
	Kicks Left	0.5	0.9	0.9
Kicker	Kicks Center	0.1	0	0.1
	Kicks Right	0.9	0.9	0.5

If we assume that decisions between the left or right side are made symmetrically (with equal probabilities), then this game can be represented by a two-by-two matrix, where $0.7 = (1/2)(0.5) + (1/2)(0.9)$.

		Goalie	
		Remains Center	Breaks side
Kicker	Kicks center	0	1
	Kicks side	0.9	0.7

The optimal strategies of the kicker and goalie are the following:

a. More often kick center and remain center.
b. More often kick side and break side.
c. Choose center and side equally often.

7. The prisoner's dilemma applied to sports. The contracts of two WNBA basketball players from the New York Liberty expire this year. The team's general manager must decide if he wants to resign one of the players or both and for how many years. To facilitate this decision, the general manager

decides to separate the two players and ask each about the other player's value to the team. The players can say that their teammate is valuable and worth resigning (cooperate), or not valuable and not worth resigning (defect). Based on their replies and the following matrix, the general manager will decide whether he will resign the players and for how many years.

		Player B	
		Cooperate	Defect
Player A	Cooperate	(3,3)	(0, 6)
	Defect	(6,0)	(1,1)

For example, if player A cooperates and player B defects, the resulting payoff of (0, 6) means that player A will not be resigned and player B will be resigned for six years.

Solution: To solve this game, suppose we are player A. We see that if player B cooperates, we will do better if we defect, resulting in a payoff of a six-year contract instead of a three-year contract. Similarly, if player B defects, we will do better if we defect. This is because we will receive a one-year contract as opposed to not being resigned if we cooperate. As a result, our strategy is to defect.

Similar analysis shows that player B's strategy is to defect. As a result, (defect, defect) is the solution to this game, even though both players would do better if they cooperated.

Game Theory and Sports

Game theory is a system for predicting how people should optimally behave in situations of conflict. As such, it is relevant to sports, since in a typical sport, both players and coaches try to outsmart their opponent by anticipating their decisions. It is a theory that has come out from games played in a modern life. In this sense, it is like probability theory. Other than sports, it is widely applied in economics, operation research, political science, etc. Game theory was developed out of games of strategy in the 1920s. As pointed out earlier, its importance in economic theory was shown by awarding of 1993–1994 Nobel Prize to three prominent game theoreticians. Game theory, the abstraction of games, is the branch of mathematics that analyzes situations of conflict and cooperation in terms of gains and losses of opposing players. Modern game theory originated in 1944 with the publication of *The Theory of Games and Economic Behavior* by John von Neumann and Oskar Morgenstern.

A mathematical game is a competition of two or more players, each of whom can choose between various strategies that have payoffs. It is assumed that the players will act in a rational manner, meaning that they attempt to choose strategies that favor their gain. Here, game means the specification of a set of players, rules for playing the game, and an assignment of payoffs to all possible endings resulting from the various actions of the players. In games of total conflict, a payoff representing a gain for one player directly corresponds to a loss for the other(s). It is natural to assume that players will seek a strategy that will maximize their gain (or minimize their total loss). A basic type of total-conflict game is the two-person zero-sum game. Such games are also called matrix games because they can be represented by a matrix. We will see game theory applies to more than just game.

Soccer's Penalty Kicks

The widespread interest in sports provides a great opportunity to catch students' attention in mathematics and statistics classes. Many students, whose eyes would normally glaze over after few minutes of number crunching, happily spend many hours analyzing data from their favorite sport.

This article describes how soccer's penalty kicks may be used as a teaching aid in mathematics and statistics classes. It presents lessons for teaching topics such as percentages, functions and their graphs, some basic algebra, geometry, and trigonometry, some elementary probability and statistics including Simpson's paradox, geometric progression, and game theory, and linear programming. It also provides hints for using penalty kicks data for teaching more advanced topics such as confidence interval, test of hypothesis, regression, time series, and Markov chains.

Introduction

The difficulties faced by educators teaching mathematics and statistics are recognized by the community of instructors involved. To help this matter, many textbooks try to motivate students by introducing varied applications. This addresses both students' apparent desire to see the relevance of their studies to the outside world and also their skepticism about the value of the mathematics and statistics. This is, of course, a great idea. However, it works mostly with students who are deeply committed to a particular academic or career field. For typical students, applied examples may fail to motivate especially if they do not find them interesting or of immediate concern to them.

So, is the prospect of motivating students entirely bleak? Despite many voids in student interests and despite many ideas that fail to motivate them, students have some common interests that we can build on as we try to teach them mathematics and statistics. Based on my experience, connecting their studies to a familiar theme with which they have interest or concern almost always works better.

In what follows, we describe a simple example of use of penalty kicks in soccer for teaching some concepts of mathematics and statistics. We start with few words about penalty kicks for reader who may not be familiar with soccer.

Penalty kicks. A penalty kick (penalty) is a free kick taken from the penalty mark thirty-six feet (twelve yards or approximately eleven meters) from the center of the goal and with no player other than the goalkeeper of the defending team between the penalty taker and the goal.

Penalty kicks occur during a normal play. They also occur in some tournaments to determine who progresses after a tie game; though similar in procedure, these are not penalty kicks and are governed by slightly different rules.

In practice, penalties are converted to goals more often than not, even against world-class goalkeepers.

A penalty kick may be awarded when a defending player commits a foul punishable by a direct free kick against an opponent or a handball, within the penalty area (commonly known as "the box" or "eighteen-yard box"). It is the location of the offence and not the position of the ball that defines whether a foul is punishable by a penalty kick or direct free kick, provided the ball is in play. The penalty taker (who does not have to be the player who was fouled) must be clearly identified to the referee.

The penalty kick is a form of direct free kick, meaning that a goal may be scored directly from it. If a goal is not scored, play continues as usual. As with all free kicks, the kicker may not play the ball a second time until it has been touched by another player even if the ball rebounds from the posts. However, a penalty kick is unusual in that, unlike general play, external interference directly after the kick has been taken may result in the kick being retaken rather than the usual dropped ball.

Lessons

Objective: To teach basic geometry and trigonometry.

Background information: Penalty kicks are normally taken from the penalty mark, which is a midline spot thirty-six feet from center of the goal. The penalty mark has the same distance from both goalposts.

Opening question: Suppose the penalty kick can be taken from any point in the field as long as its distance from the center of goal is thirty-six feet. Suppose also that the goalkeeper stands in the center of the goal and can protect the area inside a circle of radius eight feet. From which point the penalty taker has the greatest chance of scoring?

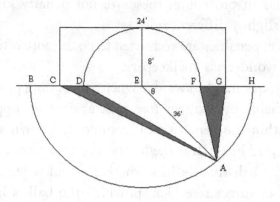

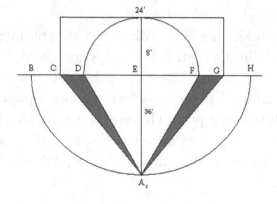

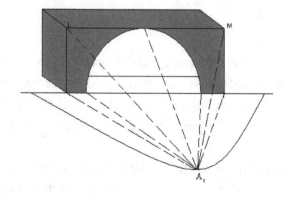

Solution

Consider a kick taken from an arbitrary point A shown in figure 1. The goal width (CG) is 24 feet, or 7.32 meters. Since we assume that the goalkeeper stands at the center of the goal and can protect 8 feet, or 2.44 meters (r) to his either side (segment DF), a goal can be scored only if the penalty taker kicks the ball inside the triangles ACD or AFG.

Let θ denote the angle $\angle AEF$, and assume, for now, that the penalty taker can only kick the ball on the ground. Then it is easy to show that the areas of triangles AEF and AEG are respectively 36 x 8 x Sin(θ)/2 ft (11 x 2.44 Sin(θ)/2 m) and 36 x 12 Sin(θ)/2 (11 x 3.66 Sin(θ)/2 m). Note the same values present the areas of the triangles ADE and ACE, respectively. Using these, the total area inside the triangles ACD and AFG is 36 x 2 Sin(θ) (11 x 0.61 Sin(θ. To determine the best point, we note that since sin(θ) is an increasing function for $0 \leq \theta \leq \pi/2$, the total area increases as the penalty taker moves the ball toward penalty mark—that is, the center of the half circle. In fact, the total area is maximum when the ball is moved to the penalty mark where $\theta = \pi/2$.

Now, for the general case, the penalty taker can kick the ball inside the volume shown in figure 1. Since for any point selected on the half circle the base of this volume is the same, the choice with the largest possible height produces the maximum volume. This, referring to the case discussed above, shows that as before, the penalty mark is the best choice.

Activity

Determine what happens to the effective scoring area if

- the penalty can be taken from a distance less than thirty-six feet from the center of the goal.
- the goalkeeper's side-to-side range changes.
- the goalkeeper moves forward out of the goal.

For penalty kick taken from the penalty mark, find the effective scoring angle (not area).

Suppose that the penalty taker can kick the ball with the speed of v.

- Could you think of a way of incorporating this dimension of the penalty kick into the problem?
- Would this change your answer to the problem of choosing the point with the highest chance of scoring?

Objective: To teach game theory and linear programming.

Background information: Conflict has been a central theme throughout human history. It arises whenever two or more individuals, with different interests, compete to control the course of actions or events. Game theory analyzes the rational choice of strategies—that is, how players select strategies to obtain preferred outcomes.

Consider the game played between the opposing goalie and a penalty taker. The penalty taker can elect to kick toward one of the two goalposts or else aim for the center of the goal. The goalie can decide to commit in advance (before the kick) to either one of the side, or else remain in the center until he sees the direction of the kick.

Opening question: Can you think of optimal strategies for the penalty taker and goalie?

Solution: This two-person zero-sum game can be represented as follows, where the payoffs are the probability of scoring a goal.

		Goalie		
		Breaks left	Remains Center	Breaks right
	Kicks Left	0.5	0.9	0.9
Kicker	Kicks Center	0.1	0	0.1
	Kicks Right	0.9	0.9	0.5

If we assume that decisions between the left or right side are made symmetrically (with equal probabilities), then this game can be represented by a two-by-two matrix, where $0.7 = (1/2)(0.5) + (1/2)(0.9)$:

| | | Goalie | |
		Remains Center	Breaks side
Kicker	Kicks center	0	1
	Kicks side	0.9	0.7

The optimal strategies of the kicker and goalie are the following:

a. More often kick center and remain center.
b. More often kick side and break side.
c. Choose center and side equally often.

CHAPTER 13

Mathematics and Alcohol

LCOHOL, A SEMI-LUXURY
beverage, is also a drug that acts
on the central nervous system. Although it had always been regarded
as either a bad habit or a sin, the realization that alcoholism is a
disease only came at the turn of the century. Today, alcoholism is one
of the world's greatest health problems. Apart from environmental,
sociocultural, and psychological influences, recent epidemiological
studies have provided clear evidence of a genetically conditioned
predisposition toward alcoholism. Moderate consumption of alcohol
decreases the risk for heart disease, ischemic stroke, and diabetes.
But its excessive use or misuse creates many problems and takes an
extremely heavy toll on families and society. Despite all the studies,
it is extremely difficult to draw the line between genetic influences
on the one hand, and social, cultural, and environmental factors such
as religious upbringing and the price and availability in the other.

In recent years, there has been a considerable amount of scientific
research on the effects of alcohol on memory, reflexes, coordination,

and depth perception as well as a host of other cognitive and psychomotor processes. Formulation of such research questions requires quantification of the amount of alcohol in the blood. This has led to the introduction of blood alcohol levels as a percentage of alcohol per volume of blood.

A blood alcohol level (BAC) of 0.10 is defined as one gram per kilogram of blood, meaning that alcohol is 0.10 percent, or one one-thousandth, of the blood. For example, a person weighing 150 pounds with thirty-four grams of alcohol in the body, the amount of alcohol from two and a half beers would have a BAC equal to 0.08. The same person with 210 grams of alcohol in the body would have a 0.50 BAC.

Quantification of blood alcohol level helps investigation of many other questions and opens the door for introduction of other concepts found in elementary mathematics and related subjects. For example, it has been demonstrated that the probability of being involved in an alcohol-related accident increases as the BAC increases. Estimates show that an individual with a BAC between 0.10 and 0.14 is forty-eight times more likely to be involved in a motor vehicle accident than an individual who has not been drinking. This can be used in discussion of probability and its definitions.

The BAC of an individual is determined by three primary factors: body weight, amount of alcohol consumed, and amount of elapsed time from first drink until a breath and/or blood sample is taken. As an example, a 180-pound man who consumes seven drinks over a three-hour period will have a BAC of around 0.10. In contrast, a 110-pound man who consumes the same seven drinks over three hours will have a BAC of around 0.20. One drink, which is defined as one ounce of liquor, five ounces of wine, or a twelve-ounce beer, consists of approximately fourteen grams of alcohol. Note that all these information and data can be used for developing lesson plans. The following is a specific example of alcohol-related information useful for teaching functions, an important concept in elementary mathematics.

After a drink, the alcohol begins to enter the bloodstream

almost immediately, and the blood-alcohol level rises rapidly. Once the person stops drinking, her natural metabolic processes slowly eliminate the alcohol, and blood-alcohol level begins to fall. So, how long must one wait after drinking for safe driving?

Experiments show that after drinking six beers, four glasses of wine, or four shots of liquor, the blood-alcohol level typically rises to 0.11 gram per deciliter of blood. Thereafter, alcohol is eliminated at the rate of 0.02 gram per hour. The graph below shows the resulting blood alcohol level over time, first rising rapidly and then falling linearly. By calculating where the graph falls below the legal limit, we can find how long one must wait. This relationship can be described by an important mathematical tool known as function. Functions have different types and can be classified according to their properties. One important classification is linear versus nonlinear. To introduce these, one can refer to the fact that many substances are eliminated from the bloodstream nonlinearly, but alcohol is eliminated linearly.

Alcohol and Mathematics

After a drink, the alcohol begins to enter the bloodstream almost immediately, and the blood-alcohol level rises rapidly. Once the person stops drinking, the body's natural metabolic processes slowly eliminate the alcohol, and blood-alcohol level begins to fall. Knowing this, we may like to know how long one must wait after drinking for safe activities such as driving.

Experiments show that after drinking six beers, four glasses of wine, or four shots of liquor, the blood alcohol level typically rises to 0.11 gram per deciliter of blood. Thereafter, alcohol is eliminated at the rate of 0.02 gram per hour. This means that over time the blood-alcohol level first rises rapidly and then falls linearly. This information can be used to introduce concepts such as

1. functions and their graphs;
2. linear functions;

3. increasing and decreasing functions; and

4. maximums and minimums (derivatives)

If used as a lesson, as a class activity, the instructor can ask students to find the length of time one needs to wait before deriving. For this, one needs to examine the graph of the function describing the blood-alcohol level to find out where it falls below the legal limit. Let us continue the discussion as a lesson.

To introduce students to multivariate functions, the instructor can pose the questions related to the effects of the amount of alcohol consumed, person's weight, etc.

Alcohol and Statistics

Many popular newspapers publish statistics regarding alcohol, especially in relation to college students. For example, in March 2002, *USA Today* published an article titled "College Drinking Kills 1,400 a Year, Study Finds." The article was based on a research by Ralph Hingson of the Boston University.

In addition to the death toll, the report included statistics such as 500,000 college students were injured while under the influence of alcohol; 600,000 were assaulted; 70,000 were the victims of sexual assault; 400,000 had unsafe sex; 25 percent had academic problems; and 150,000 have alcohol-related health problems or tried to commit suicide. The questions regarding the way these statistics were obtained can be used for teaching basic concepts of statistics such as sampling, estimation, etc., together with their use and abuse.

The use of statistics based on such research or investigations is clear. To see how they may be used to reveal the abuse of statistics, consider the following example. Hingson has obtained 1,400 deaths using the following approach. There are about 25.5 million eighteen- to twenty-four-year-olds living in the United States, according to U.S. Census data. Thirty-one percent of this age group are enrolled as full- or part-time students in two-or four-year colleges. The number of alcohol-related motor vehicle crash deaths among eighteen-to-twenty-four-year-olds during 1998 is 3,674; 31 percent of this figure is 1,138. Similarly, applying the 31 percent factor to the 991 alcohol-related, nontraffic deaths among eighteen-to-twenty-four-year-olds in 1998 results in an additional 307 deaths. Adding the 307 and 1,138 figures equals the alleged 1,445 alcohol-related deaths annually among college students.

When estimating this number, Hingson relies on a key but unsupported assumption. It does not automatically follow that college students constitute 31 percent of alcohol deaths simply because 31 percent of eighteen-to-twenty-four-year-olds are college students. This reasoning is equivalent to assuming that because women constitute about half the population, they commit half of all crimes. In fact, men commit more than 75 percent of crime. This type of examples can be used to develop lessons regarding misuses of statistics.

The definition of what constitutes an "alcohol-related" death is another problem useful for teaching concepts such as hypothesis testing. In short, there are plenty of opportunities that we hope to explore and use as teaching tool.

Hypotheses Testing Using Alcohol Statistics

Many people think that

1. the moderate use of alcohol (few drinks) does not impair person's driving abilities such as vision, reflexes, and hand-eye coordination;
2. body weight helps person's level of tolerance and sobriety; and
3. females have a lower tolerance for alcohol than males.

This information can be used to introduce students to hypothesis testing.

A) Hypothesis Testing

Hypothesis 1: Alcohol consumption makes people impaired regardless of their claims. Moreover, the more a person drinks, the more impaired they become.

Hypothesis 2: As a person's weight increases, the effect of a certain amount of alcohol decreases—that is, the more someone weighs, the less effect one drink will have on them.

Hypothesis 3: A person's sex has no effect on their ability to handle alcohol without becoming impaired.

Testing these hypotheses requires data. To collect relevant data, we need to design and implement an experiment. This can be used to introduce students to design and analysis of experiments. The following is an example of what the instructor can present.

B) Design of Experiment

Step 1: Select a random sample of individuals who are willing to participate in this experiment. It is better to have the same number of males and females (balanced design) and people with different weights.

Step 2: Test individuals in your sample using, for example, a driving simulation game such as Sega's Outrun and record their scores.

Step 3: Test them again after they consumed two beers.

Step 4: Test them once more after they consumed enough beer to raise their blood-alcohol level (content) to 0.06 percent, one percentage point above assumed legal level to declare a person impaired.

Note: It takes typical college students approximately four beers to raise their blood alcohol content to 0.06 percent.

As a class activity, the instructor can ask students to discuss the possible problems and biases of this experiment. For example, some subjects who are "video game experts" may score better on video games than others. Some might have played Sega's Outrun before, while others have never seen one. To account for this, one could test subjects before and after drinking and use the difference between the scores. This can be used to teach concepts such as dependency and the paired experiment. Next, the instructor can pose questions such as individual alcohol tolerances and its dependency on previous exposure to alcohol, weight, and gender.

Notes: We actually performed the experiment described above using a group of friends during a three-month period and drew the following conclusions. I used this to introduce t-test X^2 test, and regression.

Hypothesis 1: The single sample t-tests provided sufficient evidence to conclude that two beers and four beers have a significant effect on an individual's ability to drive. It can also be noted that the test of the effect of four beers is more significant than the effect of only two beers. This indicated that the more alcohol a person consumes, the more impaired the person becomes. I also used regression analysis to determine the effect, if any, of alcohol on an individual.

Hypothesis 2: Unfortunately, the sample size of the data was not large enough to yield any significant X^2 tests. However, the X^2 value for the effect of four beers is significantly larger than the X^2 value for the effect of two beers. This indicates that as the number of beers consumed increases, the weight of the individual consuming them becomes more significant.

Hypothesis 3: There were two tests performed to analyze the data for this hypothesis. Again, the X^2 tests were insignificant because of the small sample size. Although the paired t-tests were also insignificant, they supported our claims that gender has no effect on a person's ability to handle alcohol without being impaired. Note that the participants were divided by gender. This same division could have been achieved if the individuals were divided by weight, since the women were the lightest and the men were the heaviest. To better this experiment, a larger group of individuals that were approximately the same weight should have been taken.

Alcohol, a Teaching Tool

Teaching introductory mathematics and statistics to non-majors has always been a challenge to instructors. Here, the challenge lies on how to tie the material to students' everyday concerns or interests so that the subject matter becomes relevant and useful. Some recent research shows that the applied quantitative methods taught within a context familiar or of interest to students increase their desire to learn and their tolerance to deal with complex concepts. With these in mind, I investigated an idea for motivating students in introductory mathematics and statistics classes. The motivating theme was the quantitative methods involved in studying and analyzing alcohol-related data and information. The goal was to develop lesson plans for teaching several classical topics taught in introductory mathematics and statistics courses.

Lessons

Drinking is considered a part of the college life by many students. As such, data and information related to alcohol use and abuse may be utilized as a theme in introductory mathematics and statistics classes. For example, students often wonder how much one can drink to stay below the limit, or after consumption of a certain amount of the alcohol how long one should wait before driving. Some wonder about effect of factors such as weight, gender, race, etc., in alcohol tolerance. The answers to these questions involve quantitative methods that can be used both for teaching and learning. The following presents examples of data and information useful for developing lessons.

Methodology

The method we plan to adopt may be explained by a specific example. Alcohol is absorbed into the body primarily through the lining of the stomach. For this reason, a BAC peak is usually reached within twenty minutes of the last drink. In general, chemicals in the blood stream are either eliminated by the kidneys or being broken down by enzymes from the liver. Kidneys tend to eliminate a certain proportion of chemicals during each period. If the person has x gram of alcohol in the body, ax gram of it will be eliminated over the next hour, where a is a constant. Thus, if x is the amount of alcohol in the body at the beginning of an hour and y is the amount of alcohol eliminated during the next hour, the following relation holds,

$$y = ax.$$

This is the equation of a line that passes through the origin. Since the percent of alcohol removed from body per hour can vary greatly from person to person, this can be used to teach concepts such as slope and lines with different slopes.

The liver eliminates chemicals by breaking them down with enzymes. However, the liver may not break down a constant proportion of it each hour. Instead, the percent of the chemical being broken down can depend on the amount of it in the body, which is the case for alcohol. In fact, as the amount of alcohol in the body increases, the proportion of the alcohol the body eliminates decreases. For alcohol, the proportion z of the alcohol broken down in a given hour can be approximate using the following relation:

$$z = 10/(4.2 + x)$$

where x is the number of grams of alcohol in the body at the beginning of the hour. This is an example of what is called capacity-limited metabolism in which the amount of the chemical metabolized depends on the amount of the chemical in the body. This is an equation of specific function and can be used to teach concepts such as graphing a function, derivatives (decrease per unit of x). Note that this formula is an average. For any individual person, the numbers 10 and 4.2 may vary considerably.

The examples presented are a sample of what we plan to use for developing lessons of basic algebra. Activities of the kind presented below will be included in the lessons.

1. Find the proportion of alcohol eliminated from the body during the next hour if there are fourteen, twenty-eight, and forty-two grams of alcohol in the body at the beginning of the hour. (One drink consists of approximately fourteen grams of alcohol.)

2. Let x represent the amount of alcohol in the body at the beginning of an hour. Let y represent the amount of alcohol eliminated from the body during that hour. To find the amount of alcohol eliminated during that hour, we may multiply the proportion eliminated by the amount in the body at the beginning of the hour, giving the formula

$$y = 10x/(4.2 + x).$$

Find how much alcohol is eliminated during an hour if the body began with fourteen, twenty-eight, and forty-two grams. Compare your results to the results of problem 1. This demonstrates that as the amount of alcohol, x, in the body goes up (from fourteen to twenty-eight to forty-two grams), the proportion, z, of alcohol eliminated decreases, but the actual amount of alcohol eliminated, y, increases. This is a good example of increasing and decreasing functions, another classification used in algebra.

In summary, the process of the elimination of alcohol from the body serves as an introduction to a study of functions in general and rational functions in particular at an intermediate algebra level. The focus on lesson can be the graphs of the functions with emphasis on interpretation of the horizontal and vertical asymptotes in the context of elimination of alcohol from the body. Another lesson can focus on algebraic manipulation of the rational functions, solution of equations with rational expressions, realistic domain of a function, inverse functions, all based on formulations discussed earlier.

Questions regarding weight and gender can serve as an introduction to statistical test of hypothesis and also regression and correlation analysis. These possibilities will also be investigated.

Drinking, the Fourth Leading Cause of Preventable Death in the United States

Drinking has been an accepted part of Western culture for generations. It is associated with everything from weddings to wakes. Studies show that moderate consumption of alcohol decreases the risk for heart disease, ischemic stroke, and diabetes. However, its excessive use or misuse creates many problems and takes an extremely heavy toll on families and society.

According to the World Health Organization, alcohol misuse causes 3.3 million deaths annually and contributes to more than two hundred diseases and injuries. Globally, it is the fifth leading risk factor for premature death and disability among age group fifteen to forty-nine and the first risk factor for the age group twenty to thirty-nine.

In the United States, alcohol misuse cost approximately $250 billion annually. It is responsible for approximately eighty-eight thousand deaths (sixty-two thousand men and twenty-six thousand women) per year, making it the fourth leading preventable cause of death.

In 2013, of the 72,559 liver disease deaths in age group twelve and older, 45.8 percent were alcohol related. It was also the primary cause of almost one in three liver transplants. Drinking alcohol increases the risk of cancers of the mouth, esophagus, pharynx, larynx, liver, and breast. Moreover,

- alcohol is the number-one drug problem in America;
- there are more than twelve million alcoholics in the United States;
- in the United States, eighteen thousand people are killed in an alcohol-related car accidents every year;
- in year 2000, nearly seven million in age group twelve to twenty were binge drinkers. Binge drinking is defined as having five or more alcoholic drinks on one occasion for men, or four or more for women.

- three-fourths of all high school seniors report being drunk at least once;
- adolescents who begin drinking before the age of fifteen are four times more likely to become alcoholics than their counterparts who do not begin drinking until the age of twenty-one;
- people with a higher education are more likely to drink;
- higher income people are more likely to drink;
- excessive drinking is responsible for more than 4,300 deaths among underage youth each year;
- people aged twelve to twenty drink 11 percent of all alcohol consumed.

The statistics on alcoholics listed above do not include victims who may not even drink at all. Alcohol is responsible for 73 percent of all felonies, 73 percent of child beatings, 41 percent of rapes, 81 percent of wife battering, 72 percent of stabbings, and 83 percent of homicides. Finally, while there are twelve million alcoholics, an estimated forty million to fifty million people, such as family members, suffer the consequences of alcoholism.

Everything Has a Price

The excessive use or abuse of alcohol creates many problems and takes an extremely heavy toll on families and society. It is an example of the price societies pay for what is referred to as "freedom" of choice. The main drawback of such freedom is the price others have to pay for the choices we make. Alcoholism is one of the world's greatest health problems. Apart from environmental, sociocultural, and psychological influences, recent epidemiological studies have provided clear evidence of a genetically conditioned predisposition toward alcoholism.

Information to Develop Lesson on Functions

After a drink, the alcohol begins to enter the bloodstream almost immediately, and the blood-alcohol level rises rapidly. Once the person stops drinking, the body's natural metabolic processes slowly eliminate the alcohol, and blood-alcohol level begins to fall. So, how long one must wait after drinking for safe driving?

Experiments show that after drinking six beers, four glasses of wine, or four shots of liquor, the blood alcohol level typically rises to 0.11 gram per deciliter of blood. Thereafter alcohol is eliminated at the rate of 0.02 gram per hour. This means that over time the blood-alcohol level first rises rapidly and then falls linearly. This information can be used to introduce students to concepts such as

1. functions and their graphs;
2. linear functions;
3. increasing and decreasing functions; and
4. maximums and minimums (derivatives).

As a class activity, the instructor can ask students to find the length of time one needs to wait before driving. This requires examining the graph of the function describing the blood-alcohol level to find out where it falls below the legal limit.

To introduce students to multivariate functions, the instructor can pose the questions related to the effects of the amount of alcohol consumed, person's weight, etc.

Genetic Protection from Alcoholism

There are far fewer alcoholics in Japan, China, and Korea than there in the European countries. The reason for this is their marked inability to tolerate alcohol. Scientists from the Institute of Human Genetics, University of Hamburg, have discovered that this manifestation is of genetic origin because of an enzyme defect.

Alcohol (ethanol or ethyl alcohol), although a semi-luxury beverage, is also a drug that acts on the central nervous system to affect the essential psychomotor functions and also has a long history in every cultural circle. Intemperate indulgence had always been regarded either as a bad habit or a sin, but the realization that alcoholism is a disease only came at the turn of the century. Today, alcoholism is one of the world's greatest health problems. Apart from environmental, sociocultural, and psychological influences, recent epidemiological studies have provided clear evidence of a genetically conditioned predisposition toward alcoholism. However, it is extremely difficult to draw the line between genetic influences on the one hand, and social, cultural, and environmental factors such as religious upbringing, the price and availability of alcohol, and family drinking habits, for example, on the other. The dramatic rise in alcoholism and in its associated diseases in the postwar period, especially in Europe, is clearly attributable to the increased per capita consumption. This goes hand in hand with alcohol abuse, physical and mental dependence, and social and family problems, although severe addiction has only been observed in a small part of the population and to varying degrees of intensity in the different cultural circles.

The question arises as to whether alcoholics might differ in their personalities from the majority of the population even before the onset of the disease, whether we can talk about a behavior abnormally, or whether their biogenetic structure might be the underlying cause of their abnormal drinking habits (i.e., the development of their addiction).

There can be no doubt that alcohol metabolism is subject to a number of different influences that trigger individual variations in behavior patterns or toxic reactions after alcohol has been consumed. Studies conducted on families, twins, and adoptive relationships have revealed that genetic influence plays a major role in alcoholism. Twenty years ago, the proportion of women among the 2 or 3 percent alcoholics in the population of Germany (a total of some one million to two million people) was only 10 percent. Today, they represent some 30 percent, a rise that has brought a concomitant increase in the incidence of the so-called fetal alcohol syndrome. This has now become the commonest deformity of uniform genesis observed among newborn babies.

In the course of their studies on the problem of alcoholism, scientists in Hamburg have discovered an enzyme defect to be the underlying genetic cause of alcohol intolerance. It had long been known that a majority of the Koreans were unable to tolerate alcohol compared to those in the European countries—although the reason for this is to be sought neither in state decrees nor in strict morality. In other words, many people of Mongoloid origin who have been imbibing only small quantities of alcohol, flushing symptoms are observed accompanied by reddening of the face, accelerated pulse rate, and abdominal discomfort.

CHAPTER 14

Mathematics and Diversity

THE UNITED STATES is an amazing country. Socially, there is a wide range of people with different physical appearances, skin colors, nationalities, and ethnicities with all views from far left to far right. The society is open to many liberal ideas and behaviors and, at the same time, conservative and close to many others. It seems that in some sense, the United States is somewhere between a developed and a developing country.

Despite all the attempts to provide equal opportunities, the United States is a country with many social problems. In my view, there are several noticeable reasons for this. First, there is a great deal of unreal perceptions of life. Media is mostly responsible for furnishing and promoting fantasy life. Just few hours on the front of TV would convince one that this is really the case. Media spreads the idea that everybody is or should be fit, look good, be complete, be happy, and be entitled to have fun. Constantly watching such programs creates unreal expectations especially in young people.

Thus, facing reality often becomes a very disappointing affair leading to feel like a loser, and as such feel angry, negative, and disappointed. Such people naturally become careless who often turn against society and its rules. Even part of conflict between different groups has root in their personal disappointments.

Here is an example. I have taught in the university level for almost fifty years. In the United States, a large percentage of young people look up to actors, singers, and athletes and try to follow what they see in Hollywood movies. Every boy wants to date cheerleaders (a fantasy), and every girl wants to be with the most good-looking and popular boy. Very few, of course, get what they want. Those not good looking and having the skills or charm become disappointed, depressed, and sick in a variety of different ways. They act strange in a variety of the ways. In my view, one major cause of crimes and unrest relates to males' need and fascination with sex and lack of legalized prostitution for those who cannot have it. For people who do not have time or skill, prostitution provides a form of release. It decreases the number violent crimes against women especially in open societies such as the United States. It is puzzling that in a country classified as the leader of capitalist system, one can buy everything with money but sex. The effect is multiplied, noting that this is a country that mixes everything with sex, even sports. In fact, one may ask why a team needs cheerleaders, and if so, why should they be young and sexy women? What is the effect of such a thing on guys who are there to watch "sport"? Is it to change their focus and make them fantasize about these women while away from their not-so-fit wives? Is this a healthy activity? It is interesting that even Middle Eastern countries have recognized this and have tried to reduce the tension by asking women to cover themselves.

Of course, other sources of unhappiness include economy, discrimination, etc. These factors are studied extensively, and there is no need to discuss it further here.

Finally, I believe that the U.S. empire, like any other, will eventually collapse not because of typical historic causes but because of internal social problems.

Here is another point worth mentioning. One of the conflicts that bothers outsiders like me is the blame most men get for relationship with women. For example, consider Tiger Woods's story. Clearly, he should be blamed for what he has been doing. But what about all those women? They all knew that he was married. Shouldn't they share the blame? I always wonder, how is it that in a country where one can drive at the age of sixteen, one is considered a child when it comes to sexual activities? Yet statistics related to the number of underage teenagers sexually active are shocking.

You may find my views surprising especially if you have never lived outside America. In my view, there are two types of logic in the world, global and local. Killing is considered bad everywhere, and that is an example of global logic. Drinking before the age of twenty-one in the United States is bad and is an example of local logic or cultural logic. Driving at the age of seventeen is bad in most countries based on their cultural logic; here, it is OK.

So who should be blamed for inappropriate, let us say, sexual relationship? In most countries, women, because they are considered more mature than men of same age. Additionally, they are the ones who may become pregnant. I personally find it illogical for a woman to lead a man and then claim that she said no in the last minute.

The United States is an open society in the sense that people can openly meet each other in any setting with almost no restriction. This is not the case in many countries especially when meeting is between individuals of opposite sex. This obviously has many advantages in the sense that people can exchange their ideas, work together, collaborate, etc. However, it creates many other types of social problems regarding the relationships. As we know, the rate of divorce is very high in the United States even though people meet and even live with each other before getting married. One reason is this openness. After all, we are human beings with many weaknesses and temptations. Additionally, people usually meet each other in a pleasant setting, are often nice to each other, and even try to help and please each other. Even though the chance of being attracted to a colleague, coworker, secretary, etc., may be very small, when repeated, it may add up to something substantial. According to some reports, many companies furnish an environment that will help romance between employees, noting that this will increase the productivity. This in turn leads to relationships outside marriage and in some cases divorce.

Here is another problem I see with relationships and marriage. In the United States, most married couples are of the same age. In mid-ages, women go through menopause and become disinterested in sex; whereas, men are still active and in need. As a result, many affairs outside marriage take place. In many countries, they make sure that there is an age difference. After all, women get mature faster than men.

Varity versus Variability

If you search for the difference between the meanings of variation and variety, you will find explanations such as this: variation suggests deviation or change from something we recognize as the norm, and variety suggests a range of things within a particular group. You will also find examples to clarify such as this: Imagine that I grow roses. In my garden, I have twenty different types of roses—a wide *variety* of roses. I decide to breed a rose, which has the same structure as one particular rose in my collection, but with no thorns, because I keep pricking my fingers when I am gardening. The new rose is a *variation* of the original. When I have successfully bred it, it will add to the variety of rose types in my garden—now I will have twenty-one different types. The range is the variety. The change is the variation.

Now, to me, as a math person, this is a little confusing. After all, mathematics is a symbolic language to express the complex concepts in the shortest possible format. Here is my simple example to see the difference. Suppose that we have two classes each with ten students. I give them a true-false test with ten questions. In one class, four students score three and six students eight. In the other class, four students score five and six students seven. Looking at the grades in both classes, the variety is two, only two types of grades; whereas, the deviations from the expectation (in this case, five) measured by standard deviation are very different.

Developing a Measure for Diversity

In recent years, diversity has become an issue of great concern and importance, and many research, investigations, discussions, and publications have been devoted to it. This has been triggered by realization that most new jobs in the economy will require a postsecondary education, and women and racial/ethnic minorities will compose a majority of the workforce. As a result of this awareness, a transformation has taken place that links diversity in the student body with the development of new teaching and learning practices. Moreover, diversity in the student enrollments has led to the development of new academic support programs and the revision of education policies and curricula to reflect the diversity of human experience perspectives.

Despite these developments, researchers are still seeking to find satisfactory answers to some basic questions regarding diversity. One such a question relates to quantification of diversity of institutions such as universities. Clearly, to address issues regarding diversity, one needs to measure the diversity. This is important, noting, for example, that diversity is one of the items used for budget allocation in institutions such as the Pennsylvania State System of Higher Education, SSHE. Here are some examples.

(1) Two universities have two thousand students each. The first university consists of one thousand white, five hundred black, and five hundred Hispanic. The second university consists of eight hundred white, eight hundred black, and four hundred Hispanic. What are the diversities of each university, and which university is more diverse?

(2) A university is part of a system. The gender composition of the student body of the system as a whole is 40 percent female and

60 percent male; 30 percent and 70 percent are the corresponding proportions of this university. To what extent the gender composition of this university differs from that of the system as a whole? If another university in this system has 50 percent female and 50 percent male, which university has a gender composition closer to that for the system? If this system sets a goal regarding the gender composition, how do we decide or measure which university is closer to the defined goal?

To quantify diversity, we need to find a measure for variety. To see how we may do this, recall that asking a yes-or-no question admits an initial uncertainty about what the correct answer might be. The answer to such a question informs the questioner of the right option, thus removing the initial uncertainty.

Next, consider a university that is made up of two equally represented types such as male, female, or black and white. This can be represented by a fair coin. Now, when flipping such a coin, we are uncertain about the outcome. Here, there are only two possible outcomes, and each has a 50 percent probability of occurrence. If we guess the outcome (head or tail), then we are 50 percent uncertain about our guess. If we receive a message that the outcome was head, then the 50 percent uncertainty is removed. This is equivalent to one *bit* of information. So in this sense, information is the same as uncertainty. If we flip the coin twice (HH, HT, TH, TT), then whatever is our guess, we are 75 percent uncertain about our guess, since there are four possible outcomes (like a university that's made up of four types). To remove the uncertainty, we need *two bits* of information, one bit for each coin.

The answer to a yes-or-no question, thus, is taken to convey one *bit* of information, which constitutes a basic unit of measurement. This is, in fact, equivalent to counting options or logical alternatives by the exponents of two rather than by their number.

For one coin, the number of options is 2^1. For two coins, we have $4 = 2^2$. The amount of *information* a message, say, a, of the set of possible messages A conveys then becomes the *difference between two states of uncertainty:* the uncertainty U(A) before or without knowledge of that message, and the uncertainty U(a) after or with knowledge of that message. In this algebraically equivalent form, information is seen as a measure of *difficulty of making appropriate* (to a degree better than chance) *decisions*, and because a less expected message is more informative, information can also be interpreted as a measure of the *surprise value* of a message. So if a decision maker must pick one of the eight alternative courses of action and is given a report that shows that six of them lead to certain failure, there remains $8 - 6 = 2$ options to choose from, making the report worth two bits of information (difference between the exponents of 2^3 and 2^1). This is now equivalent to receiving the answers to yes-or-no questions. To remove the remaining uncertainty, the decision maker will have to gather one more bit of information or risk a 50 percent chance of failure. The risk is, of course, considerably less than the risk that existed before receiving the report (87.5 percent).

Entropy may be defined as a measure of *observational variety* or of actual (as opposed to logically possible) diversity. Unlike the measure of selective information, entropy takes into account that messages or categories of events may occur with unequal frequencies or probabilities. Entropy can also be defined as a measure of "uncertainty" of a random phenomenon. Suppose that some information about a random phenomenon is received. Then a quantity of uncertainty can be regarded as the quantity of transmitted information.

To see how these ideas may be related to diversity, consider two populations that are made up of two types. Suppose that one population has 80 percent of type 1 and 20 percent of type 2, and the other population has 50 percent of each type. Decision or guessing the type of a randomly selected member involves less uncertainty for 80–20 case than 50–50 case. In fact, guessing that a randomly selected member is of type 1 involves 20 percent uncertainty for 80–20 case and 50 percent uncertainty for 50–50 case.

Applications

Like many other institutions, colleges and universities have adopted a proactive commitment to diversity as their central mission. As a result, a transformation has taken place that links diversity in the student body with the development of new teaching and learning practices and new academic support programs.

To address questions regarding diversity, it is necessary to quantify it first. This section proposes investigation of a measure for diversity based on basic ideas of the information theory and entropy. Its goal is to develop a logical and scientifically based measure that can be used to measure and compare communities and universities for their diversities. I think that compared to the measures such as the percentage of certain minority groups, a measure based on the notion of variety is more appropriate and meaningful.

Diversity

The measure I investigated involves ideas from information theory and entropy. Information is a measure of the *amount of selective work* a message enables its receiver to do. Accordingly, asking a yes-or-no question admits an initial uncertainty about what the correct answer might be. The answer to such a question informs the questioner in the sense of "selecting" one of the two options he or she has in mind, thus removing the initial uncertainty.

For example, consider a university that is made up of two equally represented types such as male, female, or black and white. This can be represented by a fair coin. Now, when flipping such a coin, we are uncertain about the outcome. Here, there are only two possible outcomes, and each has a 50 percent probability of occurrence. If we guess the outcome (head or tail), then we are 50 percent uncertain about our guess. If we receive a message that the outcome was head, then the 50 percent uncertainty is removed. This is equivalent to one *bit* of information. So in this sense, information is the same as uncertainty. If we flip the coin twice, then whatever is our guess, we are 75 percent uncertain about our guess, since there are four possible outcomes (like a university that is made up of four types). To remove the uncertainty, we need *two bits* of information, one bit for each coin.

The answer to a yes-or-no question, thus, is taken to convey one *bit* of information, which constitutes a basic unit of measurement. This is, in fact, equivalent to counting options or logical alternatives by the exponents of two rather than by their number.

Information theory combines mathematical elegance with practical utility and plays an essential role in modern communication and technology.

Genetic variability is a measure of the tendency of individual genotypes in a population to vary from one another also. Variability is different from genetic diversity, which is the amount of variation seen in a particular population.

Mathematics of Diversity

A measure of the information gained from a message is obtained by finding the minimum number of yes-and-no questions required to identify the message. The answer to a yes-or-no question is taken to convey one bit of information, which constitutes a basic unit of measurement. A convenient measure in a binary system with additive property is the logarithm to base 2.

We think that compared to measures such as the percentage of certain minority groups, a measure based on the notion of variety is more appropriate and meaningful.

The measure I propose involves ideas from information theory and entropy. Information is a measure of the *amount of selective work* a message enables its receiver to do. Accordingly, asking a yes-or-no question admits an initial uncertainty about what the correct answer might be. The answer to such a question informs the questioner in the sense of "selecting" one of the two options he or she has in mind, thus removing the initial uncertainty.

For example, consider a university that is made up of two equally represented types such as male, female, or black and white. This can be represented by a fair coin. Now, when flipping such a coin, we are uncertain about the outcome. Here, there are only two possible outcomes, and each has a 50 percent probability of occurrence. If we guess the outcome (head or tail), then we are 50 percent uncertain about our guess. If we receive a message that the outcome was head, then the 50 percent uncertainty is removed. This is equivalent to one bit of information. So in this sense, information is the same as uncertainty. If we flip the coin twice, then whatever is our guess, we are 75 percent uncertain about our guess, since there are four possible outcomes (like a university that is made up of four types). To remove the uncertainty, we need two bits of information, one bit for each com.

The answer to a yes-or-no question, thus, is taken to convey one *bit* of information, which constitutes a basic unit of measurement. This is, in fact, equivalent to counting options by the exponents of two rather than by their number. Mathematically,

$$U(A) = \log_2 N_A$$

where N_A is the number of logical alternatives. For one coin, $N_A = 2$ and $\log_2 2 = 1$. For two coins, $N_A = 4$ and $\log_2 2^2 = 2$. The amount of *information* a message, say, a, of the set of

$$I(a \varepsilon A) = \text{Log}_2 N_A = \text{Log}_2 N_a = \text{Log}(N_a/N_A) = \text{Log}_2 P_a$$

possible messages A conveys then becomes the *difference between two states of uncertainty*: the uncertainty U(A) before or without knowledge of that message, and the uncertainty U(a) after or with knowledge of that message—that is, where $Pa = Na/NA$ is the logical probability of the alternatives in a relative to A. In this algebraically equivalent form, information is seen as a measure of *difficulty of making appropriate* (to a degree better than chance) *decisions,* and because a less expected message is more informative, information can also be interpreted as a measure of the *surprise value* of a message. So if a decision maker must pick one of the $NA = 8$ alternative courses of action and is given a report that shows that six of them lead to certain failure, there remain $Na = 8 - 6 = 2$ options to choose from, making the report worth

$$l(\text{Report}) = U\text{before-Darter} = \log_2 8 - \log_2 2 = 2$$

bits of information. This is equivalent to receiving the answers to yes-or-no questions. To remove the remaining uncertainty, the decision maker will have to gather one more bit of information or risk a 50 percent chance of failure. The risk is, of course, considerably less than the risk that existed before receiving the report (87.5 percent).

Entropy may be defined as a measure of *observational variety* or of actual (as opposed to logically possible) diversity. Unlike the measure of selective information, entropy takes into account that messages or categories of events may occur with unequal frequencies or probabilities. Entropy can also be defined as a measure of "uncertainty" of a random phenomenon. Suppose that some information about a random phenomenon is received. Then a quantity of uncertainty can be regarded as the quantity of transmitted information.

To see how these ideas may be related to diversity, consider two populations that are made up of two types. Suppose that one population has 80 percent of type 1 and 20 percent of type 2, and the other population has 50 percent of each type. Decision or guessing the type of a randomly selected member involves less uncertainty for 80–20 case than 50–50 case. In fact, guessing that a randomly selected member is of type 1 involves 20 percent uncertainty for 80–20 case and 50 percent uncertainty for 50–50 case.

Quantification of Diversity

Like many other colleges and universities, the State System of Higher Education (SSHE) has adopted a proactive commitment to student diversity because their central mission has to be linked with the future of a diverse society. The system is aware that, by the year 2000, most new jobs in the economy will require a postsecondary education, and women and racial/ethnic minorities will compose a majority of the workforce. As a result of this awareness, a transformation has taken place that links diversity in the student body with the development of new teaching and learning practices. Moreover, diversity in student enrollments has led to the development of new academic support programs and the revision of educational policies and curricula to reflect the diversity of human experience and perspectives.

Changes in higher education practices and curricula began nearly thirty years ago, when institutions first opened their doors to groups that previously had been excluded from higher education. When the "experiment" began, many campuses were not prepared for the changes they would undergo as a result of including more adult students, women, and racial/ethnic minorities in their student bodies. These changes in student enrollments were connected with major intellectual and social movements that raised important questions about the production and transmission of knowledge, as well as access to education.

To address questions regarding diversity, it is necessary to define diversity and, more importantly, to quantify it. In this article, a measure for diversity is introduced based on basic ideas of information theory and entropy, which, compared to the frequently used measure—namely, percentage of nonwhite members—is more appropriate and reasonable. Because comprehension of the proposed index involves notions from information and entropy, elementary concepts of these subjects are presented. This is done in a form suitable for the objective of this article.

Entropy

Entropy may be defined as a measure of *observational variety* or of actual (as opposed to logically possible) diversity. Unlike the measure of selective information, entropy takes into account that messages or categories of events that may occur with unequal frequencies or probabilities. Entropy can also be defined as a measure of "uncertainty" of a random phenomenon. Suppose that some information about a random phenomenon is received. Then a quantity of uncertainty can be regarded as the quantity of transmitted information.

To see how these ideas may be related to diversity, consider two populations that are made up of two types. Suppose that one population has 80 percent of type 1 and 20 percent of type 2, and the other population has 50 percent of each type. Decision or guessing the type of a randomly selected member involves less uncertainty for 80–20 case than 50–50 case. In fact, guessing that a randomly selected member is of type 1 involves 20 percent uncertainty for 80–20 case and 50 percent uncertainty for 50–50 case. Formulating the idea, we get the following measure for diversity D:

$$D = p \log \frac{1}{p} + (1-p) \, \text{Log} \frac{1}{1-p}$$

where p is the percentages of type1 members. For the above example, the values of D are respectively 0.87 and 1.

Summary

Information gained from a massage is obtained by finding the minimum number of yes-and-no questions required to identify the message. The answer to a yes-or-no question is taken to convey one bit of information, which constitutes a basic unit of measurement. A convenient measure in a binary system with additive property is the logarithm to base 2. This is, in fact, equivalent to counting options by the exponents of two rather than by their number. Mathematically,

$$U(A) = \log_2 N_A$$

where N_A is the number of logical alternatives. For one coin, $N_A = 2$ and $\log_2 2 = 1$. For two coins, $N_A = 4$ and $\log_2 2^2 = 2$. The amount of *information* a message, say, a, of the set of

$$I(a \epsilon A) = \log_2 N_A - \log_2 N_a = \log(N_a/N_A) = \log_2 P_a$$

possible messages A conveys then becomes the *difference between two sates of uncertainty*: the uncertainty $U(A)$ before or without knowledge of that message and the uncertainty $U(a)$ after or with knowledge of that message—that is, where $Pa = Na/NA$ is the logical probability of the alternatives in a relative to A. In this algebraically equivalent form, information is seen as a measure of *difficulty of making appropriate* (to a degree better than chance) *decisions*, and because a less expected message is more informative, information can also be interpreted as a measure of the *surprise value* of a message.

CHAPTER 15

Climate and Floods

MANY CONSIDER THE development of mathematical models for weather prediction one of the great scientific triumphs of the twentieth century. Accurate weather forecasts are now available routinely, and quality has improved to the point where occasional forecast failures evoke surprise and strong reaction among users. The story of how this came about is of great intrinsic interest. General readers, having no specialized mathematical knowledge beyond school level, will warmly welcome an accessible description of how weather forecasting and climate prediction are done. There is huge interest in weather forecasting and in climate change, as well as a demand for a well-written account of these subjects. In a book by Ian Roulstone and John Norbury, the central ideas behind modeling and the basic procedures undertaken in simulating the atmosphere are conveyed without resorting to any difficult mathematics. It opens with an account of the circulation theorem derived by the Norwegian meteorologist Vilhelm Bjerknes. This theorem follows from work of

Helmholtz and Kelvin but makes allowance for a crucial property of the atmosphere—that pressure and density surfaces do not usually coincide. This is what is meant by the term "baroclinicity." *Invisible in the Storm: The Role of Mathematics in Understanding Weather,* reviewed by Peter Lynch, professor of meteorology in the School of Mathematical Sciences at University College Dublin.

In a chapter titled "When the Wind Blows the Wind," the authors attempt to convey the ideas of nonlinearity, a phenomenon that "makes forecasting so difficult and weather so interesting." This is a key idea, and I feel that the attempt can at best be described as a qualified success.

Part 2 opens with a chapter in which the brilliant work of Carl-Gustaf Rossby is described. Rossby had the capacity to reduce a problem to its essentials and to devise conceptual models that elucidated the mechanism of atmospheric phenomena, unencumbered by extraneous details. In a landmark paper published in 1939, he explained the basic dynamics of the large wavelike disturbances in the atmosphere by using a simple model based on conservation of absolute vorticity. Linearizing this, he produced an expression for the phase speed of the waves, thereby explaining the mechanism of propagation and also providing a means of predicting the propagation of wave disturbances.

Some mathematical details of Rossby's model are presented in a tech box. Rossby's model assumed a wave disturbance of a fluid with uniform depth. When the fluid depth varies, the conserved quantity is the ratio of absolute vorticity to depth, the potential vorticity (PV) in its simplest form. This can be used to explain the effect of a mountain chain on the flow. The authors describe a flow over the Andes but do not mention that in the Southern Hemisphere the configuration of troughs and ridges is reversed. Thus, their account and their figure 5.12 are likely to be a source of confusion to readers. Rossby's formula was of limited value in practical forecasting. The atmosphere is complex, and its behavior cannot be reduced to a simple traveling wave on a uniform background flow. A much more complete understanding of how midlatitude disturbances develop from small beginnings was provided by Jule Charney when he showed that they grow through baroclinic instability. Charney's work is rightly given prominence in the book. Having explained the mechanism of wave growth, he went on to produce a system of equations that could be used for practical numerical prediction while avoiding the problems encountered by Richardson. Charney then led the team that carried out the first successful prediction on the ENIAC computer in 1950. This was the beginning of real numerical weather prediction. The story is very well told in the book. The limitations on prediction imposed by the chaotic nature of the atmosphere are then discussed.

The work of Edward Lorenz was crucial to our understanding of what can and cannot be achieved. With our growing appreciation of the inherent limitations on weather forecasting, they show how the problem can be reduced to the solution of a system of seven equations (coupled nonlinear partial differential equations) in seven variables: pressure, temperature, density, humidity, and three components of the wind. The mathematical details are very sketchy, even in the "tech boxes" (boxes separate from the running text, with additional technical details), but the overall ideas are well conveyed.

Most of the key meteorologists are recognized in the narrative. In particular, William Ferrel's work in formulating the equations on a rotating Earth is given due prominence. But a description of the important work of the scientists working in Vienna, specifically Max Margules and Felix Exner, is omitted. Margules anticipated the problems that would arise if the continuity equation was used for prediction, and Exner carried out several numerical forecasts using a highly simplified set of equations. The method of solving a complicated system of equations by reducing them to a manageable algebraic form is given good treatment. The consequences of discretization are described by a nice analogy with pixilation of a painting, Constable's *The Hay Wain*. Bjerknes's original idea was to use mathematical equations to forecast the weather. However, the complexity of this task convinced him and his team to follow a more empirical line, which turned out to be enormously fruitful. The conceptual models of warm and cold fronts and of the life cycles of frontal depressions that emerged from the Bergen School dominated synoptic meteorology for most of the twentieth century and were of great practical benefit to humankind. More quantitative methods had to await scientific and technical developments in mid-century. During the First World War, an extraordinary numerical experiment was carried out by Lewis Fry Richardson, who, using the best data set he could find, calculated changes in pressure and wind using the basic equations of motion. However, he was unaware that errors in the initial data could completely spoil the forecast, and his results were completely unrealistic. Richardson's attempt at practical forecasting by numbers was so unsuccessful and so impractical at the time that it had a deterrent effect on other meteorologists. But, of course, Richardson's approach was ultimately the right one, and the causes of the error in his forecast are now well understood and quite avoidable. The authors provide a clear description of what Richardson achieved and of the remarkable prescience of his work.

Recently, the emphasis has shifted from deterministic to probabilistic prediction, and the method of ensemble forecasting is now at the forefront of operational practice. All this is well described, including the application of probability forecasts to loss/cost models that can be used for rational decision-making with great economic benefits.

The authors have an interest in symplectic geometry, the mathematical framework underlying Hamiltonian mechanics. They include a description of the main ideas of symplectic geometry, but this is as likely to mystify as to inspire readers, especially as the link with PV is not clarified. More practical is the account of Lagrangian advection schemes, which have led to substantial increases in numerical efficiency of forecasting models. The components of climate models are also described. In general, more schematic diagrams showing, for example, the components of an earth system model and the principal physical processes parameterized in models would have been welcome.

Recognizing that many readers are strongly discouraged by the appearance of even a few mathematical equations, the authors have endeavored to elucidate the key ideas of modern weather prediction without explicit mathematical material. This is quite a difficult task. The attempt has been reasonably successful, and readers without advanced scientific knowledge but with an interest in scientific matters should get an accurate, if incomplete, impression of how modern weather forecasts are made. Readers with more extensive mathematical knowledge may be frustrated by the absence of fuller quantitative detail.

Global Warming

Global warming is the rise in average temperature of Earth's atmosphere and oceans. It is a matter that should be taken extremely serious by all of Earth's inhabitants. Since 1980, Earth's average temperature is said to have increased by approximately 0.5 degrees Celsius. Scientists are also predicting the temperature to continue increasing by another 2.4 to 6.4 degrees Celsius by the next century. In short, this may not mean a lot; but if rates continue the way they are now, the future of the human race could be in loads of trouble.

Scientists believe there are a number of reasons that could be causing this slow increase of Earth's temperature. With 90 percent certainty, scientists think the main causes are the increase of greenhouse gases and burning of fossil fuels. It can also be observed by watching the size of the ice around the North Pole. Over the past ten years, the ice has been slowly shrinking with it being at its smallest distance ever recorded.

There is a lot of controversy over whether the global warming is actually happening, or if it even matters. With the extremely high amounts of CO_2 in the atmosphere and the constant deforestation happening, many believe that nothing is going to get any better.

Statement of the Problem

In recent years, much discussion is devoted to global warming and the related issues such as the contribution of greenhouse gas emissions from human activities. Because of the complexity of the problem and uncertainty involved, some studies have reported conflicting results leading to more questions than answers. A primary source of the problem, in my view, relates to how data were compiled and, more importantly, how the statistical analysis has been carried out. For example, an analysis of the minimum/maximum temperature data in the Antarctic Peninsula shows a significant increase in the maximum temperatures but a very little change in the minimum temperatures. So analyzing minimum/maximum temperatures could lead to a different conclusion. As a result, studies that included ozone levels into the model might find it to be a significant/nonsignificant factor.

This project plans to examine data collection methods and statistical procedures applied in global warming studies. I anticipate facing questions and problems that have not been addressed before. Through this investigation, I hope to show that this and interpretation of results are partly responsible for differences one finds in publisher reports. I am planning to utilize an approach different from those one finds in the literature. If successful, the outcome could partially explain why despite many signs of global warming, global dimming, and changes to the climate, there are still some conflicting results and views about what is taking place. The findings could also serve as educational materials for teachers and instructors.

Significance of the Problem

Statistics is used in most research and investigations. Statistical methods are intended to aid the interpretation of data that are subject to appreciable haphazard variable. The methods are eclectic, and consequently it is often difficult to decide which of several ways of analyzing data is most appropriate. In fact, there is more to the correct use of statistics than knowledge of classical statistical techniques and use of statistical software. Investigations of published research in some critical disciplines have unveiled many unintentional misuses of statistical methods as well as incorrect application or interpretation of the results. Clearly, decisions based on incorrect statistical analysis could lead to serious consequences. For example, methods applied to a complete data set are not directly applicable to data sets with missing observations. This is the case for global warming problem, since it is not possible to make regular measurements during the Antarctic winter when the station is in darkness. Additionally, the available data is from only one station (Antarctic). Other factors making the inference difficult are the following:

- It is known that Antarctic air is warming faster than the rest of the world.
- The Antarctic Peninsula has experienced major warming over the last fifty years.
- It is believed that the increase in mean surface temperature at this station is mainly due to the increases in minimum temperatures.
- For the period of interest, 1951–2004, the minimum/ maximum monthly temperatures are separately available only at this station.

Considering these, I think it is beneficial to examine statistical methods used to study global warming phenomenon, especially the part that involves factors from both space and anthropogenic activities. For example, consider the sunspot number that affects global temperature. It is known that sunspot number is periodic with a period of eleven to twelve years. As a result, the analysis of different segments of data could lead to different conclusions. In fact, the periodicity of sunspot numbers may have something to do with the fact that eleven of the twelve warmest years on record have occurred in the past fourteen years.

Method

I investigated the data collection procedures and statistical methods applied to global warming and related issues. Some of the methods utilized were complex, as present climate change is caused by a collective action of several space factors, volcano activities, as well as anthropogenic factors with their own cooling and warming. Some experts believe that it is the relation between these contributions that will determine the final outcome. With this in mind, I investigated the statistical methods used both to uncover and formulate such relationships as well as issues related to the missing observations. The latter arises frequently in application. To give an example, the stratospheric ozone concentrations are recorded using a Dobson ozone spectrophotometer. This instrument tells us how much ozone there is in the atmosphere by comparing the intensities of two wavelengths of ultraviolet light from the Sun. Since it is not possible to make regular measurements of ozone during the Antarctic winter when the station is in darkness, the missing values are usually substituted by their yearly average. I plan to study the effects of this substitution.

Finally, I investigated to see whether a phenomenon such as Simpson's paradox is present here. This phenomenon happens frequently when dealing with multidimensional data. To demonstrate how this may affect the outcome, consider the following example:

Consider the death rates because of a certain condition for two groups A and B living in two neighboring cities (table below). The last column shows that the death rates for both groups are greater in the big city than in the small city. However, when you combine two groups in each city, the overall death rate for combined population is higher in the small city. Thus, using these data, officials might reach very different conclusions depending on which part of the table they use.

	Population		Number of Death		Death Rate Per 100,000	
	Big City	Small City	Big City	Small City	Big City	Small City
Group *A*	4,675,174	80,895	8,365	131	179	162
Group *B*	91,709	46,733	513	155	560	332
Combined	4,766,883	127,628	8,878	286	*186*	*224*

This example makes it clear that for complex applications with more dimensions than two, there are more possibilities, making the chance of drawing incorrect conclusions much higher. For global warming, other than known factors, it is also necessary to investigate all the possible "sudden" factors and use statistical methods to include their contribution.

Analyzing and Communicating Flood Risk

Property and business owners as well as their elected or appointed decision makers in small rural or isolated communities do not have access to the type of data, maps, or interpretive methodologies frequently cited as "best practices" by experts. They do, however, generally record and aggregate data regarding the date and height of flood water on a point-by-point basis. We propose to leverage this data, where available, by using one or more well-known statistical tools that can be sourced by project managers and others to improve understanding, collaboration, and decision-making. Application to the data derived from the Susquehanna River in the vicinity of Bloomsburg, Pennsylvania, suggests that these tools can succinctly address point-specific issues in a simple familiar framework to streamline the planning process. Alternative analyses such as these may be valuable to engineers and others with academic, professional, and humanitarian interests in developing countries or regions.

Introduction

Periodic floods have a major influence on the scope and quality of lowlands (650 feet elevation or lower) that harbor most of the world's most populous cities. To make matters worse, cities and other urban areas are even more prone to flooding than rural environments because of the relatively greater area covered by pavement and other impermeable materials that limit percolation, hasten runoff, and increase risk. Despite this, people are increasingly attracted to urban areas because of the availability of employment, education, health services, and a variety of other cultural and economic factors. In fact, it is estimated that global flood damage could exceed $1 trillion annually by 2050.

On a worldwide basis, the government has largely failed to solve this problem. Rapid development continues in flood-prone areas throughout the world, and people as well as their investments remain at risk. In many cases, the capacities of existing flood control measures such as dams, sea walls, and other barriers are already woefully inadequate given the loss of natural percolation areas that rapid urbanization brings. Even if the funding could be found to build new flood control infrastructure, the costs would be staggering.

It seems to us that the prudent and responsible alternative would be to limit development in flood-prone areas to agricultural, park land, or other uses where periodic flooding would not cause significant economic or social impacts.

However, decision makers are often at a loss to evaluate actual flood risk. Technical studies are often prohibitively expensive and take a very long time to complete, but there are alternatives. Whereas most small isolated communities do not have access to sophisticated studies, current maps, and other best practice tools (see U.S. Army Corps of Engineers "How to Communicate Risk", http://www.corpsriskanalysisgateway.us/riskcom-toolbox-communicate.cfm), they do generally collect and aggregate the date and height of floodwater (Brenner et al. 2013; Arensburg and Hutt 2007).

In this paper, we propose a few alternative methods to evaluate flood risk in smaller or more marginalized communities that do not have updated maps or more sophisticated means to analyze or interpret their records. These alternatives are based on statistical concepts that leverage simple data like water height and date of flood events. The methods proposed involve the analysis of extremes, exceedances, excesses, record heights, and ultimate heights of floodwater.

To demonstrate, methods are applied to data recorded since 1850 from the Susquehanna River at the town of Bloomsburg in central Pennsylvania. Bloomsburg offers a particularly valuable opportunity to test these tools in that the data have been accurately and meticulously recorded for over 150 years. Further, floods are a significant problem in the area, and they compromise critical infrastructure such as drinking water, sewer, transportation, and emergency services.

The river in Bloomsburg floods when the water level exceeds nineteen feet. Since 1850, there have been thirty-eight floods with 32.75 feet, establishing the highest recorded water level. We hope by using the values derived from these data, we offer a useful and "user-friendly" platform to consider flood risk to decision makers and their constituents in small towns like Bloomsburg and elsewhere in the world especially where flooding can have catastrophic consequences.

Empirical Rule

For the Bloomsburg data, the mean, median, and standard deviation are respectively 24.3 feet, 23.4 feet, and 3.4. Also, the percentages of floods with sizes to within one, two, and three standard deviations from the mean are respectively 71 percent, 92 percent, and 100 percent. These percentages agree with the empirical rule indicating the normality this data set for. Thus, the probabilities of a future flood greater than 30.9 feet or 32.1 feet are respectively 5 percent and 1 percent. Also, the probability of a record flood (greater than 32.75 feet) is 0.006. This indicates that there is a small chance that a randomly selected future flood cresting level will exceed the present record.

Exceedances

The theory of exceedances deals with the number of times a specified threshold is exceeded. Assuming independent and identically distributed events (floods), we may wish to determine the probability of r exceedances in the next n occurrences. To apply this to Bloomsburg data, we note that since 1850, there have been thirty-eight floods exceeding 19.8 feet with the largest 32.7feet. So the mean expected value, variance, and standard deviation of the number of exceedances of 32.7 feet during the next ten and one hundred years would respectively be

$$10/173 = 0.058, (10) (172) (183)/ (173)2(174) = 0.06, \text{ and } 0.246$$
$$100/173 = 0.58, (100) (172) (273)/ (173)2(174) = 0.902, \text{ and } 0.95$$

These estimates have relatively large standard deviations because of a small sample size.

Records

This theory deals with values that are strictly greater than or less than all previous values. Usually, the first value is counted as a "record." A value is a record (upper record or record high) if it exceeds or is superior to all previous values.

To predict the future records, we have developed a simple method utilizing the following results of the theory of records for an independent and identically distributed sequence of observations (McDonnell 2013).

    a.   If there is an initial sequence of n_1 observations and a batch of n_2 future observations, then the probability

that this additional batch contains a new record is $n_2/$
(n_1+n_2).

b.   For large n, $P_{r,n}$, the probability that a series of length
n contains exactly r records is

$$P_{r,n} \sim \frac{1}{(r\text{-}1)!n}(\ln{(n)}+y)^{r\text{-}1}$$

For the Bloomsburg data, we note that there have been thirty-eight floods exceeding 19.8 feet. This gives a rate of 38/166 per year. Thus, the probability estimate of a record flood (as in (a) above) during the next ten years is

$$10/176 = 0.057 \sim 0.06.$$

Also, using (b), the probability of three records in thirty-eight observations is 0.003. With the above rate, we expect two floods in the next ten years, and the probability of three records in forty observations is 0.002930091. Hence, P (no record in next 10 years) = P (3 records in 40 observations) / (3 records in the 38 observations) = P (3 records in 40 observations) / P (3 records in 38 observations) = 0.94. This gives the probability of a record in the next ten years as 0.06, or 6 percent.

Excesses

In this approach, the probabilities of future large floods are calculated by developing models for the upper tail of the distribution for height of floodwaters. Because values above an appropriate threshold carry more information about the future large floods, this approach is reasonable. Here, one usually assumes that the tail of the distribution for flood sizes belongs to a given parametric family and then proceeds to do inference using excesses—that is, the floods greater than some predetermined value. It has been shown that the most appropriate model for tail is the so-called Pareto distribution that includes models for short, medium, and long tails. We applied this approach to the Bloomsburg data and found that the best fit is a short tail based on the largest four floods as follows:

$$\bar{G}_1(x) = 1-(1+0.24327x)^{-0.15764} \qquad 0 < x < 4.1106.$$

This led to the following upper bound for floods in Bloomsburg:

$$V_{max} = 32.7 \text{ ft.}$$

This is virtually the same as the largest flood in Bloomsburg that occurred in 2011.

Ultimate Flood

In terms of predictive value, we can avoid large standard errors and provide a confidence interval for the upper bound based on the most recent large floods or record floods. Let Y and $Y_1, Y_2, \ldots, Y_n$ represent flood size for a given region where

$$Y_1 \leq Y_2 \leq \cdots \leq Y_n.$$

Assuming that the distribution function $F(y)$ has a lower endpoint and certain conditions are satisfied, a level $(1-p)$ confidence interval for the maximum of Y is (de Haan 1981)

$$\{Y_1 + (Y_2 - Y_1)/[(1-p)^{-k} - 1], \ Y_1\}.$$

de Haan (1981) has also shown that

$$\frac{\ln m(n)}{\ln [(y_{m(n)} - y_3)/(y_3 - y_2)]}$$

is a good estimate for k.

To apply this result, we need to choose an integer $m(n)$ depending on n such that $m(n)/n \to \infty$ and $m(n)/n \to 0$ as n→∞. It is shown that the following choice works well even for the worst case:

$$m(n)=\sqrt{(eT_r)}+\sqrt{(t_r)}=\sqrt{(2.718282\,T_r)}+\sqrt{(t_r)}.$$

For the Bloomsburg data, $T_r = 107$ and $T_r = 1$ and $m(n) = 18$. Since $y_1 = 32.75$ ft., $y_2 = 32.70$ft, $y_3 = 31.2$ft, $y_{18} = 23.5$ft. We have $k = 1.767$, leading to a 90 percent one-sided prediction interval for the upper bound as (32.75 feet, 33.0 feet).

Summary

The application of alternative statistical methodology demonstrated in this paper for the Susquehanna/Bloomsburg area is a useful complement to best practice methods to evaluate flood risk in that

1. they can be used to communicate point-specific risk and answer the question most commonly asked, "How likely is it that this particular area will be inundated in any given year?"
2. risk can be communicated using a familiar simple scale such as "On a scale of one to ten, this area scores eight or has an 80 percent chance of flooding in any given year";
3. threat can be evaluated quickly with relevant data and without immediate reference to maps, which may not be available; and
4. such analyses can serve as an important touchstone between decision makers, developers, and others early and effectively in the flood control process;

References

Arensburg, A. and S, Hutt. "Challenges and Issues for Water Management in Northwest China." *IDEAS,* July 9, 2007.

Brenner, R., C. Keung, B. Rosenblum, R. Soltz, and S. Wolfe. 2013. "China's Loess Plateau—a Region of Heterogeneous Environmental Communities." Proc., GASI, Rome, Italy.

Davis, R. and S. Resnick, "Tail Estimates Motivated by Extreme-Value Theory." *Annals of Statistics.* 12 (1984) 1467–1487.

de Haan, L. "Estimation of the Minimum of a Function Using Order Statistics." *Journal of the American Statistical Association.* 76 (1981) 467–469.

Department of the Army, Office of the Chief Engineer. EM 1110-2-1411, March 1965.

Federal Emergency Management Agency [FEMA], Region 10. "The 100-Year Flood Myth." Pages 1–6, undated training document.

Hill, B. M. "A Simple General Approach to Inference about the Tail of a Distribution." *Annals of Statistics.* 3 (1975) 1163–1174.

McDonnell, T. "Global Flood Damage." *Mother Jones.* August 18, 2013.

Pickands, J. "Statistical Inference Using Extreme Order Statistics." *Annals of Statistics.* 3 (1975) 131–199.

Smith, R. L. "Forecasting Records by Maximum Likelihood." *J. of the Amer. Stat. Assoc.* 83 (1988) 331–338.

Future of the World: Pollution

Our planet is dying at an alarming rate because of the way that the human race has treated it over the past few centuries. Our world has become increasingly more technologically advanced, but with this advancement comes terrifying rates of pollution. Millions of people no longer have access to safe drinking water because of industrial runoff invading bodies of water, animal species are going extinct, and chronic illness rates are rising because of environmental factors.

The American Heritage Science Dictionary defines pollution as the "contamination of air, water, or soil by substances that are harmful to living organisms. Pollution can occur naturally, for example through volcanic eruptions, or as the result of human activities, such as the spilling of oil or disposal of industrial waste" (Rinkesh).

According to the World Health Organization, air pollution has grown by 8 percent in the past five years. This data was compiled from three thousand cities nationally to produce these alarming statistics. China is no longer the country with the most polluted air; however, this is not because they have cleaned their air. This is because other countries are becoming increasingly more polluted. Also, Americans make up about 5 percent of the global population but consumes 25 percent of the world's natural resources. Every year, people use more of the world's natural resources than can be produced. Soon enough, there will be no fuel to burn, water to drink, or food to eat. Because of the constant pollution of our planet, there has been an exponential rise in chronic illness like lung cancer because many people do not have access to clean air or water. Lung cancer rates have spiked in cities, leaving people with a 20 percent chance risk of getting lung cancer, while those who live in suburban or farm areas have a 6.67 percent chance. With the rise in pollution constantly growing and the high yearly depletion of natural resources, the future of our world looks dismal.

Much of the data I have presented in this were produced with a statistical inference. They were all tests of a random sample to represent the entire population. So based on these statistical inferences I have compiled, it is clear that the future of the world is not very bright because of the extremely harmful ways humans treat our planet.

Applying to what was learned in class: Conditional probability

A: the event that a normal person develops lung cancer: p (A) = 6.67%

B: The event that a person lives in a polluted city: P(B) = 70%

P (A and B) = 20% P(A and B)/P(B) = 0.20/0.70 = 28.6%

So living in a polluted city raises the chances of getting lung cancer from the average 6.67 percent to a startling 28.6 percent.

References

Do Something. "11 Facts About Pollution." DoSomething.org | Volunteer for Social Change, 25 Apr. 2015, www.dosomething.org/us/facts/11-facts-about-pollution.

Rinkesh. "51 Facts About Pollution." Conserve Energy Future, 25 Dec. 2016, www.conserve-energy-future.com/various-pollution-facts.php.

CHAPTER 16

Mathematics and Love

When we first met, you were pretty, and I was lonely.
Now we are both pretty lonely.

The Love Formula

$$x^2 + (y - \sqrt[3]{(x^2)})^2 = 1$$

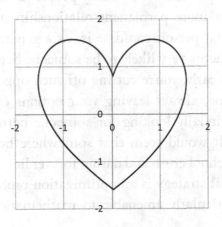

"WHY DO PEOPLE stay together in monogamous relationships? Love? Fear? Habit? Ethics? Integrity? Desperation?" Mark Colyvan asks in an essay. He considers a rather surprising answer that comes from mathematics. It turns out that cooperative behavior, such as mutually faithful marriages, can be given a firm basis in a mathematical theory known as game theory. I will suggest that faithfulness in relationships is fully accounted for by narrow self-interest in the appropriate game theory setting. This is a surprising answer because faithful behavior is usually thought to involve love, ethics, and caring about the well-being of your partner. It seems that the game theory account of faithfulness has no need for such romantic notions. He considers the philosophical upshot of the game theoretic answer and sees if it really does deliver what is required. Does the game theoretic answer miss what is important about faithful relationships, or does it help us get to the heart of the matter? Before we start looking at lasting, faithful relationships, though, let's get a feel for how mathematics might be employed to help in matters of the heart. Let's first consider how mathematics might shed light on dating to find a suitable partner.

Lover's Question

Consider the question of how many people you should date before you commit to a more permanent relationship such as marriage. Marrying the first person you date is, as a general strategy, a bad idea. After all, there's very likely to be someone better out there, but by marrying too early, you're cutting off such opportunities. But at the other extreme, always leaving your options open by endlessly dating and continually looking for someone better is not a good strategy either. It would seem that somewhere between marrying your first high-school crush and dating forever lies the ideal strategy. Finding this ideal strategy is an optimization problem and, believe it or not, is particularly amenable to mathematical treatment. In fact, if we add a couple of constraints to the problem, we have the classic mathematical problem known as the *secretary problem*. The mathematical version of the problem is presented as one of finding the best secretary (which is just a thin disguise for finding the best mate) by interviewing (i.e., dating) a number of applicants. In the standard formulation, you have a finite and known number of applicants, and you must interview these n candidates sequentially. Most importantly, you must decide whether to accept or reject each applicant immediately after interviewing him or her; you cannot call back a previously interviewed applicant. This makes little sense in the job-search context but is very natural in the dating context; typically, boyfriends and girlfriends do not take kindly to being passed over for someone else and are not usually open to the possibility of a recall. The question, then, is, how many of the n possible candidates should you interview before making an appointment? Or in the dating version of the problem, the question is, how many people should you date before you marry? It can be shown mathematically that the optimal strategy for a large applicant pool (i.e., when n is large) is to pass over the first ne (where e is the transcendental number from elementary calculus—the base of the natural logarithm, approximately 2.718) applicants and accept the next applicant who's better than all those previously seen. This gives a probability of finding the best secretary (mate) at $1e$, or approximately 0.37. For example, suppose that there are one hundred eligible partners in your village, tribe, or social network; this strategy advises you to sample the population by dating the first thirty-seven and then choose the first after that who's better than all who came before. Of course, you might be unlucky in a number of ways. For example, the perfect mate might be in the first thirty-seven and get passed over during the sampling phase. In this case, you continue dating the rest but find no one suitable and grow old alone, dreaming of what might have been.

Another way you might be unlucky is if you have a run of really weak candidates in the first thirty-seven. If the next few are also weak but there's one who's better than the first thirty-seven, you commit to that one and find yourself in a suboptimal marriage. But the mathematics shows that even though things can go wrong in these ways, the strategy outlined here is still the best you can do. The news gets worse, though—even if you stringently following this best strategy, you still only have a bit better than a one in three chance of finding your best mate.[1]

This problem and its mathematical treatment are instructive in a number of ways. Here, I want to draw attention to the various idealizations and assumptions of this way of setting things up. Notice that we started with a more general problem of how many people you should date before you marry, but in the mathematical treatment, we stipulate that the population of eligible partners is fixed and known. It's interesting that the size of this population does not change the strategy or your chances of finding your perfect partner—the strategy is as I just described, and so long as the population is large, the probability of success remains at 0.37. The size of the population just affects the number of people in the initial sample. But, still, stipulating that the population is fixed is an idealization. Most pools of eligible partners are not fixed in this way—we meet new people, and others who were previously in relationships later become available, while others who were previously available enter new relationships and become unavailable. In reality, the population of eligible candidates is not fixed but is open-ended and in flux. The mathematical treatment also assumes that the aim is to marry the best candidate. This, in turn, has two further assumptions. First, it assumes that it is in fact possible to rank candidates in the required way and that you will be able to arrive at this ranking via one date with each. We can have ties between candidates, but we are not permitted to have cases where we cannot compare candidates. The mathematical treatment also assumes that we're after the *best* candidate, and anything less than this is a failure. For instance, if you have more modest goals and are only interested in finding someone who'll meet a minimum standard, you need to set things up in a completely different way—it then becomes a satisficing problem and is approached quite differently.

Another idealization of the mathematical treatment—and this is the one I am most interested in—is that finding a partner is assumed to be one-sided. The treatment we're considering here assumes that it is an employer's market. It assumes, in effect, that when you decide that you want to date someone, he or she will agree, and that when you decide to enter a relationship with someone, again they will agree. This mathematical equivalent of wishful thinking makes the problem more tractable but is, as we all know, very unrealistic.

A natural way to get around this last idealization is to stop thinking about your candidate pool as a row of wallflowers at a debutante's ball, and instead think of your potential partners as active agents engaged in their own search for the perfect partner. The problem, thus construed, becomes much more dynamic and much more interesting. It becomes one of coordinating strategies. There is no use setting your sights on a partner who will not reciprocate. For everyone to find someone to reciprocate their interest, a certain amount of coordination between parties is required. This brings us to game theory.

The Game of Love

Game theory is the study of decisions where one person's decision depends on the decisions of other people.[2] Think of games like chess or tennis, where your move is determined, at least in part, by what you think the other player's response will be. It is important to note that games do not have to be fun and are not, in general, mere diversions. The cold-war arms race can be construed as a "game" (in this technical sense of game) between military powers, each second-guessing what the other would do in response to their "moves." Indeed, the cold war was the stage for one of the original and most important applications of game theory. The basic idea of game theory is quite simple and should be very familiar—a number of players are making decisions, each of whom depends on the decisions of the other players. It's probably best to illustrate game theory via an example. Let's start with the *stag hunt*. This game originates in a story by the eighteenth-century political philosopher Jean-Jacques

Rousseau of cooperative hunting.[3] In its simplest form, the game consists of two people setting out to hunt a stag. It will take the cooperation of both to succeed in the hunt, and the payoff for a successful stag hunt is a feast for all. But each hunter will be tempted by lesser prey—a hare, for example. If one of the hunters defects from the stag hunt and opportunistically hunts a passing hare, the defector will be rewarded, but the stag hunt will fail so the non-defector will not be rewarded. In decreasing order of preference, the rewards are stag, hare, and nothing. So the cooperative outcome (both hunt stag) has the maximum payoff for each of the hunters, but it is unstable in light of the ever-present temptation for each hunter to defect and hunt hare instead. Indeed, hunting hare is the safer option. In the jargon of game theory, the cooperative solution of hunting stag is *Pareto optimal* (i.e., there is no outcome that is better for both hunters), while the mutual-defect solution is *risk dominant* (in that it does not leave you empty-handed if your fellow hunter decides to defect and hunt hare), but it is not Pareto optimal.

That is, the cooperative solution is best for both hunters, and given that the other party cooperates in the stag hunt, then you should too. But if the other party defects and hunts hare, then so should you. Most importantly, both these outcomes are stable, since neither party will unilaterally change from cooperation to defection or from defection to cooperation (again, in the jargon of game theory, the mutual defect and cooperation solutions are *Nash equilibria*). So, in particular, if you both play it safe and hunt hares, there seems no easy way to get to the mutually preferable cooperative solution of stag hunting. Cooperation seems both hard to achieve and somewhat fragile. This game is important because it is a good model of many forms of cooperative behaviour.[5]

Consider another example, just to get a feel for game theory: the *prisoner's dilemma*. The scenario here is one where two suspects are questioned separately by the police, and each suspect is invited to confess to a crime the two have jointly committed. But there is not sufficient evidence for a conviction, so each suspect is offered the following deal: if one confesses, that suspect will go free while the other serves the maximum sentence; if they both confess, they will both serve something less than the maximum sentence; if neither confesses, they will both be charged with minor offenses and receive sentences less than any of those previously mentioned. In order of preference, then, each suspect would prefer the following: (1) confess, while the other does not confess; (2) neither confesses; (3) both confess; and (4) not confess, while the other confesses. Put it like this: it is clear what you should do—you should confess to the crime. Why? Because, irrespective of what the other suspect does, you will be better off if you confess. But here's the problem: if both suspects think this way, as surely they should, they will both end up with the second worst outcome, option 3. As a pair, their best outcome is option 2—this is Pareto optimal, since neither can do better than this without the other doing worse. But the stable solution is option 3, where both defect—this is the Nash equilibrium, since given that one confesses, the other should too. Group rationality and individual rationality seem to come apart. Individual rationality recommends both confessing, even though this is worse for both parties than neither confessing. Again, we see that defection (this time from any prearranged agreement between the suspects to not confess) is rewarded and cooperation is fragile.[6]

What have hunting stags and police interrogations got to do with dating—crude metaphors aside? First, these two games demonstrate how important it is to consider the decisions of others when making your own decisions. What you do is determined in part at least by what the other players in the game do, and vice versa. So, too, with relationships. In fact, the stag hunt is a very good model of cooperation in a relationship. Think of cooperatively hunting stag as staying faithful in a monogamous relationship. All going well, this holds great benefits for both parties. But there is always the temptation for one partner to opportunistically defect from the relationship to have an affair. This is the "hunting hare" option. If both partners do this, we have mutual defection where both parties defect from the relationship in favor of affairs. This game theoretic way of looking at things gives us a very useful framework for thinking about our original question of why people stay in monogamous relationships.

Where Did Our Love Go?

We are now in a position to see one account of how monogamous relationships are able to persist. Sometimes it will simply be the lack of opportunity for outside affairs. After all, there's no problem seeing why people cooperate in hunting stags when there are no alternatives. The more interesting case is when there *are* other opportunities. According to the game theory account we are interested in here, an ongoing monogamous relationship is a kind of social contract and is akin to the agreement to mutually hunt stag.

But what binds one to abiding by this contract when there are short-term unilateral gains for defecting? Indeed, it seems that game theory suggests defection as a reasonable course of action in such situations. If the chances of catching a stag (or seeing the benefits of a lasting monogamous relationship) are slim, defecting by opportunistically catching a hare (or having an affair) seems hard to avoid, perhaps even prudent. But we must remember that the games in question are not isolated one-off situations, and this is the key.

While defection in the prisoner's dilemma or the stag hunt may be a reasonable course of action if the situation in question is not repeated, in cases where the game is played on a regular basis, there are much better long-term strategies. For instance, both players will see the folly of defecting in the first game if they know that they will be repeatedly playing the same player. A better strategy is to cooperate at first and retaliate with a defection if the other player defects. Such so-called tit-for-tat strategies do very well in achieving cooperation. If both players are known to be playing this strategy, they are more inclined to cooperate indefinitely. There are other good strategies that encourage cooperation in these repeated games, but the tit-for-tat strategy illustrates the point. In short, cooperation is easier to secure when the games in question are repeated, and the reason is quite simple: the long-term rewards are maximized by cooperating, even though there is the temptation of a short-term reward for defection. It's the prospect of future games that ensures cooperation now. Robert Axelrod calls this "the shadow of the future"[7] hanging over the decision. This shadow changes the relevant rewards in a way that ensures cooperation.

We can make the cooperative outcome even more likely and more stable by sending out signals about our intentions to retaliate if we ever encounter a defector. In the stag hunt, we might make it very clear that defection by the other party will result in never cooperating with them again in a stag hunt. (Translated into the monogamous relationship version, this amounts to divorce or sleeping on the couch for the rest of your life.) We might even make such agreements binding by making the social contract in question public and inviting public scorn on defectors. All this amounts to a change in the payoffs for the game so that defection carries with it some serious costs—costs not present in the simple one-off presentation with which we started.

It is interesting to notice that this is pretty much what goes on in the relationship case. We have public weddings to announce to our friends and the world the new social contract in place (thus increasing the cost of a possible defection); we, as a society, frown on extramarital affairs (unless they are by mutual consent); and most important of all, we are aware of the long-term payoffs of a good, secure, long-term, monogamous relationship (if indeed that is what is wanted). Now, it seems we have the makings of an explanation of such relationships in terms of self-interest. While cooperation might look as though it has to do with love, respect, ethics, loyalty, integrity, and the like, the game theory story is that it's all just narrow self-interest. It's not narrow in the sense of being shortsighted, but in the sense that there's no need to consider the interests of others, except insofar as they impact on oneself. As David Hume puts it, "I learn to do service to another, without bearing him any real kindness; because I foresee that he will return my service, in expectation of another of the same kind."[8] In particular, there seems to be no place for love (and acting out of love) in the account outlined here.

Love Is Strange

If all I've said so far is right, it looks as though we can explain faithful relationships in terms of narrow self-interest. It's a case of "This is good for me, who cares about you." According to the game theory story, a faithful relationship is just a particular form of social cooperation. And all that is needed to keep the cooperation in place is mutual self-interest. It has nothing to do with right or wrong, or caring for your partner. It's all in the game and the focus on payoffs to the individual—or at least payoffs to the individual plus the shadow of the future. We might still frown upon noncooperation, but not for the reasons usually assumed. We, as a society, frown on defectors because that's also part of the game and it's an important part of what is required to keep cooperation alive in the society at large.

You might be skeptical of all this. You might think that people fall in love and enter a relationship not because they can get something out of it, but . . . well, why? If you're not getting something out of it, surely you're doing it wrong! OK, perhaps you get something out of it, but you stay committed through the hard times, through the arguments, through your partner's bad moods, not *purely* because it's good for you; you stick with them because they need you and you're a good person, right? It might help if that's what you believe, but one take-home message from the account I'm offering here is that there's no need for anything outside the game. We don't need to entertain anything other than self-interest as a motivation for monogamous relationships. It may well be that it's useful to believe in such things as loyalty, goodness, and perhaps even altruism, but they might all be just useful fictions—a kind of make-believe that's important, perhaps even indispensable, but make-believe all the same.

Let's look at these issues in terms of ethics. The game theory account not only leaves no room for love and romance but also seems to leave ethics out of the picture. You might think that staying faithful is *ethically right* and engaging in extramarital affairs is *unethical*. Insofar as game theory says nothing about ethics, it would seem that it cannot be the whole story. But we can take this same game theoretic approach to ethics. Ethics can be thought of as a series of cooperation problems. Thus construed, ethics is arguably explicable in the same terms.[9] The idea is that ethical behavior is just stable, mutually beneficial behavior that is the solution to typical coordination problems (basically, ethics is just a matter of "Don't hurt me and I won't hurt you"), and societies that have robust solutions to such coordination problems do better than those that don't have such solutions. As in the relationship case, it might be beneficial to engage in the pretense that some actions really are right and some really are wrong, but again such pretense will be just a further part of the game. This new twist about ethics makes your concerns about the dating and relationships case either a lot worse or a lot better, depending on your point of view. On the one hand, this broader game theoretic story about ethics allows that there is room for ethics in dating and relationships. But the ethics in question is just more game theory.

The picture of relationships I'm sketching here might seem rather different from the one we find in old love songs and elsewhere. I think the difference, though, is more of emphasis. Think of the picture offered here as a new take on those old love songs rather than a different kind of song altogether. All the usual ingredients are here but in an unfamiliar form. We have fidelity, but it's there as a vehicle for serving self-interest; ethical considerations are also there, but they, too, are not what they first seem. I suggested that all the romance and ethics might be merely a kind of make-believe, but perhaps that's overstating the case. The pretense may run very deep, and it plausibly has a biological basis. If this is right, the game theory picture can be seen as offering insight into the true nature of romantic relationships. Love, for instance, is a commitment to cooperate on a personal level with someone and licenses socially acceptable forms of retribution if defection occurs. Perhaps this conception of love doesn't sound terribly romantic and is unlikely to find its way into love songs, but to my ears, this is precisely what all the songs are about—you just need to listen to them the right way. Love is less about the meeting of souls and more about the coordination of mating strategies. If this makes love sound strange, then so be it: love is strange.

Notes

1. See Thomas S. Ferguson, "Who Solved the Secretary Problem?" (*Statistical Science*, vol. 4, 1989, pp. 282–296) for more on the secretary problem, and Clio Cresswell, *Mathematics and Sex* (Sydney: Allen and Unwin, 2003) for the many fascinating connections between mathematics and relationships.

2. For classic treatments of game theory, see John von Neumann and Oskar Morgenstern, *Theory of Games and Economic Behavior* (second edition, Princeton: Princeton University Press, 1947), and R. Duncan Luce and Howard Raiffa, *Games and Decisions: Introduction and Critical Survey* (New York: John Wiley and Sons, 1957).

3. Jean-Jacques Rousseau, *A Discourse on Inequality* (translated by M. Cranston, Penguin Books, New York, 1984, original from 1755).

4. Named after John Nash, the subject of the Sylvia Nasar book *A Beautiful Mind* (New York: Simon and Schuster 1998) and the Ron Howard movie of the same name (and based on the Nasar book).

5. Brian Skyrms, *The Stag Hunt and the Evolution of the Social Contract* (Cambridge: Cambridge University Press, 2004). This is an excellent treatment of the stag hunt and its significance for social cooperation.

6. William Poundstone, *Prisoner's Dilemma* (New York: Doubleday, 1992). This is a very accessible introduction to the prisoner's dilemma and game theory. It outlines the origins of game theory in the RAND corporation during the cold war (with a frightening application to the nuclear arms race).

7. Robert Axelrod, *The Evolution of Cooperation* (New York: Basic Books, 1984).

8. David Hume, *A Treatise of Human Nature* (ed. L. A. Selby-Bigge, Oxford: Clarendon, 1949, original from 1739) p. 521.

9. Richard Joyce, *The Evolution of Morality* (Cambridge, MA: MIT Press, 2006).

Secure Dating Protocol

At a high school party, some individuals are romantically interested in others. But because this is high school, it is absolutely central that you find out if your romantic interest is interested in you without letting on whether you yourself are interested.[1] Normal people rely on signals, hints, and elaborate ruses. Being mathematicians and computer scientists, we've got a better idea: cryptography can give us a protocol wherein you find out if the other person is interested only if you are. Here are some references that have the same material; no reason to copy good content into here.

Notes from U. C. Berkeley's CS276: Cryptography.

Notes from MIT's 6.045: Automata, Computability, and Complexity Theory.

Stable Marriage Problem

If we have n women and n men, how can we match them with each other into partnerships? Imagine that each man and woman can rank all members of the opposite sex by how attractive a partner they are. Can a matching be made so that no two people would prefer to be with each other than their partners? It turns that an algorithm exists that always produces such a "stable" matching. But there are multiple such stable matching, and it turns that the people making proposals to potential partners are the ones that get the best outcome in this algorithm, teaching us all an important lesson about not being afraid to approach that one guy/girl.

Wikipedia's description of the stable marriage algorithm, which also describes the optimality of the solution for the men and women.

Wikipedia on the stable roommate algorithm, wherein you don't have to match people by gender. This problem is much harder—for example, it need not even have a stable solution. There is also a much more complicated algorithm to find a stable matching if it exists.

A research project at the University of Glasgow, which investigated various algorithms for stable marriage and which stable matchings they choose.

The Secretary Problem

Imagine we have a long list of potential dates. We can date each for, say, a few months, after which we can rank that date relative to all the previous ones. After each date, we can either commit to that date as a partner or break things off permanently (you aren't allowed to rekindle relationships). How can one maximize their possibility of committing to their best-matched date? It turns out that a sufficiently clever approach can give one the optimal partner more than a third of the time, no matter what order the dates are dated. We'll expand on this principle to discuss why you shouldn't start looking to get married until approximately twenty-four.

Wikipedia on the secretary problem: Wikipedia has information on a variety of similar problems and on generalizations of the secretary problem.

The odds algorithm, which solves a general class of similar problems.

Stein, W. E.; Seale, D. A.; Rapoport, A. (2003). "Analysis of heuristic solutions to the best choice problem." *European Journal of Operational Research* 151: 140–152. This paper analyzed various heuristic approaches to the secretary problem, ones ordinary people might use when confronted with this problem in practice.

How a Math Genius Hacked OkCupid to Find True Love

Mathematician Chris McKinlay hacked OkCupid to find the girl of his dreams, Emily Shur. Chris McKinlay was folded into a cramped fifth-floor cubicle in UCLA's math sciences building, lit by a single bulb and the glow from his monitor. It was three in the morning, the optimal time to squeeze cycles out of the supercomputer in Colorado that he was using for his PhD dissertation (the subject: large-scale data processing and parallel numerical methods). While the computer chugged, he clicked open a second window to check his OkCupid inbox.

McKinlay, a lanky thirty-five-year-old with tousled hair, was one of about forty million Americans looking for romance through websites like Match.com, J-Date, and e-Harmony, and he'd been searching in vain since his last breakup nine months earlier. He'd sent dozens of cutesy introductory messages to women touted as potential matches by OkCupid's algorithms. Most were ignored; he'd gone on a total of six first dates.

On that early morning in June 2012, his compiler crunching out machine code in one window, his forlorn dating profile sitting idle in the other, it dawned on him that he was doing it wrong. He'd been approaching online matchmaking like any other user. Instead, he realized, he should be dating like a mathematician.

OkCupid was founded by Harvard math majors in 2004, and it first caught daters' attention because of its computational approach to matchmaking. Members answer droves of multiple-choice survey questions on everything from politics, religion, and family to love, sex, and smartphones.

On average, respondents select 350 questions from a pool of

thousands—"Which of the following is most likely to draw you to a movie?" or "How important is religion/God in your life?" For each, the user records an answer, specifies which responses they'd find acceptable in a mate, and rates how important the question is to them on a five-point scale from "irrelevant" to "mandatory." OkCupid's matching engine uses that data to calculate a couple's compatibility. The closer to 100 percent—mathematical soul mate—the better.

But mathematically, McKinlay's compatibility with women in Los Angeles was abysmal. OkCupid's algorithms use only the questions that both potential matches decide to answer, and the match questions McKinlay had chosen—more or less at random—had proven unpopular. When he scrolled through his matches, fewer than one hundred women would appear above the 90 percent compatibility mark. And that was in a city containing some two million women (approximately eighty thousand of them on OkCupid). On a site where compatibility equals visibility, he was practically a ghost.

He realized he'd have to boost that number. If, through statistical sampling, McKinlay could ascertain which questions mattered to the kind of women he liked, he could construct a new profile that honestly answered those questions and ignored the rest. He could match every woman in LA who might be right for him and none that weren't.

Chris McKinlay used Python scripts to riffle through hundreds of OkCupid survey questions. He then sorted female daters into seven clusters, like "diverse" and "mindful," each with distinct characteristics. Maurico Alejo

Even for a mathematician, McKinlay is unusual. Raised in a Boston suburb, he graduated from Middlebury College in 2001 with a degree in Chinese. In August of that year, he took a part-time job in New York translating Chinese into English for a company on the ninety-first floor of the north tower of the World Trade Center. The towers fell five weeks later. (McKinlay wasn't due at the office until two o'clock that day. He was asleep when the first plane hit the North Tower at 8:46 a.m.) "After that I asked myself what I really wanted to be doing," he says. A friend at Columbia recruited him into an offshoot of MIT's famed professional blackjack team, and he spent the next few years bouncing between New York and Las Vegas, counting cards and earning up to $60,000 a year.

The experience kindled his interest in applied math, ultimately inspiring him to earn a master's and then a PhD in the field. "They were capable of using mathematics in lots of different situations," he says. "They could see some new game—like Three Card Pai Gow Poker—then go home, write some code, and come up with a strategy to beat it."

Now, he'd do the same for love. First, he'd need data. While his dissertation work continued to run on the side, he set up twelve fake OkCupid accounts and wrote a Python script to manage them. The script would search his target demographic (heterosexual and bisexual women between the ages of twenty-five and forty-five), visit their pages, and scrape their profiles for every scrap of available information: ethnicity, height, smoker or nonsmoker, astrological sign—"all that crap," he says.

To find the survey answers, he had to do a bit of extra sleuthing. OkCupid lets users see the responses of others but only to questions they've answered themselves. McKinlay set up his bots to simply answer each question randomly—he wasn't using the dummy profiles to attract any of the women, so the answers didn't matter—and then scooped the women's answers into a database.

McKinlay watched with satisfaction as his bots purred along. Then, after about a thousand profiles were collected, he hit his first roadblock. OkCupid has a system in place to prevent exactly this kind of data harvesting: It can spot rapid-fire use easily. One by one, his bots started getting banned. He would have to train them to act human. He turned to his friend Sam Torrisi, a neuroscientist who'd recently taught McKinlay music theory in exchange for advanced math lessons. Torrisi was also on OkCupid, and he agreed to install spyware on his computer to monitor his use of the site. With the data in hand, McKinlay programmed his bots to simulate Torrisi's click-rates and typing speed. He brought in a second computer from home and plugged it into the math department's broadband line so it could run uninterrupted twenty-four hours a day.

After three weeks, he'd harvested six million questions and answers from twenty thousand women all over the country. McKinlay's dissertation was relegated to a side project as he dove into the data. He was already sleeping in his cubicle most nights. Now, he gave up his apartment entirely and moved into the dingy beige cell, laying a thin mattress across his desk when it was time to sleep.

For McKinlay's plan to work, he'd have to find a pattern in the survey data—a way to roughly group the women according to their similarities. The breakthrough came when he coded up a modified Bell Labs algorithm called K-Modes. First used in 1998 to analyze diseased soybean crops, it takes categorical data and clumps it like the colored wax swimming in a lava lamp. With some fine-tuning, he could adjust the viscosity of the results, thinning it into a slick or coagulating it into a single solid glob.

He played with the dial and found a natural resting point where the twenty thousand women clumped into seven statistically distinct clusters based on their questions and answers. "I was ecstatic," he says. "That was the high point of June."

He retasked his bots to gather another sample: five thousand women in Los Angeles and San Francisco who'd logged on to OkCupid in the past month. Another pass through K-Modes confirmed that they clustered in a similar way. His statistical sampling had worked.

Now, he just had to decide which cluster best suited him. He checked out some profiles from each. One cluster was too young, two were too old, another was too Christian. But he lingered over a cluster dominated by women in their midtwenties who looked like indie types, musicians, and artists. This was the golden cluster. The haystack in which he'd find his needle. Somewhere within, he'd find true love.

Actually, a neighboring cluster looked pretty cool too—slightly older women who held professional creative jobs, like editors and designers. He decided to go for both. He'd set up two profiles and optimize one for the A group and one for the B group.

He text-mined the two clusters to learn what interested them; teaching turned out to be a popular topic, so he wrote a bio that emphasized his work as a math professor. The important part, though, would be the survey. He picked out the five hundred questions that were most popular with both clusters. He'd already decided he would fill out his answers honestly—he didn't want to build his future relationship on a foundation of computer-generated lies. But he'd let his computer figure out how much importance to assign each question using a machine-learning algorithm called adaptive boosting to derive the best weightings.

Emily Shur (Grooming by Andrea Pezzillo/Artmix Beauty)

With that, he created two profiles, one with a photo of him rock climbing and the other of him playing guitar at a music gig. "Regardless of future plans, what's more interesting to you right now? Sex or love?" went one question. Answer: Love, obviously. But for the younger A cluster, he followed his computer's direction and rated the question "very important." For the B cluster, it was "mandatory."

When the last question was answered and ranked, he ran a search on OkCupid for women in Los Angeles sorted by match percentage. At the top: a page of women matched at 99 percent. He scrolled down . . . and down . . . and down. Ten thousand women scrolled by from all over Los Angeles, and he was still in the nineties.

He needed one more step to get noticed. OkCupid members are notified when someone views their pages, so he wrote a new program to visit the pages of his top-rated matches, cycling by age: a thousand forty-one-year-old women on Monday, another thousand forty-year-old women on Tuesday, looping back through when he reached twenty-seven-year-olds two weeks later. Women reciprocated by visiting his profiles, some four hundred a day. And messages began to roll in.

"I haven't until now come across anyone with such winning numbers, AND I find your profile intriguing," one woman wrote. "Also, something about a rugged man who's really good with numbers . . . Thought I'd say hi."

"Hey there—your profile really struck me and I wanted to say hi," another wrote. "I think we have quite a lot in common, maybe not the math but certainly a lot of other good stuff!"

"Can you really translate Chinese?" yet another asked. "I took a class briefly but it didn't go well."

The math portion of McKinlay's search was done. Only one thing remained. He'd have to leave his cubicle and take his research into the field. He'd have to go on dates.

On June 30, McKinlay showered at the UCLA gym and drove his beat-up Nissan across town for his first data-mined date. Sheila was a web designer from the A cluster of young artist types. They met for lunch at a cafe in Echo Park. "It was scary," McKinlay says. "Up until this point it had almost been an academic exercise."

By the end of his date with Sheila, it was clear to both that the attraction wasn't there. He went on his second date the next day—an attractive blog editor from the B cluster. He'd planned a romantic walk around Echo Park Lake but found it was being dredged. She'd been reading Proust and feeling down about her life. "It was kind of depressing," he says.

Date three was also from the B group. He met Alison at a bar in Koreatown. She was a screenwriting student with a tattoo of a Fibonacci spiral on her shoulder. McKinlay got drunk on Korean beer and woke up in his cubicle the next day with a painful hangover. He sent Alison a follow-up message on OkCupid, but she didn't write back.

The rejection stung, but he was still getting twenty messages a day. Dating with his computer-endowed profiles was a completely different game. He could ignore messages consisting of bad one-liners. He responded to the ones that showed a sense of humor or displayed something interesting in their bios. Back when he was the pursuer, he'd swapped three to five messages to get a single date. Now he'd send just one reply. "You seem really cool. Want to meet?"

By date twenty, he noticed latent variables emerging. In the younger cluster, the women invariably had two or more tattoos and lived on the east side of Los Angeles. In the other, a disproportionate number owned midsize dogs that they adored.

His earliest dates were carefully planned. But as he worked feverishly through his queue, he resorted to casual afternoon meetups over lunch or coffee, often stacking two dates in a day. He developed a set of personal rules to get through his marathon love search: No more drinking, for one. End the date when it's over, don't let it trail off. And no concerts or movies. "Nothing where your attention is directed at a third object instead of each other," he says. "It is inefficient."

Final Thoughts

1. Mathematics and Myths

Here is an appealing sentence from a book on the awakening of intellectuality in Egypt and Mesopotamia, the two near-Mediterranean areas in which, by Change or Providence, inexplicably but unmistakably, our present-day "Western" civilization germinated first:

"Mathematics is a form of poetry, which transcends poetry in that it proclaims a truth; a form of reasoning, which transcends reasoning in that it wants to bring about the truth it proclaims; a form of action, of ritual behavior, which does not find fulfillment in the act, but must proclaim and elaborate a poetic form of truth. Myth is a form of poetry, which transcends poetry in that is, proclaims a truth; a form of reasoning which transcends reasoning…"

2. Mathematics and politics

Mathematics plays a role in world politics. For example, around the 1870's, Karl Marx worked to understand the definition of the derivative in infinitesimal calculus, which was then about 200 years old. Among other things, he attempted to express the process of differentiation as a dialectical one. There are claims that he independently rediscovered some results on differentiation, though it is hard to track down precisely what it is that he rediscovered.

3. Good Mathematician

Not all mathematicians are good in the same way. Some are good at really quick problems, some at problems that take like half an hour, some are good at problems that take weeks or months, and some are especially good at research programs that lasted decades. The people who are considered especially good mathematicians are the ones who do work over a period of months or years and are not always also quick with little problems and puzzles. Sense of curiosity is also very helpful.